MICROBIOLOGY AND NANOBIOLOGY

MICROBIOLOGY AND NANOBIOLOGY

— Advancing Frontiers —

by
H.D. Kumar
Professor Emeritus
Indo-Soviet Friendship College of Pharmacy
Moga, Punjab

2010
DAYA PUBLISHING HOUSE
Delhi - 110 035

ISBN 81-7035-645-8
ISBN 978-81-7035-645-5

Published by : **Daya Publishing House**
1123/74, Deva Ram Park
Tri Nagar, Delhi - 110 035
Phone: (011) 27383999
Fax: (011) 23260116
e-mail : dayabooks@vsnl.com
website : www.dayabooks.com

Showroom : 4760-61/23, Ansari Road, Darya Ganj,
New Delhi - 110 002
Phone: 23245578, 23244987

Laser Typesetting : **Classic Computer Services**
Delhi - 110 035

Printed at : **Chawla Offset Printers**
Delhi - 110 052

PRINTED IN INDIA

Dedication

The author dedicates this book, with all humility, to Shri Parveen Garg, Chairman, Indo-Soviet Friendship College of Pharmacy, Moga (Punjab) who, like some of the most talented, dynamic and enterprising Vice-Chancellors of Indian universities, especially in the decades of the 1960s and 1970s, is totally devoted and committed to skilful management and enhancement of the status and reputation of the ISF College of Pharmacy as a top-ranking institution. It was largely his encouragement, kindness and hospitality that inspired the writing of this monograph.

Preface

Because microorganisms can adapt to virtually any physico-chemical environment and are efficient metabolizers, they are found everywhere. In fact, the true masters of life on Earth are not humans but microbes. They are essential for recycling biomass and for maintaining the biosphere. No discussion of infectious diseases can be completed without the description of human peaceful coexistence with a diverse microbial flora that inhabit our skin and guts. As they may well be the most brilliant and efficient chemists on Earth, microbes also attract interest as benefactors. They produce antibiotics, they genetically engineer disease-resistant plants, and they degrade pollutants.

Despite their huge importance, the fact is that much of what we know about microbes and microbial processes is based on the work done on only a handful of species–weed-like species which can be readily grown in laboratory culture. The vast majority of microorganisms found in nature (perhaps over 95 per cent) have not so far been cultured, but earnest attempts are underway to grow as many of them as possible in culture. An elegant diffusion chamber described recently allows growth of hitherto uncultured microbes in a simulated natural environment (see Chapter 5). It seems certain that their studies will pave the way to many novel processes and metabolic capabilities.

One most striking property of many microorganisms is their highly promiscuous mode of gene transfer among unrelated or remotely related taxa. Such horizontal (lateral) gene transfers have been the subject of intense research activity in the past few years in view of their strong bearing on the applicability of the biological species concept and the Linnaean system of bionomial nomenclature to prokaryotes.

The non-cellular viruses, lying at the borderline of the living and non-living, have traditionally, though uneasily, been described in textbooks of microbiology.

Viruses have erroneously been supposed to be simple creatures; in fact their enormous diversity and genetic promiscuity have made them a most creative force in evolution. It was only with the advent of nanoscience and nanotechnology that they started attracting interest as nano-organisms–most viruses have at least one dimension about or less than 100 nanometers (1 nm = 1 billionth of a meter). Importantly, the collective behaviour of matter can generate startling emergent properties between the micro- and nanometer scales–properties that clearly point to the nexus between biology and physics. Exceptionally, some giant viruses (*e.g.* mimivirus and phycodnaviruses) have all three dimensions greater than 100 nm, so are not strictly nano-organisms but in their structure and replication resemble other viruses. Viruses are therefore best treated under the new discipline of nanobiology. However, nanobiology is by no means confined to only viruses–all living cells contain a variety of subcellular organelles which behave as bionanomachines and so the processes and reactions involving the bionanomachines also come under the umbrella of nanobiology.

In the past two decades a silent revolution has transformed microbiology and created exciting opportunities for agriculture, medicine and industry. New technologies have enabled microbiologists to identify at least 20 previously unrecognized evolutionary groups of microbial life whose diversity dwarfs the better-known plant and animal kingdoms. Some new tools and techniques that have revealed the hidden microbial diversity include gene probing, genome sequencing, hybridization and DNA fingerprinting. For viruses, electron microscopy and atomic force microscopy have revealed spectacular hidden details.

Our agenda for the coming few years should be to identify the key organizational and generic principles underlying various morphogenetic processes in both prokaryotes and eukaryotes, while also pursuing the laborious elucidation of the molecular networks that control the organization of living matter. The extremely high complexity of living matter has compelled biologists to focus on a few 'model' organisms. This has no doubt been a productive approach but has led to a rather biased view–behind the apparent complexity, fundamental organizational principles based on physico-chemical laws are at work. The need of the hour is to critically examine these principles so as to understand how their ever-evolving combination leads to the dazzling diversity of patterns seen in the fascinating microbial world.

Besides the above, a few other currently active dimensions of microbiology and nanobiology that are briefly outlined in this monograph are cognition and natural genetic engineering, hyperthermophilic and other archaea, magnetosomes, metagenome libraries, microbial fuel cell, a bacterial time bomb, screening of microbes, microbial mats, medical biofilms, bilingual bacteria, and adaptive responses of microbes to toxic agents.

I am indebted to all those whose work has enhanced our knowledge of microbes and viruses and without whose critical investigations this attempted synthesis could not have been contemplated. I thank: (1) the Indian National Science Academy, New Delhi, for providing recent literature, (2) the United Nations Environment Programme, Nairobi, for giving me recurring opportunities, as Member of the UNEP Panel on Effects of Ozone Depletion, to visit several countries and interact with leading biologists, (3) to all those who kindly permitted me to reproduce their figures or illustrations; (4) to my former and present friends, colleagues, and associates, notably Profs. T.R.K. Reddy, S.P. Vyas, Ashok Kumar, O.N. Srivastava and S.P. Adhikary; and (5) to Prof. Vinay Sharma and the staff of the Department of Biosciences and Biotechnology, Banasthali Vidyapeeth (Rajasthan) for their friendly support and academic interactions in the past few years.

Several drafts of the manuscript were ably, patiently and efficiently composed by Rang Nath Singh.

H.D. Kumar

Contents

Chapter 1
General Microbiology

Understanding Microbes

The term *Microbe*, or microorganism refers to any unicellular organism. Two of the three superkingdoms of life, *Bacteria* and *Archaea*, are microbial, as also are single-celled eukaryotes (protists and yeasts). Viruses are neither self-maintaining nor self-reproducing, nor even cells, yet have traditionally (though uneasily) been regarded as microbes. Viruses are in-fact *nanoorganisms* as at least one of the dimensions of the vast majority (but not all) viruses is 100 nm or less (1 nm = 1 billionth of a meter). *Macrobes* is an equally loose catchall term for non-microbial multicellular life. They comprise the remainder of the Eukarya (the Animalia, Plantae). The characteristics generally supposed to constitute multicellularity are cellular differentiation and functional complementarity of different cell types. This is why macrobes are usually described as complex and 'higher'. Microbes are simpler and 'lower'. The border between macrobes and microbes is diffuse and vague. Many social single-celled prokaryotic and eukaryotic organisms *e.g.*, myxobacteria and cellular slime moulds, aggregate and behave in recognizably multicellular ways.

Studies of microbes are inseparable from our most basic understandings of life, from health and disease to genetics, biochemistry, evolution, geochemistry and environment. There is nothing a human being can do from one minute to the next that does not involve cooperative arrangements with our microbial partners. Our respiration, digestion, wellbeing and overall survival all depend on microbes. All our food is produced in alliance with complex communities of microbes, as also are many other life-sustaining materials.

Historically, the concept of prokaryote emphasized only negative organizational characteristics (the lack of nucleus and other membrane-bound organelles). This is why biologists regarded the microbes as deserving less esteem than eukaryotes. More recently, however, it is strongly believed that microbes in general and some specific groups in particular are extremely interesting for their positive features.

The use of DNA sequencing has revolutionized the classification of microbes. Sequencing technologies brought out the ubiquity of lateral gene transfer (even between the new superkingdoms of bacteria and archaea) and the confirmation of the endosymbiont hypotheses for mitochondria as well as chloroplasts, pointing to the incompleteness of a perspective focused on mutation and intraspecific recombination. An understanding of symbiosis implies that lineage isolation and inter-individual competition cannot account for evolutionary transformation–interspecific gene exchange and a wide range of very intimate modes of cooperation need to be incorporated more effectively (Sapp, 2005).

Microbial genetics initiated the molecular biology revolution; in fact, microbiology played an even earlier role in the configuration of contemporary biology. Müller-Wille (2005) pointed to the links between Martinus W. Beijerinck, the founder of the influential Delft School of microbiology, and botanist Hugo de Vries, in the rediscovery of Mendel, also between yeast microbiologist Emil Hansen and one of the founders of genetics, Wilhelm Johannsen. If we consider the central role of microbes in the development of early 20th century biology, then we need to reappraise the history of biology and displace Darwin from its centre!

Shapiro (1988, 1998, 2005) thought of bacterial communities as multicellular organisms and as natural genetic engineering systems, emphasizing bacterial cognition within a paradigm-shifting informatic approach in contrast to the conventional mechanistic one. He uses the terms 'cognition', 'sentience' and 'communication' for a broad range of bacterial activities; most cognitive scientists reserve such language for only a few animal species. Sophisticated cognitive information processing may be widespread in bacteria (both individual cells and communities). Many of these processes may even be more complex than those found in humans. Bacterial populations and communities must be understood as systems within this framework–systems that are sometimes embedded within broader organismal systems of interconnected and communicating cells (O'Malley and Dupre, 2007).

The 'anomalies' microbes present to prevailing species concepts are well-known. O'Malley and Dupre (2007) focused on *metagenomics* and its implications, most notably the concept of organism and its 'one genome, one organism' assumption. The distinguishing feature of *metagenomics* is that it uses environmental community DNA as the basis for DNA analyses, rather than DNA from isolated and cultured individual organisms. On this basis, strikingly different perspectives on microbial diversity and ecosystem interactions are emerging. This approach reinforces a Gaia-like perspective on ecosystems and one in which considerations of microbial ecology have a central role in discussions of global warming and biodiversity conservation. Indeed, a microbial dimension could possibly not only transform traditional conservation practices (such as seed banks) but also have major implications for environmental policy.

The Three Phases of Microbiology

The First Golden Age

The first phase of microbiology may be deemed to have started near the end of the 19th century, thanks to the works of Pasteur, Koch, and their students. Microbiological research and its applications affected health, nutrition and the environment and, indeed had an all pervasive effect on society. Discoveries concerning the microbial etiology of major infectious diseases were made in the first twenty years and led in turn to accurate diagnoses and attempts at prevention and cure. Several vaccines used today owe their existence to the works of early microbiologists. No less important, early research contributed to understanding the cycles of matter in nature, and provided a rational basis for food production and preservation.

In the first few decades of the 20th century, microbiology split into discrete, often divergent branches, and specialized identities. While medical microbiologists and immunologists mainly did research on the organismic level, trying to integrate the activities of host and parasite, environmental microbiologists (*e.g.* Sergei Winogradsky, Martinus Beijerinck and Albert Kluyver) focused on chemical processes. Its emphasis was on the basic unity of the biochemical processes in all living organisms. Microbiology in this period had not yet become a cellular science, or acquired any significant cytological orientation. While experimental biology at the time was virtually overshadowed by biochemists and their remarkable achievements in elucidating central metabolism, this did little towards understanding bacteria as cells and their complex organization. Many biologists considered bacteria–and viruses–as peculiar entities off the beaten path. It was not even agreed whether bacteria were cells, let alone whether they possessed genes or not. This attitude is best illustrated by the relative indifference that greeted the stunning discovery of genetic transformation by Fred Griffith in 1928 (Schaechter, 2003).

The following is a selected list of outstanding classical reviews contributions/ milestones published in *Bacteriological Reviews, Microbiological Reviews* and *Microbiology and Molecular Biology Reviews.* All students of microbiology should read these articles.

Year	Author	Article/Topic/Milestone
1941	S. Waksman	Antagonistic relations of microorganisms
1944	C.B. van Niel	(microbial physiology)
1946	M. McCarthy	(DNA as the agent of genetic transformation)
1947	S. Luria	(bacterial genetics)
1948	R. Dubos	Cellular structures concerned in parasitism
1949	S. Cohen	(redirection of host cell activities by plant viruses)
1958	W. Ackerman	(ditto, for animal viruses)
1959	A. Lwoff	(viral infections are a '3-body' problem, *viz.*, cell and virus, virus and organism and organism and cell)
1971	D. Baltimore	(unified view of expression of viral genomes in a parsimonious diagram)
1979	R. Dulbecco	Contributions of microbiology to eukaryotic cell biology: New directions for microbiology

The Second Golden Age

The birth of molecular genetics around 1943 heralded the second Golden Age. This era was ushered in by the classic experiment of Salvador Luria and Max Delbrück using a bacterial system to address a fundamental question of whether mutations are spontaneous or induced by the selecting environment. This experiment brought out the usefulness of bacteriophages and bacteria as model systems for the study of some of the most pressing questions in biology. In 1946 Joshua Lederberg discovered conjugation in bacteria. This experiment showed that if bacteria can mate, they must be cells.

These remarkable achievements were followed by other profound discoveries in the 1960's on fundamental aspects of biological processes such as the structure of DNA, the existence and signfiicance of plasmids, the regulation of gene expression, and the elucidation of the genetic code. Microbial systems became the first choice of those working at the frontiers, and represented the cutting edge of biology. While antibiotics came into widespread use, one of the classical arms of microbiology (the study of pathogens) waned in importance. In fact, some declared that the infectious diseases were about to be conquered and, by extension, that medical microbiology would become a relic–a declaration that later turned out to be highly exaggerated. Nevertheless, between 1950 and 1970, microbiology became virtually synonymous with molecular biology (Schaechter, 2003).

Soon, the lessons learned from microbes were found to be applicable to the study of eukaryotic cells. For some, *e.g.* yeast, the transition was direct. Other systems that cannot be easily manipulated and do not grow fast, such as animal and plant cells in culture, benefited proportionally even more from DNA cloning and sequencing. There was some waning of interest in microbiology.

The Third Golden Phase

In recent years, a new golden phase started in microbiology with the following outstanding advances.

Microbial Ecology

As long as bacteria are undergoing balanced growth, their global phenotype can be predicted. This placed bacterial growth on a reproducible baseline (Schaechter, 2003).

Bacteria are living entities, and the microbial world is exceptionally broad. Many surprising discoveries are being made in microbial ecology, with organisms being found daily in unusual and unexpected habitats. The variety of interactions between microorganisms is indeed spectacular. Symbioses between prokaryotes, and with eukaryotes consistently point to elegant solutions to the problems of living in varying environments (Margulis, 1998).

Microbial ecology has moved to center stage. Cloning techniques and PCR have made it possible to go beyond the confines of model organisms and to critically study almost any microorganism. This includes microbial species that cannot yet be cultured in the laboratory, estimated to comprise over 99 per cent of the total (see Eilers *et al.*, 2001). We can tackle questions such as who lives where, who does what, and who is phylogenetically related to whom. This capability has expanded the microbial world, even embracing whole new kingdoms such as the Archaea. Microorganisms have been recorded from such unexpected niches as thermal ocean vents and fissures in deep rocks (Taylor, 1999). Such lithotrophic organisms contribute to the existence of biomes which are not ultimately dependent on the sun or atmospheric oxygen, thus expanding the scope of microbial ecology and our understanding of the role of microbes in the cycles of matter in nature (Schaechter, 2003). Microbes in nature tend to live in communities, some with their own kin, others together with other species. We can now extend the study of microorganisms from the laboratory to their habitat in nature. Pathogenic microbes *e.g.*, are increasingly being investigated in their natural habitat. Ingenious genetic manipulations have enabled identifying genes that are selectively expressed in the host. This approach has led to the identification of novel mechanisms (Buttner and Bonas, 2003).

Lateral Gene Transfer

The genomes of many more prokaryotes than of eukaryotes have been sequenced. The resulting voluminous data has stimulated a whole industry intended to "mine" information of practical use. The demonstration of the extent and impact of lateral gene transfer (Sonea and Mathieu, 2001) has been surprising and extremely interesting. Genomic islands (sets of a dozen or more genes) appear in an unrelated species, endowing it new functions such a virulence or a novel metabolic capacity. Bacteria have successfully broken the fetters of sexual recombination by a set of highly promiscuous mechanisms that allow gene transfer between unrelated species, necessitating reappraisal of the notion that evolution can be traced strictly along separated branches of a tree. It turns out that the branches of the tree are laterally connected to one another. The ancestry of an organism is not simply monophyletic but is the consequence of multiple interactions between distinct genomes. There are indications that such interactions can also jump across the prokaryotic-eukaryotic divide. All this is playing havoc with the concept of species in microbiology (Kumar and Kumar, 2000; Schaechter, 2003).

Implications of Recent Advances

Some of the conceptual advances have strong implications for society. The growing threat of bioterrorism calls for better ways of detecting pathogenic agents, preventing their activities and spread, and finding new ways to treat affected humans, animals, and plants.

Microbial activities are playing a growing role in industrial production of pharmaceuticals and other compounds. Considerable success has been achieved in mining the genomes of cultivated and uncultivated microorganisms for enzymes with desirable physical and chemical properties. Several hydrolytic enzymes with enhanced thermal stability and targeted substrate specificities have reached the marketplace.

Integrative Microbiology

Microbiology is becoming more and more unified. Microbiologists share not just a genomic database but also the realization that microbes make use of similar capabilities for diverse purposes. According to Schaechter, common mechanisms of adhesion to surfaces, quorum sensing, signal transduction, building communities, or injecting proteins directly into host cells are seen in dissimilar organisms representing a broad spectrum of microbial life. The subdisciplines of microbiology are no longer isolated fields of study. Microbiology is becoming an integrative science and one that strives to connect and coordinate diverse elements into a biological whole (Schaechter, 2003).

Microbial Tree of Life

The first microbial genome sequence–that of *Haemophilus influenzae*–was published in 1995. Many more have been sequenced since then. This roster contains a welcome broad range of microbial phenotypes including those of soil bacteria and photosynthesizers, thermophiles and halophiles, and animal and plant pathogens. At least 12 prokaryotic phyla are represented, as are a few eukaryotes–enough to allow a meaningful examination of the Tree of (microbial) Life (Charlebois *et al.*, 2003).

Genome sequences of so many different organisms have cast some doubts on some of modern biology's fundamental tenets *e.g.*, the Darwinian-Mendelian model of parent-to-offspring ('vertical') gene flow–at least for microbes. Lateral (horizontal) gene flow–in which genes are transmitted *across, rather than along*, branches in family trees–has become a model of choice for the origin of microbial biodiversity.

Vertical inheritance with tree-like speciation fails to explain why so many gene families are distributed as they are among microbial genomes–that is, in highly diverse, sparse patterns that go against the accepted taxonomy. At the same time, lateral (recombination-dependent) events seem impossible for prokaryotes that live solitary lives inside eukaryotic cells, for example, or those which reproduce for several generations by simple binary fission.

The 1980s ushered in automated technology for sequencing DNA, the polymerase chain reaction (PCR), and improved methods for inferring phylogenies. Ribosomal RNA (rRNA) genes are ubiquitous, with highly conserved termini that make amplification by PCR easy. Several thousands of rRNA sequences became available. The rRNA tree quickly became the 'gold standard' for determining microbial relationships (Charlebois *et al.*, 2003). But rRNA opens only a small window to the microbial genome: for every gene that encodes a rRNA, there may exist a thousand that encode a protein. Protein-coding genes are less universal, more difficult to amplify by PCR, and often shorter and less information-rich than rRNA genes. Trees made from individual protein-coding genes (or from the concerned proteins) usually do not match the rRNA tree.

The verticalists argue for the inadequacies of single-gene phylogenetics. For them protein-based trees disagree because their true phylogenetic signal is too often obscured by noise and bias. In contrast, the lateralists believe that trees disagree because genes really do have different histories. To a lateralist microbial genomes are more or less ephemeral entities that are maintained rather fleetingly by the vagaries of selection and chance.

Whole genome trees have now largely converged on the rRNA-sequence tree. This convergence means that lateral gene transfer has not undermined descent with modification as the default explanation for microbial biodiversity, nor has it thrown microbial classification into disarray. Lateral transfer is perhaps not both quantitatively important and directional. One of the few widely accepted instances of lateral gene transfer is the origin of chloroplasts from archaic cyanobacteria; this is clearly visible in whole-genome trees, and even more so in 'sub-genome trees' based on functional subsets of genomes (Charlebois *et al.*, 2003). According to Charlebois *et al.* (2003), the important point is not whether a given tree is right or wrong. Rather, these trees should be used as frameworks upon which to build and test hypotheses about the rate and mode of microbial evolution.

Tree of Life or Network (Web) of Life?

Prokaryotes can swap their DNA through the processes of conjugation, transformation, and transduction. They acquire foreign genes through the process of horizontal gene transfer (HGT) which contrasts with vertical inheritance, in which an organism receives DNA from his parent. The consequence of HGT is that each prokaryotic chromosome is a mosaic of both vertically inherited and horizontally acquired genes (Welch *et al.*, 2002). Sorek *et al.* (2007) reported the large-scale analysis of the transferability of genes by analyzing the introduction of about a quarter of a million genes from other eubacterial and archaebacterial species into the eubacterium *Escherichia coli.* Their analysis brings out the likely evolutionary history of prokaryotes. It indicates that interspecies gene transfer is neither restricted to special categories of genes, nor are there categories of genes which cannot be transferred (McInerney and Pisani, 2007).

The role of horizontal gene transfer (HGT) in evolution has aroused acrimonious debate about the relevance of the Tree of Life–a long-accepted representation of the evolutionary interrelatedness of living things based chiefly on the sequence of the genes that encode the small subunit of ribosomal RNA (see Woese and Fox, 1977). The issue is whether this Tree of Life needs to be replaced with a network, or Web of Life (Doolittle, 1999). The difference between a Tree and a Web is that the former reflects evolution as a process in which new species arise (branch) from specific ancestors whereas a web better portrays microbial evolution based on the generation of variation (and creation of new species) through the horizontal transfer of genetic information between distantly related species. Using phylogenetic methods, Jain *et al.* (1999) proposed under the complexity hypothesis that not all genes are equally affected by horizontal gene transfer. Two main classes of genes are: *Informational, i.e.,* those involved in DNA replication, transcription, and translation; these form part of more complex protein-interaction networks and so are less likely to be involved in HGT; and *Operational,* which are involved in day-to-day processes of cell maintenance, such as controlling energy metabolism and the biosynthesis of nucleotides and amino acids; these may be more prone to transfer. This prompted the idea that a core of nontransferred or rarely transferred genes probably kept track of the clonal history of prokaryotes. Should this notion be true, the *Tree of Life* hypothesis would hold, even in the presence of widespread HGT (Doolittle, 1999).

One serious obstacle to understanding the role of HGT in evolution has been the lack of large-scale experimental evidence which could corroborate or reject the existence of a core set of genes. Sorek *et al.* (2007) have tested this idea: among 246,045 genes from 79 different species of prokaryotes, there was no single gene that, along with all its prokaryotic homologs, resisted transfer by way of a plasmid (*i.e.*, gene transfer by transduction) into *E. coli*. However, whereas all gene families could be cloned into *E. coli*, some genes could not be transferred. Members of informational gene families represent a considerable fraction of those resilient to cloning, supporting the complexity hypothesis (Jain *et al.*, 1999). Although 1402 genes could not be transferred, favouring the existence of core genes, transferable orthologs of these genes were found, thus opposing the core hypothesis (McInerney and Pisani, 2007).

Sorek *et al.* also found that genes from prokaryotic genomes enriched in guanine (G) and cytosine (C) are more transferable, consistent with the observation that the synonymous usage pattern of codons in recently acquired genes in *E. coli* is different from that of native genes. The explanation could be that promoter regions (DNA sequences that control the transcription of a gene) in the *E. coli* genome are GC-poor; so, GC-rich promoters might not be effective and so would result in nontranscription. One important way to introduce foreign DNA into an organism seems to be by stealth: repressing the expression of the introduced gene until it is widely distributed in the host species and then allowing expression (Doyle *et al.*, 2007). The implication of the stealth model of successful HGT is that it is easier to introduce DNA from a distant relative with different nucleotide base composition and different promoter characteristics than from a closer relative. For instance, Sorek *et al.* reported that g-proteobacterial genes are more difficult to transfer into *E. coli*–also into a g-proteobacterium. This agrees with the finding that a well-supported tree probably exists only toward the tips of the Tree of Life (within bacterial phyla, there is a strong phylogenetic signal associated with vertically inherited genes) (Pisani *et al.*, 2007).

The results of Sorek *et al.* demonstrate that even if there are some barriers to HGT into *E. coli*, they are quite low. Data on more bacterial taxa should reveal whether the results are peculiar to *E. coli*. Thus the existence of a *prokaryotic* Tree of Life is an open question. Eukaryotes have a chimeric nature (Pisani *et al.*, 2007; Lake, 2007). If we include eukaryotes in our considerations of evolution, the phylogeny of life is better represented by a network than a tree.

Microbial Phylogeny and Evolution

The origin of cells is undoubtedly biology's ultimate riddle. It constitutes arguably the strongest gap in our understanding of biology. This deep question is crucial to the search for a self-replicating RNA molecule and for re-focusing attention on the origins of biological organization.

In their review of the status of bacterial phylogeny (see Sapp, 2005), N. Pace, L. Ludwig, and K.-H. Schleifer defended the continued utility of ribosomal RNA sequences as the preferred molecular chronometer. F. Doolittle, W. Martin, and J. Lake *et al.* point to the evidence for extensive HGT, which seriously erodes phylogenetic signals and may even make it impossible to recover the history of organisms (see Harold, 2006). C. Kurland differs, and vigorously defends the universal tree; HGT is real and sometimes a nuisance, but claims of rampant promiscuity may be exaggerated. Indeed, reports of the imminent demise of the universal tree are probably greatly overblown.

The two enduring mysteries of cell evolution are: (a) the origin of the eukaryotic cell, and of mitochondria and chloroplasts; and (b) microbial symbiosis and parasitism. Critical work on these aspects may throw light on the intriguing problem and mystery of the living cell (Harold, 2006).

Self-organization Processes in Biology

Living matter not only has its special chemistry involving molecules carrying information but also is built of self-organizing systems far from thermodynamic equilibrium. The concept of self-organization has a strong bearing on our view of living matter. Much recent light has been thrown on how reaction diffusion and collective molecular behaviours are largely responsible for the generation of cellular and subcellular dynamic patterns characteristic of living matter (Karsenti, 2007).

Besides enormous complexity of life matter, a key issue is scaling. Life unfolds over very different timescales. Whereas the evolutionary timescale is millions of years, the actual activities of cellular physics and chemistry which determine the shapes, functions and behaviour of living matter take place over milliseconds, minutes or hours. Life results from the complex collision and intermingling of these different time frames.

Some processes operate at the molecular level with extremely fast rates, involving molecular dynamics, diffusion processes, chemical and enzymatic reactions. Other events take place at the subcellular or cellular level with reaction times of only minutes to hours and spatial dimensions ranging from a few micrometres to millimeters. For large organisms, some processes (*e.g.* embryonic development, circadian rhythms and ageing) operate over timescales of hours, days, months or years.

The Eukaryotic Cell

In eukaryotes, cell sizes typically range from a few micrometres (μm) to millimeters. Some free-living eukaryotic unicellular organisms (Fig. 1.1) are found mostly in sea and fresh water and have sizes of tens to hundreds of μm. The average size of cells found in multicellular organisms is around tens of μm.

Eukaryotic cells typically contain a nucleus in which the genome is packed into chromosomes. The nucleus is separated from the cytoplasm by a nuclear envelope which has pores through which molecules can be transported between the two compartments. DNA (*i.e.* genes) is transcribed into messenger RNAs (mRNAs) in the nucleus; these mRNAs then move through the pores to the cytoplasm, where they are translated into proteins. Exchange of RNAs and proteins between the cytoplasm and the nucleus also takes place through the pores and involves a gradient of chemical energy harnessed by a small protein known as Ran (Hetzer *et al.*, 2002). This protein exists in either a low- or high-energy state. In the nucleus it occurs in a high-energy state, coupled to a small chemical molecule of GTP (guanosine triphosphate, the Ran-GTP form). In the cytoplasm, it exists in a low-energy state (Ran-GDP). This

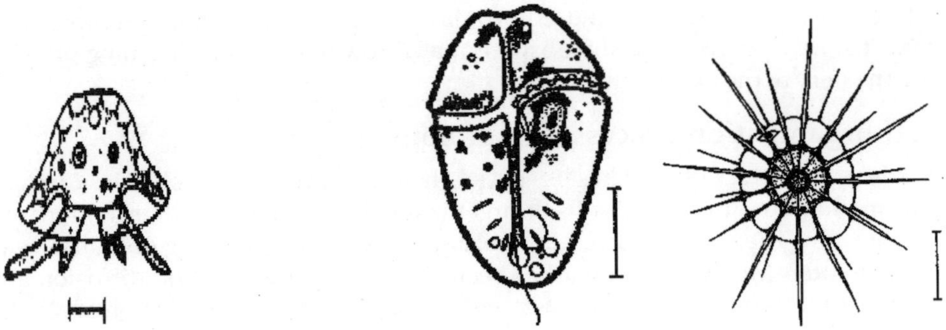

Figure 1.1: Four Examples of Free Living Unicellular Eukaryotic Organisms (protists): bars are ten micrometers. A, B, C... have cilia made of microtubules. B. *Euglena*, has a flagellum made of microtubules arranged in triplets. Cilia and flagella beat and can have various functions ranging from moving the unicellular organism to producing water flows and bringing food to be absorbed. A. *Amoeba*, which has no cilia but can crawl and change shape. C. a dinoflagellate. D. a heliozoan.

energy gradient makes up a sorting system which defines two spatial domains recognizable by cellular molecules; it generates 'fields' whose state determines how different molecules behave within them. In fact, spatial organization within the cell is determined not by stable physical interactions between molecules, but rather by the dynamic state of signaling molecules (Karsenti, 2007).

We are made of several different cell types, each type having a specific shape and function, yet all cells in our bodies have the same genome. Biologists have been greatly interested in understanding how the genome is differently expressed in different cell types. Much work has been done on the regulation of gene transcription and chromatin modifications. In contrast, much less work has been done on understanding how the different shapes corresponding to each cell type arise. These shapes are in fact determined by the cytoskeleton; so we need to learn how cytoskeleton fibres, such as the microtubules and actin microfilaments (shown in Figure 1.2) can self-organize into variable three-dimensional patterns (Jankovics and Brunner, 2006). In fact, better terms for these tubules and filaments are nanotubules and nanofilaments because at least one of their dimensions is less than 100 nm. These fibres constitute the cell's structural elements, and they are highly dynamic and versatile. They can grow and shrink at any time so that cell shapes are maintained at the constant expense of chemical energy dissipated to maintain dynamic steady states for those filaments. Cells are powered by high energy nucleotides such as ATP and GTP that

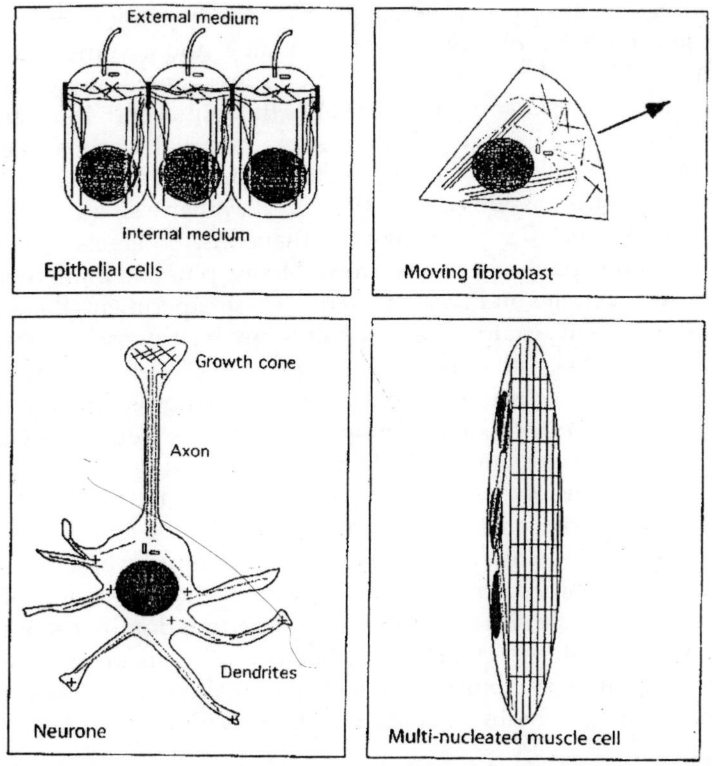

Figure 1.2: Some Cell Types Found in Multicellular Organisms. Thick bars show microtubules; thin lines indicate actin microfilaments; rectangles are centrioles and basal bodies and the grey circles and ovals are nuclei. A range of patterns is formed by microtubules and microfilaments in the various cell types. The cilia found at the apex of epithelial cells are quite similar in content and structure to those found in some ciliated protists (after Karsenti, 2007).

are hydrolyzed into low energy ADP or GDP either to promote filament dynamics or force exerted by the molecular motors which cross-link the filaments or move on them. All this happens at a scale very close to that of molecules in a highly hydrated environment. Thermal agitation and diffusion are a strong source of organization as they allow highly dynamic encounters between the different molecular components of the cytoskeleton. It is the chemical energy dissipation coupled to the specific structures and intramolecular dynamics of cytoskeleton molecules that result in vital self-organization processes. Complex and fairly large cellular patterns arise spontaneously from a mixture of thermal noise and collective molecular behaviours driven by nucleotide hydrolysis. These processes are being understood from researches on the collective behaviour of purified components of the cytoskeleton (Nedelec *et al.*, 1997, 2003) and from work done on yeast *Saccharomyces pombe* and the nematode worm *Caenorhabditis elegans* (Grill *et al.*, 2003).

Microtubules are asymmetric tubes (Figure 1.3) with a 'plus' end that actively alternates between phases of growth and shrinkage, and a 'minus' end which is rather inert and stable. Molecular motors can move directionally along microtubules, some towards the plus end and others towards the minus end. Microtubules and motors can self-organize in diverse patterns, *e.g.*, parallel/anti-parallel bundles, vortices and asters. These patterns can be generated with a limited number of molecular motors and microtubules (Nedelec *et al.*, 1997, 2003). When a single type of motor is present (minus end-oriented, *e.g.*), and if it simultaneously interacts with at least two microtubules, different patterns can be formed. Mixing plus and minus end-oriented motors create more complex and varied patterns. These experiments and simulations show how diverse steady-state patterns emerge through the self-organization of microtubules and motors with only a few components. The underlying principle is that out of thermodynamic equilibrium, steady-state patterns can be generated by oriented tubes and motors through the dissipation of chemical energy involving ATP consumption. Without motors, microtubules tend to distribute isotropically over short timescales. In the suspension an intrinsic asymmetry is created from the asymmetry of the tubulin molecule that assembles into oriented tubes. So long as motors do not work, the suspension remains symmetrical, but the symmetry breaks down when motors move and organize microtubules, so that the whole system evolves either towards a chaotic unstable situation in which no pattern dominates over time, or towards a well defined steady-state pattern. Such patterns are obtained with stable microtubules. If dynamic microtubules (microtubules alternating between growing and shrinking phases, showing dynamic instability) are used instead, greater

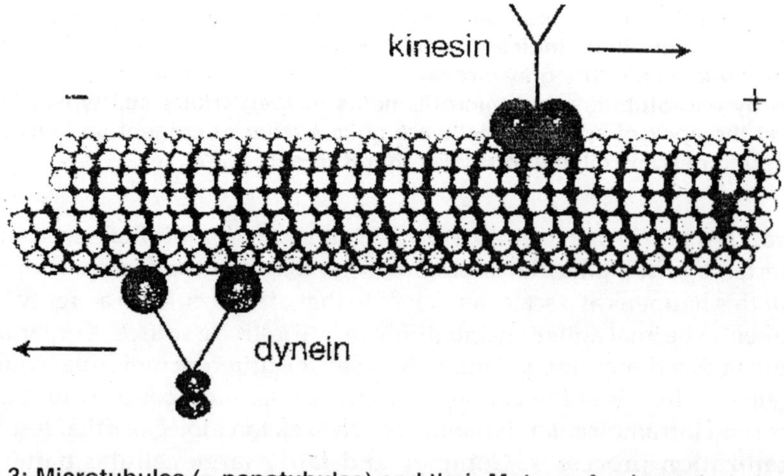

Figure 1.3: Microtubules (= nanotubules) and their Motors: A microtubule is a 25-nm diameter hollow tube made of 13 protofilaments which are built of 8-nm tubulin dimers. Motors drawn on the microtubule can walk towards the plus end of the microtubule if they are kinesins or the minus end if dyneins. Motor heads hydrolyze ATP when they move along the lattice of the microtubule. The tail of the motor (the end that does not walk on microtubules) interacts with adaptor proteins that tether the motors to various cargoes; the cargoes can range from vesicles to chromosomes or other microtubules (Karsenti, 2007).

complexity can be added to the patterns formed (Nedelec, 2002). Dynamic microtubules self-organize into patterns without motors by gravity which acts as a symmetry-breaking force in place of motors (Cortes *et al.*, 2006). Critical experiments as well as computer simulations have revealed how local regulation of microtubule dynamics and the action of motors contribute to microtubule organization and cell shape determination in *S. pombe* (Daga *et al.*, 2006).

According to Karsenti, the organization of living matter is not entirely determined by protein sequences. Interactions among proteins attaining thermodynamic equilibrium cannot lead to large-scale dynamic structures. This necessitates understanding of the actual dynamical properties of protein networks to learn about how the components of living matter create the complex patterns and behaviours observed. Such behaviours occur quite far from thermodynamic equilibrium via self-organization processes which dissipate energy through ATP and GTP consumption. In the future, a clear understanding of how complex eukaryotic cells, embryos and organisms are shaped and function, may emerge if we supplement the traditional molecular and genetic analyses with modeling and theory based on firm biological data. It is desirable for biologists to continue using molecular methods, but they should also ensure that their work is accurate enough for theorists to be able to use it with confidence.

Our agenda for the coming few years should be to identify the key organizational and generic principles involved in a variety of morphogenetic processes, while also pursuing the laborious identification of molecular networks underlying the organization of living matter. It is largely because of the high complexity of living matter that biologists have focused on the study of a few 'model' organisms. This has been a productive approach, but has led to a rather biased view–behind the apparent complexity, fundamental organizational principles based on physicochemical laws are at work. The need of the hour is to critically examine these underlying principles with a view to finding how their ever-evolving combination can lead to the spectacular diversity of patterns seen in the living world.

Cellular Microbiology

Microbiologists have traditionally studied pathogens in pure culture and cell biologists likewise studied mammalian cells in tissue culture. Both fields used rich media and conditions optimized for *in vitro* growth. These laboratory methods of growing pathogens and their hosts are quite artificial and do not reflect conditions in real life where pathogens and hosts coexist, interact, and compete in conditions that are usually not at all optimal. To better understand what happens in real life, the interaction between microbes and host cells can be studied by exploiting technological progress that allows cocultures of microbes and cells to be handled together. This area falls in the jurisdiction of the new discipline of cellular microbiology (Henderson *et al.*, 1999; Cossart *et al.*, 2000; Rappuoli, 2000).

A recently discovered new technique describes how, instead of studying one parameter at a time, microchips may be used to study, in a single experiment, all of the host genes and those of the bacteria whose expression is modified during host-pathogen interaction (Belcher *et al.*, 2000).

Figure 1.4 illustrates a typical microarray experiment. RNA is prepared from bacteria grown in standard laboratory conditions from host cells grown under optimal conditions in tissue culture, and from cells that have been infected with bacteria. A probe is then prepared from each of the RNAs by reverse transcription, usually by using oligo(dT) to probe host-cell genes and random oligonucleotide primers to anneal to bacterial genes. A balanced mixture of oligos specific for each gene present in the microarray can also be used. To probe printed microarrays, RNA is converted into a fluorescent cDNA probe by incorporating different fluorochromes during the reverse-transcription reaction (by using Cy3- and Cy5 dCTP). The probes are then used to

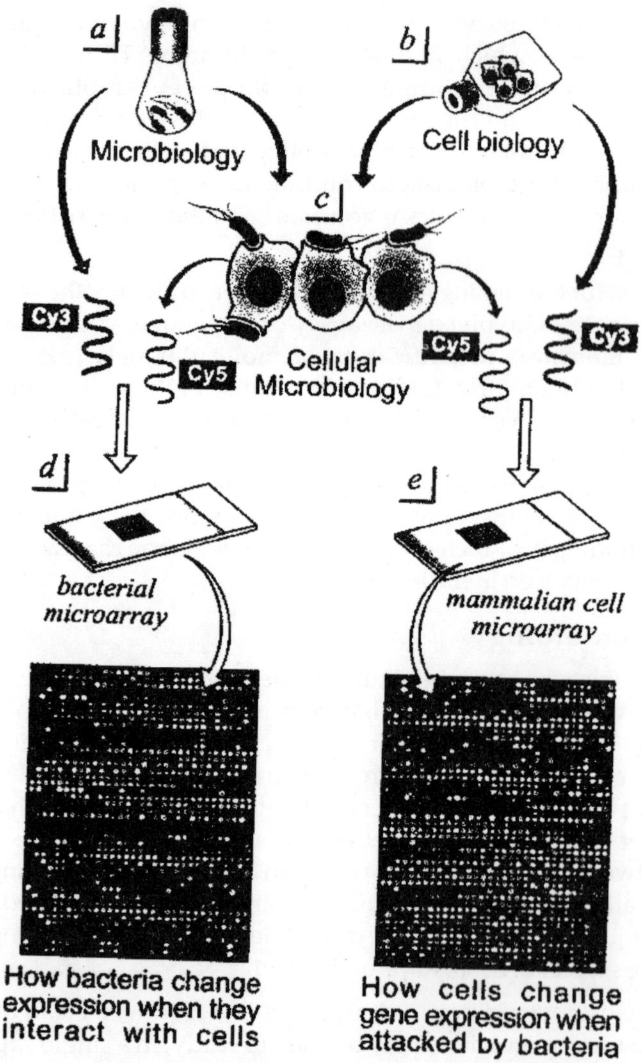

Figure 1.4: Use of Microarrays to Study Host-Cell-Pathogen Interactions (After Rappuoli, 2000). [Copyright (2000) National Academy of Sciences, U.S.A.].

hybridize microarrays containing bacterial genes (Figure 1.4d) or eukaryotic genes (e). High-density genome-wide microarrays of eukaryotic cells are commercially available (see Rappuoli, 2000). Chips representing the whole genome of a pathogen are now becoming available as soon as the genomic sequence of the bacterium is published. Usually chips are printed by using each of the bacterial genes, which have been previously amplified by PCR. Microarrays containing all genes of the sequenced bacterial genomes are either available or are being prepared, and include *Helicobacter pylori, Neisseria meningitidis, Mycobacterium tuberculosis* and *Escherichia coli.*

In the experiment shown in Figure 1.4e, the RNA prepared in b and labeled with Cy3 is mixed with the RNA prepared in c and labeled with Cy5. The two fluorescent cDNAs are then mixed together and make up the probe, which is hybridized to the chip.

Genome-wide changes in bacterial gene expression after contact with eukaryotic cells can also be obtained (Figure 1.4d) by hybridizing bacterial chips with a mixture of the probes in Figures 1.4a and c. DNA arrays provide a measure of the average mRNA present (steady-state abundance) and are not just a measure of the rate of transcription (Rappuoli, 2000).

The Inflammatory Response to Bacterial Infection

Several studies describing changes in host-cell transcription after infection with bacteria have been published in 2000 (see Rappuoli, 2000). Many studies and improved computer algorithms will be needed before we can gain a general idea of which genes are induced in mammalian cells by most pathogens and which are specifically induced by individual pathogens. However, initial studies already give a preliminary idea of the main reaction of a mammalian cell to bacterial attack. The inflammatory response is the first and main reaction of mammalian cells to infection. Chemokines and cytokines, which are directly responsible for various inflammations and transcription factors involved in the inflammatory response are affected.

Pathology, Pathogenesis, and Pharmacogenomics

Microarrays can also be used to ask very specific questions about the clinical manifestation of a disease and the role in pathogenesis of individual virulence factors and to predict the clinical outcome of specific drugs (Belcher *et al.*, 2000). They can also be used in pharmacogenomic studies to determine how a drug can modify the transcriptional response to infection. By adding the antiinflammatory drug dexamethasone to the experiment in Figure 1.4e, these microarrays could monitor how a drug recommended for clinical use can modify the response of a single cell (Rappuoli, 2000).

The Complexity Pyramid

Cells and microorganisms can efficiently adjust their intracellular machinery in response to environmental changes, food availability, and developmental state. They also correct internal errors¾battling the effects of mutations or misfolded proteins. Living systems are in fact extremely complex, versatile and robust (Oltvai and Barbasi,

2002). The detailed inventory of genes, proteins, and metabolites is not sufficient to understand the cell's complexity (Kitano, 2002). Lee *et al.* (2002) considered the cell as a network of genes and proteins and discussed the complexity of living systems. DNA is the ultimate depository of biological complexity. Information storage, information processing, and the execution of various cellular programs occurs at distinct levels of organization : the cell's genome, transcriptome, proteome, and metabolome. However, the distinctness of these organizational levels has recently been questioned. For instance, although long-term information is stored almost exclusively in the genome, the proteome is important for short-term information storage (Bhalla and Iyengar, 1999), and transcription factor-controlled information retrieval is greatly affected by the metabolome. This integration of different organizational levels means that cellular functions are distributed among groups of heterogeneous components that all interact within large networks (Jeong *et al.*, 2000). There is definite evidence for the existence of such cellular networks–the proteome organizes itself into a protein interaction network and metabolites are interconverted through an intricate metabolic web. Surprisingly, the structures of these networks are governed by the same principles; this offers a new perspective on cellular organization.

The complexity pyramid is composed of the various molecular components of the cell¾genes, RNAs, proteins, and metabolites. These various building blocks organize themselves into small recurrent patterns (pathways in metabolism and motifs in genetic-regulatory networks). In turn motifs and pathways are intimately integrated to from functional modules¾groups of nodes (*e.g.*, proteins and metabolites) that perform discrete cellular functions. The hierarchical nesting of these modules defines the cell's large-scale functional organization (Ravasz *et al.*, 2002).

One important feature of the complexity pyramid is the gradual transition from the particular (at the bottom level) to the universal (at the apex). The precise repertoire of genes, metabolites and proteins is, of course, unique to each organism. Jeong *et al.* (2000) reported that 43 organisms for which fairly complete metabolic information is available, share only ~ 4 per cent of their metabolites. Key metabolic pathways are usually shared, however, and so are some of the motifs. Conceivably key properties of functional modules may be shared across most species.

The Irreducible Nature of Eukaryotic Cells

Recent work on comparative genomics and proteomics has supported the idea that modern eukaryote and prokaryote cells have long evolved on separate evolutionary trajectories. In view of their cells appearing simpler, prokaryotes have usually been viewed as ancestors of eukaryotes (Knoll, 1992; Baldauf, 2003). However, comparative genomics has confirmed a lesson in paleontology, that evolution seldom proceeds monotonically from the simpler to the more complex (Simpson, 1944; Klasson and Andersson, 2004). Kurland *et al.* (2006) reviewed recent data from proteomics and genome sequences that points to the possibility that eukaryotes are a unique primordial lineage.

Large-scale comparative genomics and proteomics has substantiated several basic features of eukaryote cellular evolution. The evolutionary trajectory of modern eukaryotes is quite distinct from that of prokaryotes. There is no direct evidence that eukaryotes evolved by genome fusion between archaea and bacteria. Comparative genomics shows that under certain ecological settings, sequence loss and cellular simplification occur commonly in evolution. Subcellular architecture of eukaryote cells is partly a physical-chemical consequence of molecular crowding; subcellular compartmentation with specialized proteomes is essential for the efficient functioning of proteins.

Mitochondria, mitosomes, and hydrogenosomes (see Gabaldon and Huynen, 2003) constitute a related family of organelles which distinguish eukaryotes from all prokaryotes (Embley and Martin, 2006). Early eukaryotes had many introns (Roy and Gilbert, 2005), as well as RNAs and proteins found in modern spliceosomes (Collins and Penny, 2005). According to Lynch (2002), intron numbers may be affected by life-history parameters. Further, "molecular crowding" is an important physical-chemical factor that contributes to the compartmentation of even the earliest eukaryote cells (Pielak, 2005).

Nuclei, nueleoli, Golgi apparatus, centrioles, and endoplasmic reticulum exemplify *cellular signature structures (CSSs)* which distinguish eukaryote cells from archaea and bacteria. Comparative genomics together with proteomics of CSSs such as the mitochondria, nucleoli, and spliceosomes, have revealed many proteins with no orthologs evident in the genomes of prokaryotes; these are the eukaryotic signature proteins (ESPs) (Hartman and Fedorov, 2002). The ESPs within the subcellular structures of eukaryote cells serve as sign posts to trace the trajectory of eukaryote genomes from their origins. In contrast, hypotheses that attribute eukaryote origins to genome fusion between archaea and bacteria (Rivera and Lake, 2004) surprisingly do not reveal anything about the emergence of the cellular and genomic signatures of eukaryotes (CSSs and ESPs). Genome fusion fails to directly explain any characteristic feature of the eukaryote cell.

Regardless of whether we use gene content, protein-fold families, or RNA sequences (Woese *et al.*, 1990; Yang *et al.*, 2005), the unrooted tree of life can be divided into archaea, bacteria, and eukaryotes (Figure 1.5). On such unrooted trees, the three domains diverge from a population that can be called the last universal common ancestor (LUCA). But as LUCA has meant different things to different people, a better term may be a common ancestor; in this case it is the hypothetical node at which the three domains coalesce in unrooted trees. It appears that the common ancestor of eukaryotes, bacteria, and archaea was a community of organisms containing the following: autotrophs that produced organic compounds from CO_2 either photosynthetically or by inorganic chemical reactions; heterotrophs that obtained organics by leakage from other organisms; saprotrophs that absorbed nutrients from decaying organisms; and phagotrophs that were complex enough to envelop and digest prey (Kurland *et al.*, 2006)

Proteomics of Cell Compartments

Comparative genomics and proteomics uncover phylogenetic relationships among proteins making up eukaryote subcellular features and those found in prokaryotes. Three chief phylogenetic classes can be distinguished: (a) proteins (ESPs) that are unique to eukaryotes. The ESPs can be placed in three subclasses: those arising *de novo* in eukaryotes; proteins that are so divergent to homologs of other domains that their relationship cannot be traced; or finally, descendants of proteins that are lost from other domains and survive only as ESPs in eukaryotes.

The class (b) has interdomain horizontal gene transfers; these are proteins occurring in two domains with the lineage of one domain rooted within their homologs in a second domain (Canback *et al.*, 2002). The class (c) contains homologs found in at least two domains. but the proteins of one domain are not rooted within other domain(s); rather, the homologs may descend from the common ancestor (Figure 1.5). Most eukaryote proteins shared by prokaryotes are distant, not close, relatives. Thus, proteins shared between domains probably descended from the common ancestor; none seems to result from interdomain lateral gene transfer (Woese *et al.*, 1990).

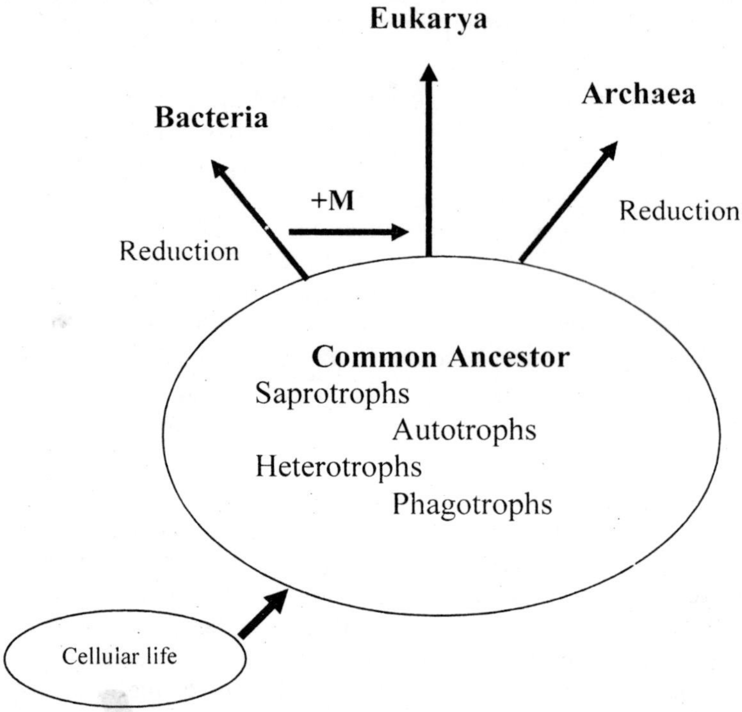

Figure 1.5: The Unrooted Tree of Life. (+M: endosymbiosis of mitochondrial ancestor) (after Kurland *et al.*, 2006).

The genomes of mitochondria have clearly evolved from ancestral α-proteobacteria (Gray *et al.*, 1999). But proteomics and comparative genomics identify only a few proteins in yeast and human mitochondria to have descended from the ancestral bacterium (Yang *et al.*, 2005). Hundreds of genes have been transferred from the ancestral bacterium to the nuclear genome, but most proteins from the original endosymbiont were lost. According to Kurland *et al.* (2006), the yeast and human mitochondrial proteomes emerged largely as products of the eukaryotic nuclear genome (85 per cent) and only to a lesser degree (15 per cent) as direct descendants of endosymbionts.

There are substantial numbers of ESPs in CSSs. For the proteome of the reduced anaerobic parasite *Giardia lamblia*; searches of 2136 proteins found in each of *Saccharomyces cerevisiae*, *Drosophila melanogaster*, *Caenorhabditis elegans*, and *Arabidopsis thaliana* yielded 347 ESPs for *G. lamblia*. This number was reduced to about 300 by rigorous screening, with ESPs distributed between nuclear and cytoplasmic compartments (see Kurland *et al.*, 2006). The ubiquity of the ESPs and the absence of archaeal descendants cannot be explained by a prokaryote genome fusion model. Most probably the host for the endosymbiont/mitochondrial lineage was an ancestral eukaryote (Kurland *et al.*, 2006).

The spliceosome is a unique molecular machine. Its function is to eliminate introns from eukaryote mRNAs (Jurica and Moore, 2003). About half of the 78 spliceosomal proteins likely to be present in the ancestral spliceosome are ESPs, whereas the other half containing the Sm/LSm proteins have homologs in bacteria and archaea (Collins and Penny, 2005). These distributions of ESPs and probable descendants of the common ancestor suggest that many components of modern spliceosomes were present in the common ancestor (Poole *et al.*, 1999).

A typical feature of eukaryotic cells is that the subdivision into subcellular compartments (CSSs) with characteristic proteomes restricts proteins to volumes much smaller than the whole cell. Concentrations of macromolecules in cells are quite high, usually about 20 to 30 per cent of weight or volume. Such densities generate "molecular crowding" that is conducive to macromolecular associations, large complexes, and networks of proteins that support biological functions (Ellis, 2001).

High densities improve the association kinetics of small molecules with proteins because the excluded volumes of the proteins decrease the effective volume through which small molecules can diffuse. The net outcome is that the high macromolecular densities within CSSs enhance the kinetic efficiencies of proteins. The same principles are applicable to the smaller prokaryotic cells, but the effects become marked in larger cells. Subdividing high densities of proteins into fairly distinct compartments containing functionally interactive macromolecules was probably an early feature of the eukaryote lineage. Cell fusion is not needed to account for the large number of eukaryote cell compartments (Kurland *et al.*, 2006).

There is no doubt that recent advances in genomics and proteomics have materially enhanced our awareness of the uniqueness of eukaryote cells. This, together with much better understanding of molecular crowding, as well as the dynamic and reductive nature of genome evolution, provide a fresh view of the origin of eukaryote cells. The eukaryotic CSSs define a unique cell type that cannot be deconstructed into features inherited directly from archaea and bacteria. No more than about 15 per cent of α-proteobacterial descendants in this proteome confirms the phylogenetic distinction between direct descent from genes transferred to the host from the bacterial endosymbiont, as opposed to descent from a hypothetical common ancestor (Kurland *et al.*, 2006).

ESPs serve as valuable markers of the novel evolutionary trajectory of modern eukaryotes. In contrast, most proteins are found in more than one domain, and most of these could have been derived from the common ancestor. The relative abundance of signature proteins among eukaryotes indicates that their genomes typically have a greater coding capacity than those of prokaryotes. What is still not known is which ESPs have been lost from prokaryotes and which were acquired by eukaryotes during their evolution.

According to Kurland *et al.*, a primitive eukaryote took up the endosymbiont/ mitochondrion by phagocytosis (Cavalier-Smith, 1987). A unicellular predator with a larger, more complex cell structure than that of present-day prokaryotes may have been the host of the ancestral endosymbiont. This scenario is not contradicted by new data derived from comparative genomics and proteomics and may be a good starting point for future work.

Eukaryote Evolution

It is generally thought that eukaryotes evolved via a merger of cells from the other two domains of archaea and bacteria. The mitochondrion, powerhouse of the eukaryote cell, probably evolved from an engulfed bacterium. What is not known is whether it was an archaeon that engulfed a bacterium? Or, did a bacterium, bacterial consortium, or RNA cell engulf first an archaeon (which became the nucleus) and then the mitochondrial ancestor? It is also possible that nuclei emerged in a virus-infected archaeon, which then engulfed mitochondria. We know little about which, if any of these hypotheses, is right (Poole and Penny, 2007).

In the mid-1990s, a rather pedestrian view of eukaryotic origins the 'archezoa hypothesis' was popular. According to this, a nucleated proteukaryote engulfed the mitochondrial ancestor. This view was supported by 'archezoa', anaerobic eukaryotes lacking mitochondria. Archezoa apparently populated the oldest branches of the eukaryote tree, suggesting that eukaryotes started diversifying before mitochondria entered the scene.

The archezoa hypothesis is actually composed of two independent hypotheses: (a) that a proteukaryote host (PEH) engulfed the mitochondrial ancestor, and (b) that modern archezoa are 'missing links' that never had mitochondria. This second hypothesis had been unanimously rejected–every archezoan examined bears vestigial mitochondria, or genes inherited from mitochondria. It is clear that all modern eukaryotes evolved from a mitochondrion-bearing ancestor.

Unfortunately, as per (b), the baby was thrown out with the bath water. The hypothesis (a) was also rejected, and since eukaryotes and archaea share several similar genes, the deposed PEH was replaced with archaea. The result was that incorporation of the mitochondrion–not the origin of the nucleus–was hailed as the crucial event in eukaryotic origins. This generated much speculation and numerous new hypotheses. None is, however, supported by observation: no archaea reside within bacteria, no bacteria reside within archaea, viruses have virtually no similarities to the nucleus, and no RNA cells exist (Poole and Penny, 2007).

The question is: if archezoa are not missing links, has evidence for the PEH theory also disappeared? It seems, no. First, while phagocytosis (one cell engulfing another) is widespread among eukaryotes, it is unknown in bacteria and archaea. No doubt eukaryotes engulf bacteria as food, and some engulfed bacteria can evade digestion. There are conditional endosymbioses in which one cell lives within another but escapes under adverse conditions, and stable endosymbioses, where host and guest become locked up into the same evolutionary trajectory. Predatory eukaryotes can even appropriate chloroplasts from their hosts. Second, every other organelle that evolved from a once free-living cell was initially engulfed by a eukaryote. Indeed, this is how the chloroplast evolved.

The origin of mitochondria under the PEH theory is not some bizarre happening that calls for special explanation; mitochondria just descended from an engulfed cell co-opted as an organelle. This was clearly an evolutionary success–bona fide archezoa no longer exist. In contrast, archaea do not engulf bacteria (nor *vice versa*). So the alternatives to the PEH hypothesis imply that bacterial or archaeal ancestors 'invented' engulfment, only to lose it again without trace–but this is an assertion that cannot be disproved. In 2006, Norman Pace rightly suggested (*Nature* 441: 289, 2006) that eukaryotes most probably did not evolve from archaea; they only share a common ancestor. This fits the PEH theory–the host was a direct ancestor of modern eukaryotes. Should this host have been an archaeon, eukaryotes would appear on the tree of life as a subdivision of archaea–they would be phylogenetically partly archaeal, and partly bacterial. The tree indeed depicts that they are part eukaryotic, part bacterial, and sister to archaea (Poole and Penny, 2007).

Microbial Diversity

Exploring microbial diversity is no different from exploring outer space, with soil representing a "final frontier" that harbors a largely unknown microbial universe. One ton of soil contains more than 10^{16} prokaryotes compared to a mere 10^{11} stars in the galaxy. Microbiologists are using mathematical inference to estimate the number of prokaryote taxa in soil. Gans *et al.* (2005) reported that the inferred diversity is staggering–higher than previously thought by almost three orders of magnitude.

The total number of bacteria in the biosphere is around 10^{31}. One important goal of microbiology is to determine how such high diversity is created and maintained. Microbial diversity can be understood in terms of the extravagance of small molecules that microbes produce–such extravagance reflects a diversity in microbial metabolism, which may be considered to have evolved as a result of the richness of the organisms' environments. But some recent publications present findings that do not support this view.

The first indication of the enormous diversity of the microbial world emerged from the use of ribosomal-RNA typing in the 1980s. Within a few years, this nurtured the new field of metagenomics which, in recent years has provided catalogues of microbial communities from various terrestrial and marine environments.

Comparison of such catalogues made by Neufeld and Mohn (2005) revealed that the apparently bare Arctic tundra exceeds fertile forest soils in phylogenetic content. Another study made by Lozupone and Knight (2007) compared information from over 100 different environments and found, rather surprisingly, that the microbial content of soils is typically less diverse than that of sediments and hypersaline environments.

Microbial communities are central to health, sustainable cities, agriculture, and most of the earth's geochemical cycles. Prokaryote communities serve as reservoirs for the discovery of new drugs and metabolic processes. Like any other reservoir, its size is crucial, but measuring the reservoir of prokaryotic diversity is a challenging task but one that can be accomplished by ignoring the organisms themselves and focusing instead on their DNA. If DNA from a single organism is purified and heated, the strands of the double helix separate or "melt". If the DNA is then slowly cooled, the strand reassociate, and the rate at which this happens is affected by the size and complexity of the DNA. Big and complex DNA reassociates slowly. This fact is exploited to estimate the size and complexity of genomes from individual organisms. Torsvik *et al.* (1990) reasoned that pooled genomic DNA from a microbial community might reanneal like the DNA from a large genome. DNA extracted from soil reassociated so slowly that it resembled a genome that was 7000 times as large as the genome of a single bacterium. It follows that there could have been at least 7000 different prokaryote taxa in the sample of soil that they analyzed. This was indeed an amazing number. Fortunately, however, there is another method of estimating prokaryotic diversity in the environment. A biological community has a characteristic abundance distribution of its various species. In principle, if we know the shape of the taxa abundance distribution curve, we know the diversity. But typically, for large organisms, species abundance distributions have been estimated by assessing the abundance of almost all of the species in a sample, which means that one must already know the number of species. In the absence of this knowledge, one still can assume that a particular species distribution pertains, and then make an estimate (Curtis *et al.*, 2002). Alternatively, one may fit a curve to the data one has and so make some estimate. The latter approach requires enough data to make a sensible decision about the underlying species distribution pattern. Most microbiologists tend to estimate diversity by looking at a gene that occurs in all cellular life forms. They infer diversity from the number of different variants that can be cloned from a sample of environmental DNA. Unfortunately, the number of clones analyzed is typically small (tens to hundreds) as compared to the billions of individual microbes being analyzed (Curtis and Sloan, 2005).

Since the pattern of DNA reassociation kinetics reflects the underlying distribution of similar sequences, it probably reflects genomic diversity. Also enough data exists in published DNA reassociation curves for bacterial communities to allow discrimination between different possible species abundance curves. By appropriate mathematical treatment of data, Gans *et al.* generated abundance curves, the most plausible of which suggests that there could be 10^7 distinct prokaryote taxa in 10 grams of pristine (free of chemical contaminant) soil. Furthermore, it is the rare organisms that comprise most of this diversity. Most of these rare organisms could be wiped out by heavy metal pollution of the soil (Curtis and Sloan, 2005).

Power law distributions describe exponentially increasing numbers of species at exponentially decreasing abundances. On the other hand, lognormal distributions suggest that at lower abundances, the number of rare species tends to decrease. Not surprisingly, therefore, wiping out rare species, as in the case with heavy metal soil pollution, reduces the ability to distinguish between the two different situations.

Exponentially increasing numbers of taxa at exponentially decreasing abundances means that, in random samples of diversity, a few abundant taxa tend to turn up repeatedly. It needs to be resolved whether the recurrence of abundant taxa is an artifact of sampling or a real reflection of low diversity.

Biodiversity of Bacteria–Taxa-area Relationship

Whereas a positive power-law relationship between the number of species in an area and the size of that area is generally applicable to plant and animal communities, no such species-area relationship has been reported for bacteria (see Rosenzweig, 1995; Horner-Devine *et al.*, 2004). In 1922, Gleason had generalized the empirical relationship between the number of species and area as a power-law, $S = cA^z$, where S is the number of species, A is the area sampled and c is the intercept in log-log space. The species-area exponent, z, denotes the rate of change of the slope with increasing area, that is, the rate of turnover of species across space. Variation in the values for c, and z indicates that different processes underlie the species-area relationship at different spatial scales (Losos, 2000).

Bacteria are certainly some of the most abundant and diverse groups of organisms on earth (Whitman *et al.*, 1998). They are involved in several important ecosystem processes, *e.g.*, trace gas emissions, decomposition and nitrogen cycling. Horner-Devine *et al.* (2004) described a taxa-area relationship for bacteria over a scale of centimeters to hundreds of metres in salt marsh sediments. Bacterial communities located close together were more similar in composition than those located farther apart; the reduction of community similarity with distance was used to show that bacteria do show a taxa-area relationship, driven primarily by environmental heterogeneity rather than geographic distance or plant composition.

It is well known that many microbes cannot be cultured with available laboratory techniques. Bacterial taxa have therefore to be identified from the sequences of indicator genes extracted from environmental samples (Stackebrandt and Goebel, 1994). The bacterial community composition of salt marsh samples was determined by Horner-Devine *et al.*, by amplifying via the polymerase chain reaction (PCR), cloning and sequencing a region of 16S ribosomal DNA (rDNA)–the most commonly used indicator gene for bacterial biodiversity. Because the bacterial diversity of salt marshes is often very high, PCR primers targeting a subset of the bacterial biota (the β-proteobacteria and relatives) were employed. In view of the absence of any single acceptable definition of 'species', by using this sequencing approach, taxa are usually defined as operational taxonomic units (OTUs) based on sequence similarity groupings (see Rossello-Mora and Amann, 2001). Horner-Devine *et al.* actually used the three most commonly used groupings–90 per cent, 97 per cent and 99 per cent sequence similarity–to define OTUs. Use of multiple OTU definitions is generally analogous to comparing different taxonomic resolutions (for example, comparing genus, species and sub-species) (see Stackebrandt and Goebel, 1994).

Sequencing a total of 945 partial 16S rDNA sequences revealed their grouping into 88 OTUs, using a taxon resolution of 97 per cent sequence similarity. Approximately 34 per cent of the OTUs were singletons while 15 OTUs were represented by more than ten sequences. Five-hundred and twenty-three sequences were members of the β-proteobacteria; the remainder were members of the γ-proteobacteria and the δ-proteobacteria (Horner-Devine *et al.*, 2004). The work of Horner-Devine *et al.* (2004, 2004a) has demonstrated that prokaryotes also exhibit taxa-area relationships; this means that the taxa-area relationship is a universal law.

Novel Marine Microbes

Two recently identified groups of marine bacteria that account for about 20 per cent of bacterial life on the ocean surface convert light to energy (Pennisi, 2000). These two groups of bacteria harness light to transport electrons and power cellular processes. They help in bringing energy into the food chain. It also explains the long standing enigma of how so many bacteria can survive in the open ocean, where there is so little for them to eat.

The newly discovered bacteria have light-harnessing abilities previously reported in a fungus and in archaea that thrive in the most hostile salty environments, such as salt ponds. One of the two bacterial groups contains a protein, bacteriorhodopsin, that enables them to live by harnessing light to generate ATP. The second group uses a type of bacterial chlorophyll that, until now, no one had reported from bacteria in the open ocean.

Drs. Edward DeLong and Paul Falkowski discovered the two new groups by using novel techniques for tracking down new species. Typically biologists have assessed microbial diversity by growing environmental samples in the laboratory, but this technique misses many organisms that fail to survive long enough to be counted. To address this problem, DeLong used a molecular biology tool whereas Falkowski used a biophysical technique.

DeLong's group collected DNA directly from seawater samples, without first isolating the individual organisms. In one 130-kilobase DNA fragment they fished out this way, they found a gene that encoded the small ribosomal subunit RNA belonging to a group of marine bacteria called SAR86. No one could ever grow the bacteria in the lab, so it was known only by this one ribosomal gene.

But DeLong found that very gene and more in the fragment. When the rest of the sequence–which included some 120 genes–was compared to other genes available in the public archives, most looked unexceptional for this type of bacterium. But one stood out: a gene for bacteriorhodopsin, a rhodopsin protein that, contrary to what the name implies, had never before been found in bacteria. It revealed a mechanism of photosynthesis that was totally unknown in bacteria.

DeLong termed the new protein proteorhodopsin and explored its function by introducing it into *Escherichia coli*. The modified bacteria not only made the protein, but also reacted to light by moving protons out of the cell and into the surrounding medium. That's typical of bacteriorhodopsin, whose proton-pumping activity helps to create a gradient whereby energy is generated by protons flowing back into the cell.

Unlike DeLong, Falkowski's team did not go after genes. They were looking for new types of photosynthesizing organisms that might use dim infrared light emitted from deep-sea vents. It was thought that such microbes might exist in those extreme environments, but there was no proof. So the researchers fabricated a fluorometer designed to detect any changes in the electromagnetic radiation that would be caused by microbes using light energy to move electrons.

While they failed to find any organisms by the deep-sea vents, when they scanned surface waters they surprisingly picked up many positive signals in samples taken along a 1000-km line in the eastern tropical Pacific and later off the coast of New Jersey. As only bacterial–as opposed to plant–chlorophyll makes use of light at the wavelengths detected by the fluorometer, the team had found a new habitat for phototrophic bacteria–the sea surface. On the basis of the strength of the signals, it was estimated that these aerobic phototrophic bacteria represent about 1 per cent of the surface phytoplankton. Although these bacteria and DeLong's bacteria use different mechanisms to harness light, they both use that light to generate ATP. This gives them an advantage in environments where food is sparse, and they in turn serve as nourishment for organisms higher up the food chain.

Mammalian Microbiota

Mammals are metagenomic in the sense that they are made of not only their own gene complements but also those of all of their associated microbes. Thus, our "metagenome" is a composite of *Homo sapiens* genes and genes present in the genomes of the trillions of microbes that colonize our adult bodies (Turnbaugh *et al.*, 2007). The vast majority of these microbes live in our distal guts. "Our" microbial genomes (microbiomes) encode metabolic functions that we have not had to evolve wholly on our own, including the ability to extract energy and nutrients from our diet.

To understand the coevolution of the mammals and their indigenous microbial communities, a network-based analysis of bacterial 16S ribosomal RNA gene sequences was conducted by Ley *et al.* (2008) from the fecal microbiota of humans and about 60 other mammalian species living in two zoos and in the wild. The results indicate that host diet and phylogeny both influence bacterial diversity, which increases from carnivory to omnivory to herbivory; that bacterial communities codiversified with their hosts; and that the gut microbiota of humans living a modern life-style is typical of omnivorous primates (Ley *et al.*, 2008).

The acquisition of a new diet is a fundamental driver for the evolution of new species. Coevolution refers to the reciprocal adaptations occurring between interacting species. Most modern species of mammals arose during the Quaternary (1.8 Ma to the present), when there was extensive expansion of C4 grasslands dominated by plants that perform photosynthesis through the Hatch-Slack cycle rather than the Calvin cycle typical of C3 plants. The expansion occurred in response to a fall in atmospheric CO_2 levels and/or climate changes. The switch to a C4 plant-dominated diet led to selection for herbivores with longer gut retention times necessary for the digestion of lower-quality forage (Stevens and Hume, 2004). Indeed, the community of microbes in the gut constitutes a potentially critical but unexplored component of diet-driven speciation (Ley *et al.*, 2008). To infer past evolutionary processes from comparative analyses of extant mammalian gut microbial communities, Ley *et al.* analyzed the fecal microbial communities of over 100 individual mammals representing 60 species from 13 taxonomic orders, including 17 nonhuman primates. The majority of the nonhuman primate species studied were omnivores (12 of 17), but some were leaf-eating species, including a langur. The herbivorous giant panda and red panda were included from the Carnivora. Most animals were housed at two American zoos. Others were examined in the wild or domesticated . A data set of >20,000 16S rRNA gene sequences was generated. For comparison of the human, primate, and nonprimate mammalian gut microbiotas, the samples also included published fecal bacterial 16S rRNA sequences (>3000) from wild African gorilla, Holstein cattle, Wistar rats, and healthy human males and females ranging in age from 27 to 94, living on three continents and including a strict vegetarian.

Network-based analyses were used to map gut microbial community composition and structure onto mammalian phylogeny and diet, thereby complementing phylogeny-based microbial community comparisons.

The ensemble of sequences studied by Ley *et al.* provides a general view of the mammal gut microbiota: They detected members of 17 phyla (divisions) of Bacteria. The majority of sequences belong to the Firmicutes (65.7 per cent of 19,548 classified sequences) and to the Bacteroidetes (16.3 per cent). Some other phyla represented were the Proteobacteria (8.8 per cent of all sequences collected; 85 per cent in the Gamma subdivision), Actinobacteria (4.7 per cent) and Verrucomicrobia (2.2 per cent).

The network-based analyses revealed that overall, the fecal microbial communities of same-species (conspecific) hosts were more similar to each other than to those of different host species.

Because heavy isotopes of carbon and nitrogen are known to bioaccumulate in the food chain (Koch *et al.*, 1994), to obtain a more objective marker of diet, Ley *et al.* measured stable isotope ratios of carbon and nitrogen, ^{13}C and δ^{15}N, in the feces. Heavy isotopes were found to be enriched in the order herbivore <omnivore < carnivore.

Detailed critical analyses of the results supported an association between microbial community membership and diet, and an independently validated dietary clustering observed in the network diagrams that is free of bias in assigning hosts to one of the three diet categories (Ley *et al.*, 2008).

Underlying the correlation between bacterial community composition and diet is the partitioning of bacterial phyla among hosts according to diet. Herbivore microbiotas contained the largest number of phyla (14), carnivores contained the fewest (6), and omnivores were intermediate (12). Herbivores had the highest genus-level richness, followed by omnivores and carnivores.

Gut (Intestinal) Bacteria

Considerable and valuable healthy information lies hidden in the human body and its intestinal flora. The microbes found in the intestines produce several chemicals that cause disease. Treating disease means treating the bacteria which produce these chemicals.

In 1981, the young biochemist Jeremy Nicholson was using nuclear magnetic resonance spectroscopy (NMR) (NMR can identify chemicals based on the magnetic properties of atomic nuclei) to study how red blood cells absorb cadmium, a carcinogen. Prompted by the thought that he would achieve the best results if he could mimic the cells' natural environment, he added a few drops of blood to the cells and ran the test. A huge variety of strange signals and spectra appeared. A sample of blood or urine contains thousands of metabolites–signatures (biomarkers) of all the chemical reactions occurring in the body at a given time. Identifying those chemical signatures and their significance can enable not only to better understand different diseases–based on chemical reactions that had gone awry–but also to identify early warning signs and potential interventions. Nicholson's work has led to the creation of two new fields: *metabolomics*, which studies the metabolites that cellular processes leave behind, and *metabonomics*, which characterizes the metabolic changes a biological system experiences in response to stressors.

The metabolome refers to the collection of chemicals produced by human metabolism. Whereas the genome provides detailed information about a person's genetic makeup, the metabolome reveals how genes interact with the environment and reveals the state of a person's physical health. The metabolome adds value to genome information and puts it in perspective.

But deciphering it is a Herculean task. It requires the analysis of blood, urine, breath and feces within large populations. For instance, to find potential chemical biomarkers for high blood pressure, Nicholson analyzed the urine of 4,630 individuals from the U.K., the U.S. and Asia and compared the urinary metabolites with blood pressure data to determine if any consistent metabolic differences exist between individuals with hypertension and those without it (see Wenner, 2008).

Instead of making hypotheses and then devising experiments to test them, Nicholson performs experiments first and tries to decipher his results later. He sifts through the range of chemicals produced by the genes people have, the food they eat, the drugs they take, the diseases they suffer from and the intestinal bacteria they harbour.

Gut bacteria are Nicholson's chief focus. They influence how our bodies break down food and drugs and indicate why food affects people differently: some people cannot derive benefit from one of soy's components because they lack the gut microbes necessary to process it. Deciphering which metabolites come directly from our gut microbes can be difficult, but in some cases it is easy–they are the chemicals that are not produced by cells or ingested in food (Wenner, 2008).

Gut microbes certainly play a crucial role in human health and disease. They help us absorb nutrients and fight off viruses and "bad" bacteria. Disrupting intestinal colonies, *e.g.*, with antibiotics, often leads to digestive sickness. In fact, virtually every kind of disease has some gut bacterial connection.

The most well-known disease-causing gut organism is *Helicobacter pylori*, which can trigger peptic ulcer. In recent years, obesity has been linked to the relative abundance of two dominant intestinal bacterial phyla; and dysfunctional intestinal bacteria are associated with nonalcoholic fatty liver disease, inflammatory bowel disease and some types of cancer. In fact, it is likely that the organisms have a role in some neurological disorders. Messing around with the gut microbes may mean messing around with brain chemistry. Work is underway to match metabolites with specific bacteria–there may be about 1000 species and more than 10 trillion bacterial cells inside us at any given time!

Microbiologists have extracted gut bacteria from fecal samples but it has been almost impossible to culture the samples afterward because they survive only in highly acidic, oxygen-free environments. New DNA-sequencing technologies have enabled scientists to identify gut bacteria easily. The National Institutes of Health has launched its Human Microbiome Project in Dec. 2007 with the goal of fully characterizing the human gut flora.

Correlating metabolites with health could facilitate designing of urine sticks similar to those used in pregnancy tests to regularly check the fitness of our gut flora. In fact, some companies already sell food products that help keeping these bacterial populations in check. This is done with live beneficial bacteria (*probiotics*) or compounds that help these grow (*prebiotics*), or combinations of the two (*synbiotics*).

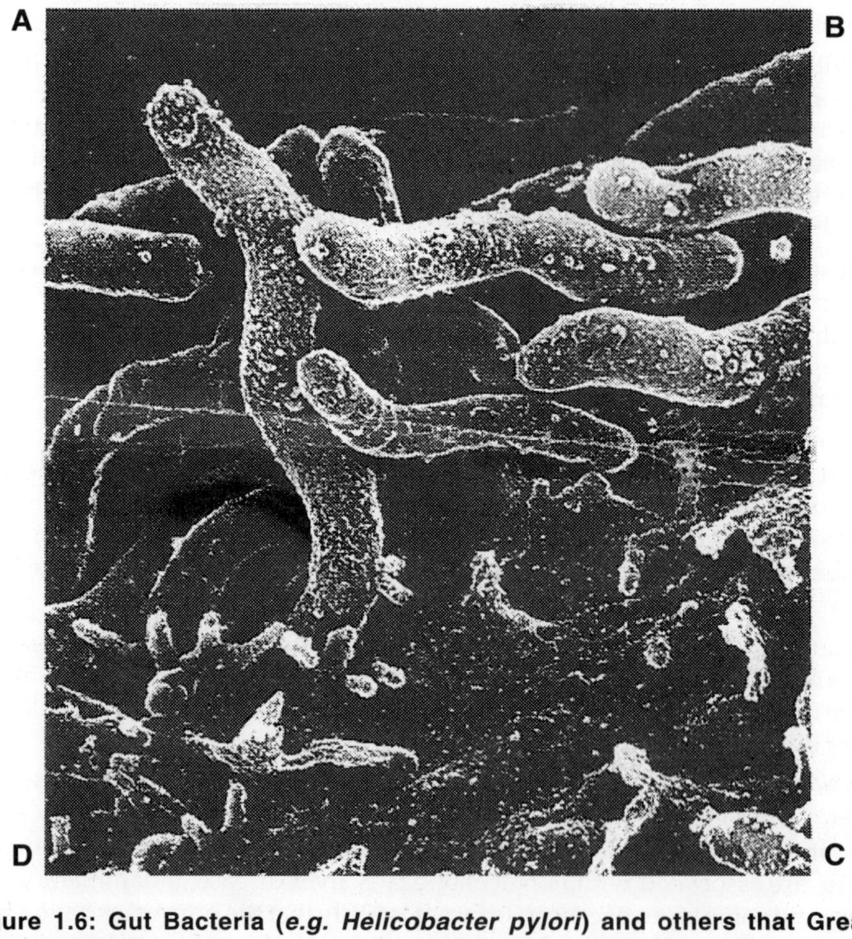

Figure 1.6: Gut Bacteria (*e.g. Helicobacter pylori*) and others that Greatly Influence Health.

These medications are "functional foods" and are rarely, if ever, tested in clinical trials. One notable exception is VSL #3, a combination of eight bacterial species sold in packet form by the Gaithersburg, Md. based VSL Pharmaceuticals. VSL#3 has effectively treated ulcerative colitis and irritable bowel syndrome (Wenner, 2008).

According to pharma giant Pfizer, the human genome offers only about 3,000 potential drug targets, because just a subset of genes produces proteins that can be bound and modified by druglike molecules. But there may be 100 times as many genes in the microbial pool, according to Nicholson.

Genes provide only limited information about a person's risk for disease. Nicholson's dream is when physicians will provide personalized health care on the metabolome. Simple blood or urine tests would detect the risk of cancer or heart disease early enough to begin preventive therapy; drugs would be tailored to each person's metabolic profile–in many cases, they would not target our organs but our bacteria.

The Human Microbiome

Trillions of bacteria, fungi and viruses colonize the human body. The human gut may contain at least a kilogram of bacteria alone. Several ambitious projects have been launched to characterize the human microbiota in its entirely. Not only microbiologists but also food and pharmaceutical industries are greatly excited by this opportunity. When it comes to useful applications to human health, the microbiome might offer several clear advantages over the more famous genome–human genes are extremely difficult and risky to tamper with. But, in principle at least, the microbiome should be easier to change by the selective addition or removal of bacterial species, or by altering their genetic composition. Some antibiotics and 'probiotic' foods calm inflammatory bowel disease in some cases. Intestinal inflammation can be reduced by a single molecule produced by a gut bacterium (see *Nature* pp. 602 and 620, 29 May 2008). It is also emerging that certain foods, or the bacteria contained in them, can change gut microbiota in ways that improve general health.

All people contain exotic environments that support microbial communities as diverse as any rainforest. There is a unique ecological perspective on food itself, and the effects that different foodstuffs, such as processed versus unprocessed ones, have on these environments.

How important to humans are the microbes that eke out a living in our intestines? Jeffrey Gordon *et al.* (St. Louis, Missouri) compared particular genetic sequences from gut microbes found in the faeces of 106 mammals, some wild and others from American zoos.

Among the 60 species represented in this sample, carnivores had a less diverse internal flora than omnivores, and omnivores had less diverse than herbivores. Modern humans support gut microbes that are broadly similar to those of other omnivores. This is surprising in view of the importance of agriculture and cookery in human ecological history (see *Science* doi: 10.1126/science.1155725, 2008).

Human Microbiome Project (HMP)

The human body is teeming with microbes, with about 10 trillion microbial cells living in the gut. These microbial cells exceed the number of human cells by 10 to 1. Between them, they harbour millions of genes, compared with no more than about 25,000 estimated in the human genome. In fact, some biologists regard a human as a 'super-organism'–a human being represents a community that adds up to more than the sum of its parts.

The term microbiome means collection of microorganisms living in and on the human body. Huge sums of money in the USA and the European Commission have been allotted to identify and characterize the human microbiome. Given the multifaceted nature of the microbiome, it is being studied by a large and varied community. Whereas the NIH's five year HMP will identify which bacteria are lodged where in the body, followed by compiling a reference set of their genetic sequences, the European Commission's project named Metagenomics of the Human Intestinal Tract (MetaHIT), will focus on microbial inhabitants of the gut, and how they can lead to obesity and inflammatory bowel disease.

Today it is possible to process millions of base pairs in only a few hours. This abolishes the need to grow bacteria; the human microbiome can be explored by studying genes *en masse*, without studying the organisms themselves. Sequencing technology can facilitate this task easily.

Genes, age, diet, lifestyle and geography all affect which bacteria live in a person's body. But what is not very clear is: how many bacterial species colonize the entire body, not just the gut or mouth? Another issue is about which ones everyone shares, and what is the size of the core microbiome?

In the first year of the project, the researchers plan to collect samples of faeces plus swabs from various other organs from 250 volunteers. They will sequence short, variable stretches of DNA that code for components of ribosomes so as to roughly identify which bacteria are present in each person and how many the volunteers have in common. When the diversity is known, the researchers will use shotgun sequencing to analyze many short fragments of DNA from all over the microbes' genome and show which genes are present (Mullard, 2008).

Some have misgivings about the project. One challenge could be that the core microbiome might be incredibly small. Even if two people may each have 1,000 types of bacteria in their guts, they might have only 10 species in common, for instance. Also the commonalities might not lie in the genes, but rather in bacteria that have the same metabolic and physiological role. Another reservation is that the major focus of the Project is on sequencing, not on actual application to disease.

But MetaHIT differs from HMP–it will focus chiefly on the role of the gut microbiota in obesity and inflammatory bowel disease. Unlike the HMP which is comparing people's microbiota on a species level, MetaHIT detects differences in microbial genes and the proteins they express without necessarily worrying about which species they came from.

In fact, Ley *et al.* (2006) reported that obese and lean individuals have markedly different profiles of bacteria in their guts. When the obese volunteers went on a one-year diet and lost about a quarter of their body weight, their bacterial profiles became more like those of the lean people. Conceivably, part of the tendency to gain weight could lie in 'obesity-causing' bacteria in the gut which release more calories from food than those found in lean people (Turnbaugh *et al.*, 2006). To explore this, in Denmark Oluf Pedersen is collecting faecal samples from 120 obese volunteers and 60 controls to determine specific microbial genes that might be involved in obesity. In Spain, the microbiotas of patients with inflammatory bowel disease are being compared with those of genetically matched controls so as to explore the effect of drugs.

One major challenge is: how to study such a vast and unknown community? Both groups will use shotgun sequencing, which generates scraps of sequence from the many different genes and species. For this approach, a reference sequence will be needed against which to compare and identify these scraps–currently, there are not enough known bacterial gene sequences to match the fragments against.

Both the HMP and MetaHIT intend to sequence the complete genomes of hundreds of bacterial species, initially using shotgun sequencing of a few select species that can already be grown, and piece together their whole genomes from overlapping fragments. The HMP hopes to provide 600 reference genomes and MetaHIT will provide another 100. With a broad enough reference database, the genetic capabilities of some of those recalcitrant, unculturable species may be predicted on the basis of similarities with known genes.

The Creativity of Symbiosis

Symbiotic associations such as parasitism may not merely be one influence in evolution. Instead they could be among the most important creative forces in the process. According to Lynn Margulis, through symbiosis, organisms pool their genetic resources to accomplish more than what either partner could achieve separately. Complex cells may have evolved through a series of symbioses between simpler organisms.

Margulis maintains that by relying on mathematical models based on competition between organisms, most biologists ignore the effects of other biological influences. They claim that natural selection is responsible for the creative novelty one sees in evolution. While natural selection may select the symbionts over the individual partners or an animal with one genotype over another, it is an intrinsic part of the evolutionary process. But natural selection is the editor, not the author.

The best illustration of this idea is the serial endosymbiotic origin of eukaryotic cells: that eukaryotic cells, those with internal organelles, descended from symbiotic unions of simpler, unnucleated prokaryotic cells.

As Margulis predicted, organelles such as mitochondria and chloroplasts do contain genes unrelated to those in the nucleus of eukaryotic cells but similar to those in some bacteria. This finding has much evolutionary significance–eukaryotic cells did not have to reinvent respiration and photosynthesis through genetic trial and error. The first eukaryotes simply acquired the skills from unrelated bacteria by incorporating those cells as endosymbionts (internal symbiotic partners).

A series of endosymbiotic events has added to the complexity of many organisms. Such endosymbioses are not restricted to prokaryotes. Susan E. Douglas and others at the Institute of Marine Biosciences in Halifax have shown that two eukaryotes had joined symbiotically to create a species of freshwater alga called a cryptomonad. Cryptomonads harbour an unusual membranous structure known as a nucleosome, which contains both DNA and RNA. It was suggested that the nucleosome is the remnant of a red alga that parasitized a eukaryotic host long ago. Douglas *et al.* showed that the DNA in the nucleosome is more similar to that of red algae than to that in the cryptomonad's nucleus. Apparently, the ancestral cryptomonad became photosynthetic by incorporating red algae containing chloroplasts.

Symbiosis is clearly integral to most organisms, not just the early eukaryotes and simpler organisms such as lichens. Some 90 per cent of all land plants are mycorrhizal– that is, fungi associated with their roots are essential to the plants' ability to draw nutrients from the soil. Most herbivorous animals and insects depend on their gut microorganisms to digest the cellulose they eat.

Peter W. Price of Northern Arizona University proposed an evolutionary scheme about how herbivores could arise from symbioses between animals and microscopic plant parasites. In this system, the parasite would have already evolved enzymes for digesting the materials of its host plant.

If an animal host were to establish a symbiosis with the parasite, it could thereafter share in the nutrients contained in the vegetable matter. The herbivorous partnership could then spread out to wherever the plant grew. Its success in finding new food resources would favour the natural selection of more animals capable of forming similar symbioses.

It has long been known that communities of soil microorganisms perform many important functions such as controlling climate, enhancing crop production, nitrogen-processing in the soil, decomposing organic matter and remediation of environmental contamination (see Dechesne *et al.*, 2008; Leibold *et al.*, 2004). Microbial communities in the mouth and the gut have attracted strong biomedical interest. *The interactions within some symbiotic communities have been thoroughly well characterized but in general, not much is known about how this balance is achieved within many microbial communities.* Understanding precisely the function of these communities is quite difficult: artificial microbial communities grown in the laboratory become unstable because of competition among species. Kim *et al.* (2008) developed a community of three different species of wild-type soil bacteria with syntrophic interactions using a microfluidic device to control spatial structure and chemical communication. Critical study of this community suggested that defined microscale spatial structure is both necessary and sufficient for the stable coexistence of interacting bacterial species in the synthetic community. A suitable spatial structure balances the competition and positive interactions within the community, even when the rates of production and consumption of nutrients by species are mismatched. These findings provide evidence for a class of communities that require microscale spatial structure for stability, and

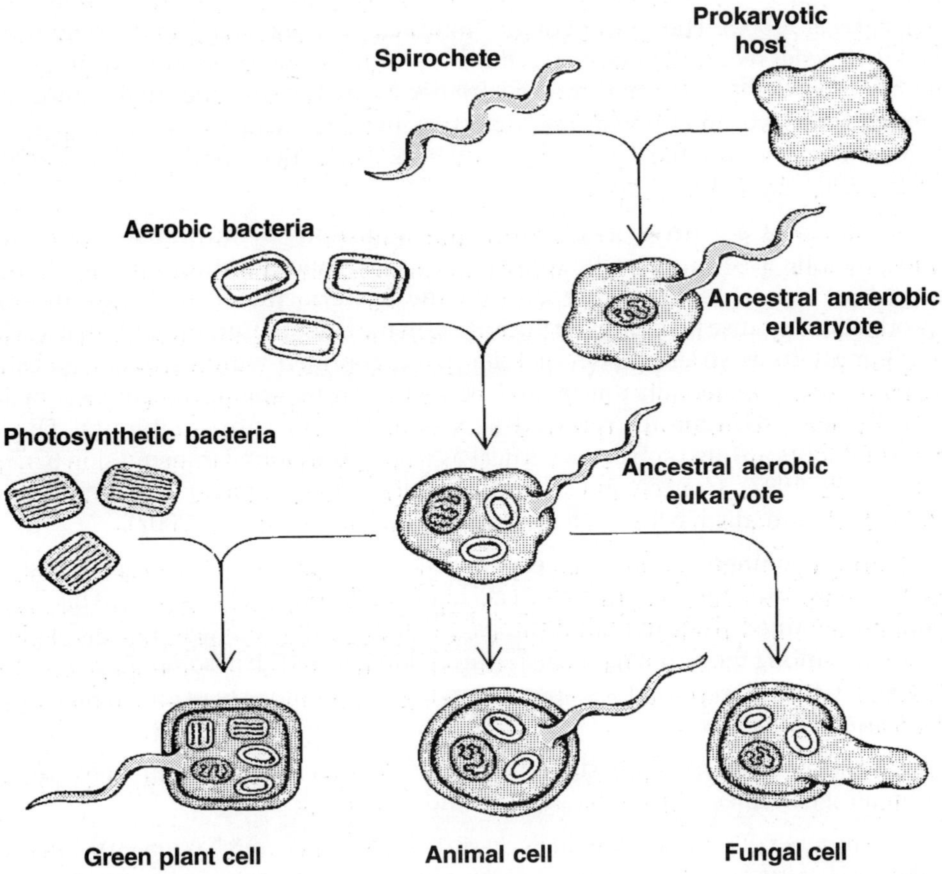

Figure 1.7 Probable evolution of complex cells through a series of symbioses between simpler organisms (*Sci. Amer.* Jan. 1992, p. 131).

these results predict that controlling spatial structure may facilitate harnessing the function of natural and synthetic multispecies communities in the laboratory (Kim *et al.*, 2008).

Minimal Genome Needed for Symbiosis

Many pathogenic and mutualistic intracellular bacterial species have much smaller genomes than their free living relatives (Anderson and Kurland, 1998). The reduction of genome size goes hand in hand with the loss of many genes as an adaptation to their life conditions, in which it becomes possible to obtain many molecules from the host. The smallest bacterial genome reported so far has been that of *Mycoplasma genitalium*, an intracellular parasite of epithelial cells, and comprises a circular chromosome of 580 kb with only 470 coding genes (Fraser *et al.*, 1995). The recently sequenced genome of *Buchnera* sp. APS, primary (P) endosymbiont of the aphid *Acyrthosiphon pisum*, is also extremely reduced; it has one circular chromosome of 641 kb, with only 564 coding genes, plus two small plasmids (see Shigenobu *et al.*, 2000; Gil *et al.*, 2002). The concept of the "minimal genome" refers to the minimum number of genes required to support cellular life. Intracellular symbionts are of special interest in this context because they do not need to maintain genes that pathogens require for survival and to evade host detection. In the sequence of *Buchnera* sp. APS, Shigenobu *et al.* found that it has only five (0.9 per cent genes without an homologue in *Escherichia coli* (Gil *et al.*, 2002).

Buchnera is a γ proteobacterium that maintains a mutualistic obligate endosymbiotic association with aphids: *Buchnera* cannot be cultured outside the aphid host, whereas aphids depend on the bacteria for normal growth and reproduction because they provide the host with the essential amino acids that are in short supply in its strict phloem diet. *Buchnera*, confined within specialized cells (bacteriocytes), is maternally transmitted. Besides *Buchnera*, aphids sometimes contain other bacteria, commonly referred as secondary (S) endosymbionts. These endosymbionts are also subject to vertical as well as horizontal transmission (Chen and Purcell, 1997). Conceivably, S-endosymbionts may interact and modify the established mutualism between the aphid and *Buchnera* (Gil *et al.*, 2002).

During evolution, *Buchnera* has experienced a dramatic decrease of genome size, retaining only essential genes for its specialized lifestyle. Physical mapping of *Buchnera* genomes obtained from five aphid lineages showed that the genome size is not conserved among them, but has been reduced down to 450 kb in some species (Gil *et al.*, 2002). Gil *et al.* reported six species having a genome size even smaller than *Mycoplasma genitalium*.

Buchnera species from aphids of the genus *Cinara* (subfamily Lachiniae) possess the smallest genomes, indeed the smallest known genomes.

The number of genes present in *Buchnera* spp. chromosomes have been estimated on the basis of the fact that the known sequence of the *Buchnera* sp. APS chromosome contains 564 protein coding genes in 641 kb, meaning that the genome of *Buchnera* sp. CCE, the smallest genome found in the study of Gil *et al.*, should have about 396 protein coding genes, which is still higher than the minimal genome estimated both from computational analysis (327-256 genes or even less (see Mushegian and Koonin, 1996; Koonin, 2000) and transposon mutagenesis (350-265 genes) (Hutchinson *et al.*, 1999). However, the minimal genome is a hypothetical situation, perhaps not easily attainable in a natural environment. Nevertheless, the reduced genome size does indicate that 395 genes can sustain independent cellular life, which is the minimum set reported so far (Gil *et al.*, 2002).

The Species Concept for Prokaryotes

A bacterial species is essentially a collection of strains characterized by at least one diagnostic phenotypic trait; purified DNA molecules of the strains show 70 per cent or higher reassociation values, as per the recommendations of Wayne *et al.* (1987). This species definition is universally applicable within the prokaryotic world (Brenner *et al.*, 2000; Stackebrandt *et al.*, 2002), but is difficult to implement because of technological limitations in identifying diagnostic traits and in conducting the pairwise DNA-DNA reassociation experiments. It also does not generally help in predicting phenotype (Staley, 1997, 2004). This definition is much broader and not covered by any of the eukaryotic species definitions (Staley, 1997, 2004). If we applied this standard to eukaryotic species we will have to include members of many taxonomic tribes in the same species (Sibley *et al.*, 1990). Furthermore, some strains showing >70 per cent DNA-DNA reassociation values are classified into different species, even different genera, usually on the basis of pathogenicity or host range as, for instance, strains of *Escherichia coli* and *Shigella* spp., making the current prokaryotic classification inconsistent (Konstantinidis and Tiedje, 2005).

To understand these issues and to help advance the species definition for prokaryotes, Konstantinidis and Tiedje compared the gene content of 70 closely related and fully sequenced bacterial genomes with a view to identifying whether species boundaries exist, and to learn the role of the organism's ecology on its shared gene content. The average nucleotide identity (ANI) of the shared genes between two strains turned out to be a robust means to compare genetic relatedness among strains, and ANI values of ~94 per cent corresponded to the traditional 70 per cent DNA-DNA reassociation standard of the current species definition. At the 94 per cent ANI cutoff, current species include only moderately homogeneous strains, *e.g.*, most of the >4-Mb genomes share only 65-90 per cent of their genes, probably as a consequence of the strains having evolved in different ecological settings. Furthermore, diagnostic genetic signatures (boundaries) can be found between groups of strains of the same species, and the intergroup genetic similarity is often as high as 98-99 per cent ANI, suggesting that justifiable species might be found even among organisms that are nearly identical at the nucleotide level. Importantly, a large fraction, sometimes up to 65 per cent, of the differences in gene content within species is associated with bacteriophage and transposase elements, pointing to a critical role of these elements during bacterial speciation. These findings seem consistent with a definition for species that would include a more homogeneous set of strains than provided by the current definition and one that considers the ecology of the strains in addition to their evolutionary distance (Konstantinidis and Tiedje, 2005).

Konstantinidis and Tiedje found that strains of the same species sometime vary up to 30 per cent in gene content. This prompts questions about whether they should belong to the same species, and that these intraspecies differences are presumably driven by differences in the ecology of the strains, thereby supporting a more ecological and stringent definition for prokaryotic species.

The work of Konstantinidis and Tiedje suggests that a more stringent and natural definition for prokaryotic species that is flexible enough to accommodate the ecological distinctiveness of the organisms, is much better than the current definition because such a definition would more reliably predict the phenotype and ecological potential of species.

Modern Concept of Bacterial Species

A great deal of uncertainty surrounds the issue of what constitutes a species for bacteria–apparently rampant horizontal gene exchange takes place between microorganisms and seems to blend microbial populations. Fraser *et al.* (2007)

considered the distribution and effect of mutation and recombination frequencies and, by combining a modeling approach with a review of the available data, uncovered the conditions under which bacterial speciation may occur.

Whereas higher organisms can be classified into taxonomic groups on the basis of their ability to interbreed, there is no definite criterion for bacterial species demarcation other than definitions based on human interests such as disease-associated bacteria. Currently, a prokaryotic species is usually defined as 'a genomic coherent group of strains which share a high degree of similarity in independent features' (Stackebrandt *et al.*, 2002). Practically, a prokaryotic species is thought to be a group of strains with DNA-DNA hybridization (DDH) ratios >70 per cent, and >97 per cent 16S rRNA gene-sequence identity (Vandamme *et al.*, 1996). Since the 1970s, prokaryotic species delineations have been based wholly on data generated by DDH experiments–bacteria in which genomic DDH ratios exceed 70 per cent have not been reported to show less than 97 per cent 16S rRNA gene-sequence identity. In this modern approach, assessment of the genetic similarity among isolates rests on the degree to which their genomes hybridize under standardized conditions.

However, even though DDH experiments are the touchstone of current bacterial taxonomy, they do suffer from some shortcomings. Firstly, there are some problems associated with the performance and reproducibility of DDH experiments. Second, experimentally determined genomic similarities do not represent actual sequence identities because DNA heteroduplexes are formed when the two strands show over 80 per cent sequence identity (Kauffman and Levin, 1987). This means that a difference of 20 per cent in sequence identity can sometime lead to DNA reassociation value between 0 per cent and 100 per cent (Coenye *et al.*, 2005). Kostantinidis and Tiedje (2005) who compared the average nucleotide identity (ANI) of the shared genes in two strains with their DDH ratios, reported that ANI is linearly related to DNA-DNA reassociation values. Also, Henz *et al.* (2005), by using a novel strategy called 'genome blast distance phylogeny (GBDP)', derived phylogenies based on the entire genomic information of organisms. These genomics-based approaches can potentially remove the shortcomings of DDH and can make DDH experiments an acceptable method for bacterial species demarcation (Bae and Park, 2006).

Microbial Forensics

The intrinsic adaptive strategy for replication and survival within a host enables microbial pathogens to cause disease. Microbial-related disease is also sometimes the result of a forced interaction between microbe and host or manipulation of the microbe's genome by unscrupulous or evil persons. In the case of both naturally occurring "emerging" infectious diseases, and disease caused intentionally by terrorists or militants, it is crucial to establish "attribution", such as the source of a pathogen, and its origin and relatedness to other strains and species; this can reveal mechanisms by which virulence arises and the host-microbial equilibrium is disrupted (Cummings and Relman, 2002). For microbial diseases, these issues can be addressed by using molecular microbial signatures. The study of emerging infectious diseases and new pathogens (Relman, 1999), and the criminal justice system have evolved similarly: Both have moved away from dependence on biological phenotypes of the suspected perpetrator, such as fingerprints, and shifted towards more dependable and quantifiable molecular markers, such as polymorphisms (variations) in the DNA sequence (Cummings and Relman, 2002).

Comparative genome sequencing facilitates analysis of genetic variation and relatedness within and between species, also for resolving differences between strains that superficially look identical. However, the rate of accumulation of genetic variations during evolution for some microbial species, *e.g. Bacillus anthracis*, the causative organism of anthrax, is quite low; for these organisms, the ability to discriminate between strains has been limited by the limited number of known genetic polymorphisms.

Read *et al.* (2002) reported the detailed identification of genetic polymorphisms in two related strains of *B. anthracis* by comparative full-genome sequencing. They also suggested a statistical model that distinguishes between true genetic polymorphisms and random sequencing errors. The discriminating power of these polymorphism markers has been shown by the typing of closely related *B. anthracis* strain. Their successful polymorphism detection and analysis in such a genetically homogeneous bacterium is a valuable contribution to the molecular typing field and establishes a methodology for the comprehensive identification of sequence polymorphisms and their depoloyment as typing markers (Cummings and Relman, 2002).

Microbial forensics can be defined as the detection during evolution of molecular variations between related microbial strains and their use to determine the origin, relationships, or transmission route of a particular isolate. These variations or markers can include genome sequence polymorphisms that are detected by direct sequencing or by hybridization-based methods; genomewide patterns of gene expression, which are measured with DNA microarrays; and differences in protein or small-molecule patterns, which may be detected by spectroscopic or other methods. All these techniques can also be used to study population structure, species evolution, and acquisition of virulence (Israel *et al.*, 2001) .

The application of molecular markers in forensic studies has already resulted in striking discoveries. For example, by using multiple-locus variable number tandem repeat (VNTR) analysis, the Aum Shinrikyo *B. anthracis* bioterror strain could be identified as the veterinary vaccine strain, Sterne 34F2 (Keim *et al.*, 2001). Indeed, cultivation of an organism is sometimes not necessary for genotyping because random or targeted genome amplification from picogram quantities of DNA can suffice for microbial forensic analysis of micromanipulated single cells, and for direct analysis of clinical specimens (Dean *et al.*, 2002).

Green Fluorescent Protein as a Molecular Lantern

The green fluorescent protein (GFP) is an excellent reporter molecule in cell biology. GFP fluorescence is now widely used as an effective tool to monitor gene expression, localization, mobility, traffic, and interaction of various membrane and cytoplasmic proteins. The chief merit of GFP as a fluorophore is its cofactor-independent and highly stable fluorescence; the fluorescence remains stable not only in the presence of denaturants but also persists over a wide range of pH (Tsien, 1998). GFP-tagged proteins are easily visualized (seen) by fluorescence microscopy and also can be monitored by various techniques including fluorescence resonance energy transfer (FRET), fluorescence recovery after photobleaching (FRAP) and fluorescence correlation spectroscopy (FCS). Indeed, GFP is so important that the Nobel Prize in chemistry was awarded in 2008 for the discovery and development of GFP to Osamu Shimomura, Martin Chalfie and Roger Tsien. GFP was first discovered in jellyfish *Aequorea victoria* (Figure 1.8) by Shimomura *et al.* (1962). Prasher *et al.* (1992) cloned the gene that encodes GFP. Two years later Chalfie *et al.* (1994) reported that GFP could be expressed in heterologous systems. It was this remarkable finding that boosted the popularity of GFP. The structural and molecular details of GFP were carefully worked out by Ormö *et al.* (1996).

GFP is a compact, barrel-shaped protein consisting of 238 amino acids arranged in 11 β strands and an α-helix runs through the central axis of the cylindrical structure, giving it a *monolithic cylindrical symmetry* that looks like an elegant *molecular lantern* (Figures 1.8, 1.9). The fluorophore of GFP is (*p*-hydroxybenzylide-neimidozolidinone). It is responsible for the post-translational modification upon folding of the polypeptide chain by internal cyclization followed by oxidation of the residues Ser65-Tyr66-Gly67. The uniqueness of GFP stems from the fact that the fluorescence of GFP uniquely does not depend on the idiosyncrasies (any cofactor or specific enzyme) of the jellyfish–*A victoria*–it only needs some molecular oxygen to fluoresce. The fluorophore is located in a highly constrained environment being protected from the bulk solvent by the surrounding rigid β strands (Haldar and Chattopadhyay, 2007, 2009).

While wild-type GFP shows two absorption/excitation maxima (at 398 and 478 nm at 25°C) and has only one emission maximum, mutations in the fluorophore component and in the surrounding β barrel structure can change its photophysical and spectral properties. By means of an elegant protein engineering approach involving random and site-directed mutagenesis, Tsien (1998) and Patterson (2004) could generate several variants of GFP, *e.g.* the blue fluorescent protein (BFP), cyan fluorescent protein (CFP) and enhanced yellow fluorescent protein (EYFP). These variants show improved spectral properties and photostability, making GFP an ideal system for studying directed protein evolution (Haldar and Chattopadhyay, 2009).

Because secondary structure is essential for its fluorescence, GFP is an excellent model for protein folding. Unlike tryptophans in proteins, GFP does not fluoresce when denatured, and the green fluorescence is regained only when the protein refolds.

Most studies of protein folding have typically been done with α-helical globular proteins. As GFP has a β-barrel structure, it could conceivably be a valuable tool for advancing our knowledge of the folding of β sheets/barrel proteins. Another notable feature of GFP is that despite its large size (27 kDa), most (but not all) proteins retain their biochemical characteristics upon fusion with GFP (Pucadyil *et al.*, 2004). Hu *et al.* (2002) showed that GFP can be dissected into two non-fluorescent fragments

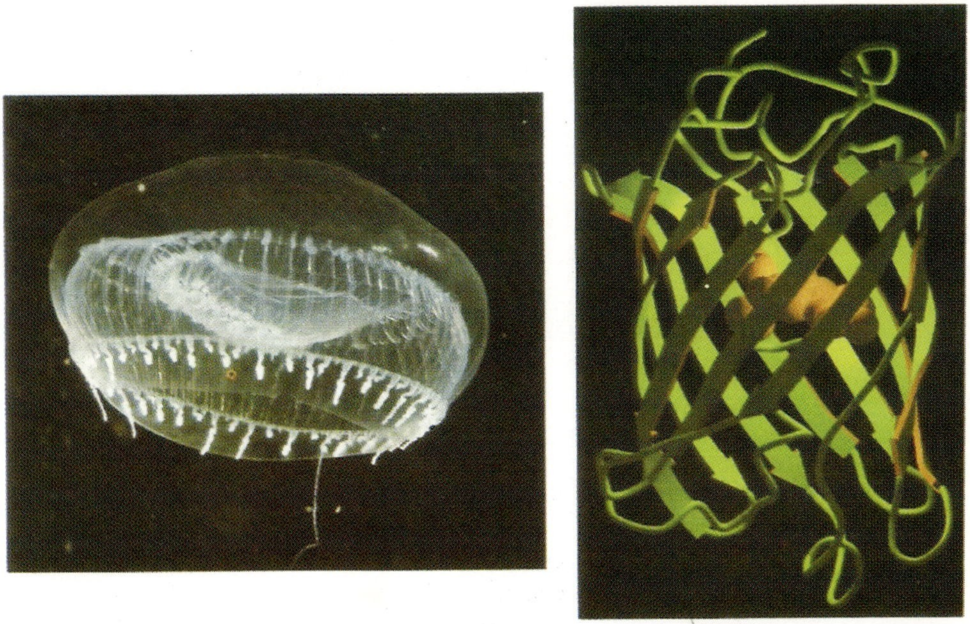

Figure 1.8: The Jellyfish *Aequorea victoria* (Left) as a Source of GFP (Right) (After Haldar and Chattopadhyay, 2009).

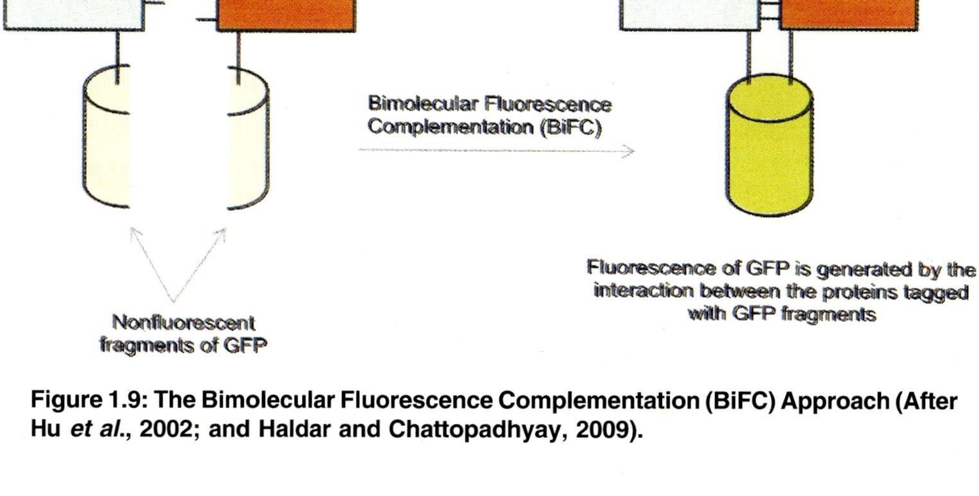

Figure 1.9: The Bimolecular Fluorescence Complementation (BiFC) Approach (After Hu *et al.*, 2002; and Haldar and Chattopadhyay, 2009).

which can be brought together by interactions between proteins fused to each fragment, causing GFP fluorescence (Figures 1.8, 1.9). This approach is known as bimolecular fluorescence complementation (BiFC) and can be used for monitoring protein-protein interactions in live cells. Indeed, GFP acts like a powerful lantern that throws light on some of the dark mysteries in the interior of living cells.

Chapter 2
Gene Transfer and Gene Swapping in Bacteria

Introduction

Real species are defined by the ability of their constituents to exchange genes. This sexual activity explains the maintenance of species as cohesive units whose members are closely related and are of similar genetic architecture.

The problems associated with classifying groups of asexual organisms into discrete species have long been known, and the situation with bacteria is even worse. Bacteria reproduce asexually, yet they can also obtain genes from other organisms, even those of different kingdoms. Moreover, the amounts, types, and sources of imported genes can vary among lineages, allowing gene transfer to blur the boundaries of bacterial groups at every taxonomic level.

The classification of bacteria into discrete taxonomic units is obstructed by the potential for lateral gene transfer (LGT) among lineages at virtually all phylogenetic levels. In most bacterial genomes, large proportions of genes are introduced by LGT, which occurs between very distantly related organisms.

Bacteria can acquire genes by conjugal transfer, by phage-mediated insertions and by the uptake of native DNA from the outside sources (Ochman *et al.*, 2000; Ochman *et al.*, 2005). But in the face of the diversity of mechanisms that can plant virtually any gene in virtually any organism, bacterial genomes remain small (on the order of 500-10,000 kb) and are not simply arbitrary assortments of genes of mixed heritage.

The completion of each bacterial genome sequence prompts a search for horizontally acquired genes. This research usually proceeds by scanning the genome sequence for regions of atypical base composition, a surprisingly accurate method for identifying one class of recently acquired genes. The rationale for this approach has its foundations in research performed half a century ago; base composition [usually expressed as the relative proportion of guanine and cytosine (G+C) residues, per cent G+C] had been determined for hundreds of bacterial genomes, leading to the general observations: (i) that the diversity (ranging from 20 per cent to 80 per cent G+C) is much greater than that in eukaryotes, (ii) that despite this variation, the base composition within an organism is fairly consistent over the entire chromosome, and (iii) that closely related organisms have similar G+C contents (see Ochman *et al.*, 2005). The observed heterogeneity among genomes, coupled with the compositional homogeneity within genomes, implicates gene transfer between organisms of different G+C contents as a source of intragenomic variation in base composition and codon usage patterns (Lawrence and Ochman, 1998).

LGT is pervasive and an ongoing process within bacterial genomes. Genes with sporadic distributions and atypical sequence features arise by LGT, and gene transfer sometimes can occur at all taxonomic levels, even among the genes common to all life forms. With the potential for the LGT of any gene and among all organisms, bacterial species and other taxonomic groupings might not be definable entities. This creates a need to determine whether LGT is resorting the genes in bacterial genomes, eradicating the vestiges of bacterial species, and confounding attempts at phylogenetic classification.

By adopting a whole-genome approach, which examined the history of every gene in numerous bacterial genomes, Ochman *et al.* showed that LGT does not hamper phylogenetic reconstruction at many of the shallower taxonomic levels. Despite the high levels of gene acquisition, the only taxonomic group for which appreciable amounts of homologous recombination were detected was within bacterial species. Taken as a whole, the results derived from the analysis of complete gene sequence inventories support several of the current means to recognize and define bacterial species.

While gene acquisition is frequent, gene replacement is relatively rare, resulting in fundamentally two classes of protein-coding sequences within bacterial genomes: (i) the orthologs that are conserved among taxa and not prone to gene transfer and exchange among species; (ii) the acquired genes, which are generally unique to a genome and, unlike orthologs, encode proteins of uncharacterized functions. This means that despite high levels of LGT, bacteria probably form coherent groups at the shallower taxonomic levels because LGT is concentrated in a class of genes that are not suitable candidates for phylogenetic analysis (Daubin *et al.*, 2003).

Despite the massive influx of new genes into bacterial genomes, the only taxonomic group for which appreciable amounts of homologous recombination have been detected is within bacterial species–a finding very similar to the concept of species applied to sexually reproducing eukaryotes, *i.e.*, groups of organisms that exchange genes.

If certain mechanisms that limit homologous recombination are operating in several bacterial taxa, then bacterial species can be viewed as assemblages of lineages that are sufficiently closely related to potentially exchange shared genes, and the practice of delimiting bacterial species on the basis of some prescribed level of sequence divergence seems justified.

Why Species?

One overarching issue emanates from the fact that many of the genetic and genomic properties were characterized in groups of lineages that were already designated as distinct bacterial species. *E. coli* and *Salmonella typhimurium* were each discovered more than a century ago, and their classification is founded on schemes devised before there was any knowledge of genes or genetics.

Bacterial species are commonly recognized according to their cellular properties and metabolic capabilities. A good example is that of *E. coli*, a mammalian commensal, which ferments lactose but not citrate, but *Salmonella enterica*, a mammalian pathogen, is lactose negative and citrate positive. Examining the genetic basis of these traits, the lac operon is a G+C-rich region unique to *E. coli*, whereas the citrate utilization (as well as many *Salmonella* virulence determinants) is conferred by low G+C genes present only in *Salmonella*. Therefore, assignment of isolates to each of these bacterial species seems to have been largely based on traits that were introduced by LGT. Consequently, the species so defined have turned out to be discrete biological entities that, because of genetic and mechanistic reasons, rarely exchange homologs (Ochman *et al.*, 2005).

Speciation Attributable to Gene Acquisition

Strangely, despite massive amounts of LGT, bacteria seem to form fairly cohesive groups at many taxonomic levels (Daubin *et al.*, 2003; Kurland *et al.*, 2003). These groupings are the result of a nonarbitrary process of gene acquisition in which divergent organisms serve as a persistent source of novel genes in a genome, and the levels of recombinational exchange among homologs shared by related species are quite low (Ochman *et al.*, 2005). As such, LGT could probably be considered as an agent that promotes and maintains bacterial species (Lawrence and Ochman, 1998). Acquired genes have strong role in bacterial diversification by supplying previously unavailable traits, which can allow the rapid exploitation of new environments. Once acquired, these capabilities are strictly vertically transmitted and they help to subdivide the population and allow the phenotypically distinct lineages to diverge at the sequence level to the point where there is a recombinational barrier to gene exchange.

Gene Swapping

One major revelation of microbial genome sequencing has been that many organisms are swapping genes from strain to strain, even across species. For example, about 25 per cent of the genome of *Escherichia coli* has been acquired from other species.

Gene swapping, or horizontal gene transfer is a widespread phenomenon, having very broad implications. Borrowed genes can spread antibiotic resistance from one pathogen to another or help an organism survive new or stressful conditions.

This can happen often enough to change the dynamics of microbial communities. For systematists trying to figure out the relationships between different organisms, however, gene swapping causes big headache as it blurs the boundaries between species, making it difficult to determine where organisms belong on the family tree (Pennisi, 2004).

Gene transfer can occur in different ways. Sometimes a dying bacterium spits out its DNA, and other bacteria retrieve and discard it or incorporate segments into their own chromosomes. Conjugation can also lead to genetic exchange when two bacteria come in contact with each other. Viruses that infect a cell sometimes pick up host DNA as they replicate, carrying it to the next bacterium they infect. Finally, independent pieces of bacterial DNA (plasmids) can enter foreign cells and–if they survive the cell's defense mechanisms–set up residence separate from the host's genome.

Microbes from similar environments are more likely to swap DNA. Similarly, big genomes exchange genes with other big genomes. Organisms seem to limit gene exchange to microbes on nearby branches of the family tree, probably because their chromosomes share certain characteristics. Genes are exchanged between species with similar chromosomal structures; the exact point where replication stops on a particular species' chromosomes, for example, can limit what genes can be incorporated into that genome (Pennisi, 2004).

Sometimes, external factors facilitate gene transfer. Conjugation works best in dense microbial communities. And plasmids can only get into a target cell that has the proper proteins on its surface. When environmental conditions become harsh, DNA can be scavenged from the neighbours. Horizontal gene transfer facilitates rapid microbial adaptation to stress. Higher-than-suspected transfer rates have been seen among microbes living in nutrient-poor environments, where sharing genes may be crucial to survival.

The convergence of fresh theoretical ideas in evolution and the imminent explosion and avalanche of genomic data is sure to drastically change our understanding of the biosphere. This can even lead to revision of concepts such as species, organism and evolution. According to Goldenfeld and Woese (2007), the molecular reductionism that dominated twentieth-century biology will be superseded by an interdisciplinary approach that embraces collective phenomena. In this transformation, the most critical role will be played by horizontal gene transfer (HGT) involving the non-genealogical transfer of genetic material from one organism to another–*e.g.*, from one bacterium to another or from viruses to bacteria. HGT is pervasive and powerful among many prokaryotes in which it accelerates the spread of antibiotic resistance. In fact, the prevalence of HGT argues against regarding microbes as organisms dominated by individual characteristics; rather their communications by genetic or quorum-sensing channels point to the fact that microbial behaviour should be understood as predominantly cooperative.

In the wild, microbes form communities, invade biochemical niches and drive biogeochemical cycles. There are strong indications that microbes absorb and discard genes as needed, in response to their environment. Rather than discrete genomes, there is a continuum of genomic possibilities, which throws doubt on the applicability of the concept of a 'species' to microbes (Goldenfeld and Woese, 2007). One recent example of this idea emerged in the study of genomes recovered from natural samples as opposed to clonal cultures–studies of the spatial distribution of rhodopsin genes in marine microbes suggest such genes to be 'cosmopolitan', wandering among bacteria (or archaea) as environmental pressures dictate. Another exciting finding is that viruses have a fundamental role in the biosphere, in both immediate and long-term evolutionary senses. They are an important repository and memory of a community's genetic information, contributing to the system's evolutionary dynamics and stability. This is suggested, for example, by prophage induction, in which viruses lying latent in cells can become activated by environmental influences; the resulting destruction of the cell and viral replication is an effective mechanism for the dispersal of host and viral genes.

Microorganisms can efficiently reconstruct their genomes in the face of dire environmental stresses, and in some cases their collective interactions with viruses seem to be crucial to this. This situation questions the validity of the concept of an organism in isolation. Rather there may be a continuity of energy flux and informational transfer from the genome up through cells, community, virosphere and environment. Goldenfeld and Woese (2007) have even suggested that a defining characteristic of life is the strong dependence on flux from the environment, whether it be of energy, chemicals, metabolites or genes (see Frigaard *et al.*, 2006; Vetsigian *et al.*, 2006).

The implications of collective phenomena mediated by HGT (= LGT) are most pervasive and important in evolution. The crucial role of translation, represented by the genetic code, is manifested by the well documented optimization of the code. Its special role in any form of life leads from its striking interface that early life evolved in a Lamarckian way, with vertical descent marginalized by the highly robust early forms of HGT.

Refinement through the horizontal sharing of genetic innovations probably triggered an explosion of genetic novelty, until the level of complexity warranted a transition to the current era of vertical evolution (see Frigaard *et al.*, 2006; Vetsigian *et al.*, 2006).

Gene Transfer Among Bacteria in Natural Environments

Gene transfer among bacteria in nature can occur by three chief mechanisms: (1) transformation, in which extracellular DNA is taken up by recipient bacteria; (2) conjugation, in which genetic material is transferred from one bacterium to another by cell-to-cell contact: and (3) transduction, in which the transfer of genetic information between bacteria is mediated by bacteriophages. Other possible mechanisms include capsduction by a small phage–like structure (Joset and Guespin–Michel, 1993), protoplast fusion (Matsushima and Baltz, 1986), and transposition (Berg, 1989).

Transformation

Transformation can be divided into natural transformation and artificial transformation. In natural transformation, recipient bacteria develop a physiological state of competence and actively take up and incorporate extracellular DNA into their genetic information. This seems to occur only in bacteria, but recombinant DNA technology (artificial transformation) can be used with both prokaryotic and eukaryotic cells.

Natural transformation (usually referred to as "transformation") can be distinguished experimentally from conjugation and transduction by its sensitivity to nucleases of DNA (*e.g.*, DNase). In as much as gene transfer by transformation occurs via extracellular DNA, DNase can degrade the DNA and prevent transformation. In contrast, during transduction, transducing DNA is protected by the protein coat (capsid) of the phage and is never exposed extracellularly to DNase. In conjugation, DNA is transferred by cell–to–cell contact, and the transfer of DNA from donor to recipient cells also occurs without exposure of DNA to DNase. Therefore, the addition of DNase can be used to eliminate the possibility of transformation in experiments on transduction and conjugation (Yin and Stotzky, 1997).

Transformation in the environment consists of the following sequence of major steps (Lorenz and Wackernagel, 1994).

1. Release of DNA from cells.
2. Dispersal of the released DNA into the environment.
3. Persistence of the released DNA in the environment.
4. Development of competence in potential bacterial recipients.
5. Interaction of recipient cells with the released DNA.
6. Uptake of exogenous DNA and the processing of the entering DNA in cells.
7. Integration of the exogenous DNA into the endogenote and the expression of its gene(s).

Plasmid Transformation

Plasmid transformation is defined as that which is mediated by a plasmid (an extrachromosomal, double–stranded, circular molecule of DNA capable of self–replication). Usually transformation by plasmids is less efficient than by chromosomal DNA. This may be a result of the covalently closed circular structure of plasmids. For plasmid transformation of Gram–positive bacteria, double–stranded DNA has first to be converted to single–stranded DNA outside the cells. After the single–stranded DNA enters the recipient, it is usually impossible for the linear single strand of a monomeric plasmid to recircularize and replicate. But the separate entry of two complementary single strands of a monomeric plasmid that may reassemble into a complete plasmid molecule, can occur sometime.

Fair Allocation of Plasmids in Daughter Cells

Cell division sometimes proves hazardous for a cell that needs to segregate the two copies of its genome into daughter cells. Prokaryotic cells often contain extrachromosomal elements such as plasmids which also need to be allocated fairly. The process of partitioning of the R1 multidrug resistance plasmid has been examined in *Escherichia coli*; it is powered by the actin-like ATPase ParM. ParM shows structural homology to actin, but in filamentous form, it shows some dynamic instability reminiscent of microtubules. Time-lapse fluorescence microscopy has facilitated tracking of plasmid segregation in *E. coli*. ParM filaments form spindles that can capture a plasmid at each end and then push them to opposite ends of the cell at about 50 nm/s. At first, spindles do not always align with the long axis of the cell, but alignment probably does occur during elongation. After reaching the poles, the filaments undergo heavy depolymerization. Interestingly, the relocated plasmids do not remain fixed at the poles; they sometimes diffuse and suffer several rounds of segregation before cell division (see *J. Cell Biol.* 179: 1059, 2007).

Transformation by Cell Contact

This is a special form of transformation, and needs cell–to–cell contact, but the transfer is partially sensitive to DNase I (*i.e.*, not completely inhibited) and is bidirectional. The mechanisms involved in this process are not well known, but it has been observed to occur in some species of *Bacillus, Haemophilus, Pseudomonas, Streptococcus, Acinetobacter, Vibrio,* and *Escherichia* (see Yin and Stotzky, 1997). This form of transformation may be significant in gene transfer between intact cells in natural environments.

Transfection

This is an unusual variant of transformation. The extracellular DNA is the naked DNA of bacteriophages, and the result is a virus–infected recipient cell (Birge, 1994). Although transfection has basic similarities with transformation, such as the requirement for competent cells, it differs from transformation, at least in the DNA–uptake process. The importance of transfection in the natural environment is not known.

Transformation *in situ*

Even though transformation was the first described mechanism of gene transfer in bacteria, transformation of only a limited number of bacteria has been studied, in only a few environmental settings.

Griffith (1928) first demonstrated transformation by conducting experiments in mice, using two variants of *S. pneumoniae* (then called *Pneumococcus* or *Diplococcus pneumoniae*). Griffith's work constitutes a milestone in biology, because it led to the discovery that the "transforming principle" was DNA, the carrier of genetic information.

Graham and Istock (1978) were the first to report the exchange of chromosomal genes by transformation in *Bacillus subtilis* in sterile soil. Transformation of chromosomal markers between *B. subtilis* and *B. licheniformis* also occurs in sterile soil, showing that interspecific transformation may occur in natural environments. Transformation of chromosomal and plasmid genes coding for amino acid biosynthesis and antibiotic resistance among strains of *B. subtilis* has also been observed in sterile and nonsterile soils (Lee and Stotzky, 1990), but the transformation frequencies in sterile and nonsterile soils are significantly lower than in pure culture.

Aquatic Habitats

The existence of high molecular weight extracellular DNA in marine environments suggests that transformation probably occurs in this habitat, *e.g.* in *Vibrio* spp. Transformation is more likely to occur in water columns than in marine sediments in nature. Paul *et al.* (1992) also reported intrageneric and intergeneric transformation by plasmid pQSR50 among marine strains of *Vibrio* and between *Escherichia coli* and *Vibrio* species in sterile and nonsterile estuarine water. Frischer *et al.* (1994) demonstrated the transfer of extracellular plasmid DNA to indigenous marine bacteria by transformation. Transformants included members of the genera *Vibrio* and *Pseudomonas,* as well as members of unknown genera.

On the basis of many recent studies, it is concluded that transformation is quite important in the spread of genetic information through natural bacterial populations.

Conjugation

Conjugation in bacteria was first discovered in *Escherichia coli* by Lederberg and Tatum (1946), but the mechanisms involved in conjugation especially in the environment, are still not completely known. It is supposed to be primarily a plasmid–associated process in which conjugative plasmids carry transfer (*tra*) genes that not only mediate self–transfer but can also mobilize nonself–transmissible plasmids (*tra*$^-$) and chromosomal genes to transfer unidirectionally from donor to recipient cell via cell contact. But another type of nonplasmid conjugative element, the conjugative transposon, has also been discovered (Franke and Clewell, 1981; Salyers *et al.*, 1995). Further, another type of conjugation, called retrotransfer, has been described (Mergeay *et al.*, 1987), in which there is reciprocal genetic exchange, that is, donor bacteria can also obtain genes from recipient bacteria via the movement of nonconjugative plasmids and chromosomal genes into cells containing some types of conjugative plasmids. Note that this direction of gene transfer is opposite to that in classical conjugation and mobilization (Yin and Stotzky, 1997).

Mechanisms of Conjugation

Self-Transfer of Conjugative Plasmids

In Gram–negative bacteria, conjugation occurs in two stages. In the first, a specific cytoplasmic bridge is formed between the mating cells, whereby the plasmid–containing donor and the recipient contact each other through extracellular conjugative pili, which have been broadly divided into two morphological groups: (1) long flexible pili and (2) short rigid pili. The tip of the F-pilus (encoded by an extrachromosomal fertility (F) factor) initiates contact with the recipient cell and a stable cellular mating aggregate is formed between the two cells. A DNA–transport pore extends through the cell envelopes of the paired donor and recipient and connects the cytoplasm of the mating cells. Stage 2 involves the transfer and processing of DNA; one strand of the plasmid DNA in the donor is nicked at the specific site of the origin of transfer (*oriT*), and a mobilization protein, Mob, is covalently attached to one end (5' terminus) of the nicked DNA strand and interacts with the proteins of the transport pore to anchor this end of the transferring DNA strand to the pore (Yin and Stotzky, 1997). At the *oriT* site, DNA replication occurs by the rolling–circle mode and produces a single–stranded copy of the plasmid DNA which goes through the pore to the recipient cell. When the end (3' terminus) of the transferring DNA strand reaches the pore, it joins covalently with the leading end (5' terminus) of the same DNA strand, which is anchored there by the Mob protein–pore complex. So, a covalently closed single–stranded copy of the plasmid is moved into the cytoplasm of the recipient, and its complementary strand is generated by DNA replication in the recipient. The double–stranded plasmid then either replicates further or integrates into the recipient chromosome (Clewell, 1993; Salyers and Shoemaker, 1994).

When a plasmid becomes integrated into the donor chromosome, it also becomes able to transfer chromosomal genes to a recipient cell. Thus, the self–transmissible F–factor in *E. coli* is an episome, which can either exist autonomously in the cell or integrate into the bacterial chromosome at several different locations by recombination between homologous insertion sequences present on both the plasmid and the host chromosome. After integration, the *tra* operon of the F–factor is still functional, and the integrated F–factor can direct the synthesis of pili, rolling–circle replication, and transfer of part of the F–factor (leading or starting region of the transferred DNA) and part of the chromosome into an F^- recipient cell. Such a donor is called a "high frequency of recombination" (Hfr) cell.

In Gram–positive bacteria, no pili are found, also no specific bridge between mating pairs of bacteria via a conjugative pilus is formed. At least two categories of conjugation that are mediated by self–transferable plasmids have been found. The third category involves conjugative transposons (*vide infra*). Some Gram–positive bacteria form pheromone–responding plasmids. These plasmids are confined to the enterococci and their close relatives. The plasmids are larger than 25 kb. During the transfer the recipient cells excrete proteinaceous sex pheromones (CIAs–clumping inducing agents) which trigger the plasmid–carrying donor cell to produce surface proteins that promote cell clumping. This promotes cell–to–cell contact between donor and recipient cells, leading to conjugal transfer of plasmids. These pheromone–responding plasmids often contain genes encoding haemolysins, bacteriocins, or resistance to antibiotics (Brantl *et al.*, 1990). Another system of plasmid transfer in Gram–positive bacteria occurs in streptococci and staphylococci (Clewell, 1993). Here, the transfer of plasmids occurs at frequencies ranging from 10^{-6} to 10^{-3} per recipient), depending on the plasmid type and the genotype of the mating pair, and requires that donor and recipient cells be co–cultivated on a solid surface (*i.e.*, direct contact between donors and recipients). These plasmids do not respond to pheromone signals. Conjugative plasmids of streptococci can mobilize non-conjugative plasmids. Staphylococcal conjugative plasmids encode resistance to gentamicin, and have a very narrow host range.

Mobilization

Mobilization is a process by which smaller nonconjugative plasmids that cannot transfer themselves because they do not encode a self–conjugative system, are transferred from donor to recipient with the help of self–transmissible plasmids. This can occur in two ways. (1) the nonconjugative plasmids can use the transfer system of a co–resident conjugal plasmid to move from the host cell into the recipient; here nonconjugal plasmids first cause the production of proteins to initiate the transfer process by making a single–stranded nick at their own *oriT* site. The resulting DNA–protein complex then interacts with proteins of the mating pore encoded by the conjugative plasmid and continues the conjugal transfer process. The nonconjugative plasmids can also be mobilized by *conduction* (Clarke and Warren, 1979): the plasmid forms a co–integrate with a conjugal plasmid and enters the recipient cell as part of the conjugal plasmid. Subsequently, the co–integrate plasmid can separate into the original two plasmids. Mobilization has been reported in both Gram–negative and Gram–positive bacteria.

Broad-Host-Range Plasmids

Although the host range of a conjugative plasmid may be limited to a relatively few closely related organisms, many plasmids have a very broad host range. Indeed, some plasmids of Gram–positive bacteria can also be transferred to Gram–negative bacteria and *vice versa*. Moreover, plasmids can be transferred from *E. coli* to the eukaryotic yeast *Saccharomyces cerevisiae*. Interkingdom transfer is best exemplified by the Ti plasmid of the bacterium *Agrobacterium tumefaciens* which can transfer a segment of its DNA (T–DNA) to plants.

Conjugative Transposons

A transposon (Tn) is a DNA segment that carries genes required for transposition and can move around on the chromosome and plasmid (Prescott *et al.*, 1996). Conjugative transposons are discrete DNA elements that are normally integrated into a bacterial genome. These transposons were first discovered in *Enterococcus faecalis* (formerly called *Streptococcus faecalis*) and *Streptococcus pneumoniae.* Since then they have been found in numerous genera of Gram–positive and Gram–negative bacteria and are widespread in the bacterial world. Conjugative transposons are similar to (1) other transposons, as they excise from and integrate into chromosomal and plasmid DNA, but they appear to have a different method of excision and integration; (2) plasmids, as they have a covalently closed circular transfer intermediate and are transferred by conjugation, but the intermediate does not replicate; and (3) bacteriophages, as their excision and integration resemble those of temperate phages, but they do not form viral particles and are not transferred by phage transduction. Some properties of conjugative transposons are summarized below.

1. Size can range from 10 to over 150 kb among different bacterial strains. 2. DNA sequence differs between bacterial strains. 3. Mechanism of transposition: (a) the conjugative transposon is excised from the chromosome to form a covalently closed double–stranded circle of DNA, which is the transposition intermediate; and (b) the circular form can integrate either back into the chromosome or into a plasmid in the same cell (intracellular transposition), or it can transfer itself by conjugation and integrate into the genome of the recipient (intercellular transposition). However, the integration mechanisms are quite different among the various DNA elements involved,

although they also share some common features: (1) their ends do not cross–hybridize with each other, which is different from standard transposons; (2) the excised circular form is the transfer intermediate, and it is transferred similarly to a plasmid from an internal *oriT* site, but the regenerated double–stranded form of the circle can integrate into the chromosome of both the recipient and the donor; and (3) they can mobilize plasmids by two means: the conjugative transposon can either integrate into the plasmid and produce a self–transmissible chimera, or it can mobilize plasmids by providing the mating pore, much as self–transmissible plasmids mobilize nonconjugative plasmids (Salyers and Shoemaker, 1994; Yin and Stotzky, 1997).

Conjugation *In Situ*

Conjugation has been generally thought to be the major mode of gene transfer in natural environments and so most studies on gene transfer focussed on this mechanism. It has been demonstrated in many environmental settings such as animal systems, conjugal transfer of different plasmids between bacteria of the intestinal tract and associated with the respiratory and/or urogenital tract and association of various Gram–positive and Gram–negative bacteria with insects. Plants (*e.g.* gene transfer by conjugation *in planta* in *Erwinia, Klebsiella, Pseudomonas, Rhizobium,* and *Xanthomonas, Agrobacterium*; soil conjugal transfer of chromosomal genes from prototrophic to auxotrophic strains of *E. coli* in sterile soil, (Weinberg and Stotzky, 1972). Rhizosphere (plasmid transfer by conjugation in the rhizosphere among the rhizosphere bacteria *Agrobacterium, Pseudomonas,* and *Rhizobium* (Lynch, 1990); Fresh Water (eg. in lakes, ponds, and rivers, mainly in *E. coli*); Activated Sludge (probably a good site for gene transfer by conjugation, as in this relatively nutrient–rich ecosystem the autochthonous microflora is likely to be well adapted to grow and survive and to act as recipients); Wastewater Treatment (the transfer of an antibiotic–resistance, R–plasmid, between *E. coli* strains in some sewage–treatment facilities (see Yin and Stotzky, 1997); and Epilithon (conjugal transfer of plasmid DNA between strains of *P. aeruginosa* on the surface of stones in rivers, as well as the transfer of mercury–resistance genes from indigenous epilithon organisms to a restriction–deficient strain of *P. putida*).

Transduction

Bacterial viruses or bacteriophages (phages) are the key elements in this process of gene transfer. Phages are predators or obligate intracellular parasites of host bacteria and have no independent metabolic ability. Their genetic material is packaged within an outer coat (capsid). Capsid is made mainly or entirely of protein subunits caused capsomeres. Most phages have only DNA (usually double–stranded) as their genetic material, but some phages (which are non–transducing) may contain only RNA. Phages lyse bacterial lawns on agar plates to produce plaques, which serve as an index of the concentration (titer) of the phages. This titer is expressed as the number of plaque–forming units (PFUs) in a given volume (Yin and Stotzky, 1997).

Phages are classified into the following morphological groups (Joset and Guespin–Michel, 1993) (1) tail-less icosahedral; (2) contractile tail–containing phages; (3) noncontractile tail–containing phages; (4) filamentous, and (5) enveloped phages may also be classified into three types according to their infection mechanism: (1) virulent; always lyse the infected bacterial cell to release their progeny; (2) temperate; can either enter the lytic cycle as virulent phages or enter the lysogenic cycle in which the phage genome is retained as a replicable entity (prophage) for a long time within the host; and (3) filamentous; attach to the sex pili of F^+ Gram–negative bacteria, mediating production of progeny phage particles from the growing cell instead of lysing the cells, as the generated viruses are secreted through the cell wall (Duckworth, 1987). The most successful transductions are mediated by temperate phages. The F^+ specific filamentous phages can also mediate transduction of plasmid DNA.

A. Lytic Phages

The life cycle has 4 steps: (1) Adsorption of phage particles (virions) on bacterial cells at specific receptor sites; this limits the host range of the phage to only a few species of bacteria. (2) Injection: the nucleic acid of the phage enters the host cell. In some phages, the tail contracts acting as a micro–syringe to inject phage nucleic acid into the cytoplasm of the host. The capsid remains outside the cell. Phage tails may also have certain enzymes that weaken the cell envelope, facilitating the penetration process. (3) The latent period: phage proteins and nucleic acid are synthesized and assembled into nascent phage particles within the bacterial cell. In the early latent (eclipse) period, no virions are present within cells. The rate of increase in the number of phages in the latent period is linear, not exponential. (4) The rise or burst period: infected cells rapidly lyse and release mature virions. The average number of viral particles released per infected cell is called the "burst size" and is characteristic of the phage.

B. Temperate Phages

These can develop in one of two directions: (1) the lytic mode of development, as above, and (2) lysogenization, in which a temperate phage maintains its presence in a host cell without killing the cell and producing free infectious particles. Lysogenization is a latent infectious state, during which the relationship between the phage and its host is called lysogeny. Kokjohn (1989) defined lysogeny as the indefinite persistence of the phage DNA as a prophage in the host cell without the production of phage particles. Bacteria having the potential to produce phage particles are called lysogenic bacteria. The latent form of the phage genome that remains in the host without killing it is called a prophage. The prophage is usually integrated into the bacterial genome, but in some cases it exists independently. Lysogenic bacteria are immune to superinfection by the same (and sometimes another closely related) phage. Usually, lysogeny is controlled by both the bacterium and the phage: if the cell is starving, the lysogenic state will be favored. If the cell is active, induction and the lytic cycle will be favoured (Yin and Stotzky, 1997).

Lysogenic induction involves the initiation of phage reproduction in a lysogenized culture. Some environmental factors (*e.g.*, ultraviolet radiation and other mutagens) induce a prophage to move from the temperate cycle to the lytic cycle.

Lysogenic Conversion

Sometime a change in the phenotype of the host bacterium (the lysogen) is caused by a temperate phage. Such acquisition by a lysogenic bacterium of novel phage–

encoded, phenotypic properties is called lysogenic conversion (or phage conversion). Lysogenic conversion is important in the ecology of both a phage and its host and may have some effect on the potential for the transfer of genes by transduction in the environment.

Types of Transduction

During the life cycle of a phage, bacterial genes are sometime incorporated into a phage capsid and injected into another bacterium, resulting in transduction of bacterial genes. There are essentially two types of transduction: generalized, and specialized. Another type, capsduction, may belong to another system of gene transfer.

1. Generalized

In generalized or unrestricted transduction, any DNA fragment of a bacterial genome or plasmid can be transferred when the fragment is packaged within a phage capsid. The transducing phages can inject the DNA from the bacterial chromosome or plasmids into another bacterial cell, but they cannot initiate a lytic cycle in the recipients, and the transductants will not show immunity to superinfection, as they do not contain phage genes (Yin and Stotzky, 1997).

Cotransduction occurs when two genetic loci are packaged in the same phage and transduced together into the recipient. It is used to measure the relative distance between the two genetic loci by the probability of their coinheritance. When the distance between the two genes is short, the probability of their coinheritance is high.

2. Specialized

In specialized or restricted transduction, a specific set of bacterial genes, because of imprecise excision of the prophage during induction of the lytic cycle, is introduced into another bacterium by a temperate phage. When the DNA of a temperate phage has been injected into the potential donor cell, it is integrated into the bacterial chromosome at a specific region. The phage DNA then replicates with the host DNA and remains in the progeny cells of the lysogen as a prophage until induction occurs. After being induced to the lytic stage, the phage DNA normally is correctly excised from the bacterial chromosome, and the exact initial viral DNA molecule is reproduced. Occasionally, however, imprecise excision occurs resulting in a transducing particle that possesses most of the phage genome as well as a portion of the bacterial genome immediately adjacent to the integration site. Only temperate phages seem to be capable of specialized transduction, and only genes close to the integration site of the phage are packaged and transduced. The best known example of specialized transduction is phage lambda of *E. coli*.

During specialized transduction, sometime *abortive* transduction can occur. Transduced DNA fragments from a lysed lysogen that are not integrated into the endogenote of the new recipient persist in the recipient but do not replicate, although the genes on the transduced DNA fragment are expressed. In colonies that develop from such transduced cells, only one of the daughter cells receives the transduced DNA in each cell division, resulting in a clone in which only some of the cells contain the transduced DNA fragment. The nonintegrated DNA fragment tends to be protected from degradation and can persist within the clone for many generations.

3. Capsduction

In *Rhodobacter capsulatus*, the size and structure of the *gene transfer agent (GTA)* is similar to that of a small virulent phage. At least five major and three minor proteins comprise the capsid, which contains linear double–stranded DNA. GTA particles are synthesized and released in one or two synchronized waves during the growth of GTA–producing cells. Because no phage–like functions (*e.g.*, cell–killing, plaque-formation) and no plasmid–like structures have been found in the various GTA–producing strains, the GTA–encoding genes are probably located on the host chromosome. The modes of production and extrusion of this particle are not known but the GTA often mediates transduction through the steps of adsorption, injection, and homologous recombination reminiscent of a phage.

Factors Influencing Transduction in the Environment

Transduction appears to be a widespread mechanism of gene transfer among bacteria. The packaging of genetic material in phage particles has many advantages under natural conditions, as it not only protects the DNA from nucleases and adverse environmental conditions, but may also be a good survival strategy for bacterial genes. Therefore, transduction may also be an important method for gene transfer in natural environments.

As transduction involves an interaction between transducing phages and host bacteria (donor and recipient, respectively), high concentrations of bacteria and phages are probably required. The numbers of phages and bacteria that have been detected in different types of soil, water (fresh, marine, coastal, estuarine, and waste), wet sludge solids, and hospital settings are more than sufficient to ensure transduction, indicating that transduction can occur in the environment, probably in the following stages (Stotzky, 1989): (1) induction of a prophage to the lytic stage or successful infection by a lytic phage; (2) incorporation of bacterial gene(s) into the new virion; (3) release of the virion containing the bacterial gene into the ambient environment; (4) infection of another bacterial cell with this virion; (5) incorporation, propagation, and expression of the original bacterial gene in the infected bacterium; (6) transfer of the new genetic information to the progeny of the infected bacterium; and (7) repeat of the lytic–prophage cycle to spread the new genetic information to other bacteria in the habitat (Yin and Stotzky, 1997). Some biological and physicochemical environmental factors that influence transduction in the environment include phage host range, restriction modification systems, immunity to superinfection; lysogeny, molecular size of transduced DNA, survival of phages in the environment, multiplicity of infection, concentration of phages and bacteria, type of recipient cells, metabolic state of the host, temperature, ionic composition and concentration, pH, duration of interaction between bacteria and phages, surfaces and solar radiation (Yin and Stotzky, 1997).

In summary, transduction does occur in the environment and may be as important, if not more so, than conjugation and transformation as a means of gene transfer in some environments, especially in structured ones with a high solid: liquid ratio, such as soil, where cell–to–cell contact may not occur often and free DNA may not survive for long periods. It has been shown that the survival of phages is enhanced by their adsorption on clay minerals and other particles in soil (see Yin and Stotzky 1997) so the cryptic genes carried by the phages might be expressed after they infect appropriate hosts.

The following overall conclusions and implications may be made about the three major mechanisms of gene transfer (*i.e.* transformation, conjugation, and transduction) in natural environments.

1. Gene transfer in two directions has been shown by: (1) cell–to–cell contact transformation; (2) retrotransfer in conjugation; and (3) reciprocal chromosomal transduction. This means that gene transfer by all these mechanisms could occur bidirectionally in the environment.
2. In some bacteria, there are additional mechanisms for gene transfer in nature. For example, streptomycetes can fuse their hyphae (anastomosis), resulting in gene exchange. Transposition, in which specific mobile molecules of DNA change position within or among a bacterial genome and plasmids, may also be important in gene transfer in the environment, as indicated by reports of transposon–mediated conjugation. Lightning in natural environments may possibly electroporate indigenous cells in nature.
3. The traditional definitions of the three major mechanisms are gradually becoming more and more ambiguous as the result of such discoveries as cell-to-cell transformation, capsduction, transposon–mediated conjugation, and retrotransfer.
4. Interactions and synergy among and between the mechanisms of gene transfer are possible in the environment. Each mechanism is not entirely independent but may interact with others and may be used alternatively by the same bacteria or phages, depending on the environmental conditions. Bacteria in the environment live as mixtures and in a community rather than as single species or strains, as on an agar plate in the laboratory. Therefore, several mechanisms of gene transfer may be acting simultaneously in a mixed bacterial community, and ambient environmental conditions may favor one or another mechanism. When the environmental conditions change, the mechanisms of gene transfer probably also change (see Yin and Stotzky, 1997).

Far-Flung Gene Transfers in Nature

Many prokaryotes in nature very frequently pass genetic information back and forth, by very diverse means.

Not only plasmid-borne genes but chromosomal genes can also be transferred by a conjugative process, and this latter mode may be as common as plasmid transfer.

Conjugative transposons are another variation on this theme. In streptomycetes, the genetic elements may well be integrated plasmids. In some respects, however, they are unlike other transposons; as they are not flanked by insertion sequences, their locations along chromosomes are not random, and they usually carry transfer-determining genes. Conjugative transposons are common in Gram-positive cocci and in anaerobes such as bacteroides, where they can contain as many as 200,000 base pairs.

Far-flung gene transfers seem to be quite common in many prokaryotes. One reported case is that between a particular isolate of *Prevotella ruminicola*, an anaerobe that resides within the rumen of cattle, and *Bacteroides uniformis*, which resides in the human intestine (A. Salyers, Univ. of Illinois, see *ASM*, 57 (a) p. 450, 1991). The transferred gene specifies resistance to the antibiotic tetracycline, and the copies of the gene obtained from the two types of bacteria were more than 97 per cent identical by DNA sequencing.

This example illustrates the surprising distances these genes can move across genera, and the observations raise the question of transfers across highly disparate mammalian hosts and their bacterial flora.

Recombination and Clonality

Bacteria such as *Escherichia coli* have been generally regarded as being primarily clonal organisms and so the genetic variation within clones has been viewed to be almost exclusively the result of the mutational process.

Genetic exchange in bacteria was first demonstrated in 1928 by Frederick Griffith. Later, the debate centered on the nature of the transforming principle. Work of Avery *et al.* and Hershey and Chase showed that the genetic material was in fact DNA. More-recent debates on this issue have related to the role that genetic exchange and recombination play in the evolution of bacterial species and clones. This issue is important for understanding fundamental evolutionary processes, and has also enormous applied importance. Two areas of research that are influenced by these ideas are the evolutionary dynamics of pathogenic and infectious microorganisms, and the potential spread of recombinant microbes or genes into the natural environment (Guttman, 1997).

The concept of the bacterial clone has been loosely used and is thought of as a group of bacteria descended through clonal (asexual) propagation from a common ancestor. This implies that all genetic transmission has occurred vertically, from parent cell to daughter cells. For most species of bacteria, this is too restrictive since genetic information can also be transferred horizontally between cell lines by different recombination mechanisms. Commonly genetic exchange in bacteria involves the non-reciprocal transfer of segments or pieces of genetic material. These recombinant segments typically constitute only a small fraction of the genome. Their insertion into the clonal frame (Milkman and Bridges, 1990) creates a mosaic genome.

The pathogenic *Neisseria* strains are good examples of both highly clonal and freely recombining populations.

For many years after the discovery of transformation, transduction and conjugation, bacteria were believed to exchange genetic material frequently and promiscuously, as observed by rapid, horizontal spread of antibiotic resistance genes between species and even genera. But general opinion changed after it was found that serotypically indistinguishable pathogenic strains of *E. coli* could be isolated from geographically and temporally widespread hosts (Orskov and Orskov, 1983). These findings helped popularize the clonal nature of bacterial populations which emphasized the mainly vertical, non-sexual transmission of genetic material between individual strains or clones of a species and de-emphasized a role for recombination in bacterial evolution (Guttman, 1997).

DNA Sequences and Intragenic Recombination

Comparative nucleotide sequence studies of *E. coli* loci in the mid 1980s cast some doubt on the clonal concept. Sequence data were examined for patterns consistent with intragenic recombination. Intragenic recombination produces novel mosaic alleles by reshuffling parts of a locus between individuals, and is identified in DNA sequence data by the statistical clustering of polymorphic sites among specific groups of taxa. These studies proved the occurrence of intragenic recombination. The observation of extensive intragenic recombination in several loci questioned the generality of the clonal concept.

It appears that recombination may be very localized and of minimal long-term importance in the evolution of *E. coli*.

Gene Genealogies and Intergenic Recombination

Comparisons of gene genealogical histories from multiple loci constitute an alternative means of evaluating the importance of recombination. This approach is based on the rationale that under the assumption of strict clonality the genome will be maintained as a single linkage unit. Consequently, every locus in the genome will have passed through identical genealogical histories and, therefore, genealogical relationships between strains and clones will be identical at all loci. Thus if some particular mutant allele of a gene is positively selected, the strain carrying that allele will increase in frequency in the population. With strict clonality, the entire genome linked to the selected allele will also increase in frequency. So the alleles at these linked loci will tend to predominate in the population to the same degree as the selected allele. Besides linked neutral alleles, deleterious alleles will also face high frequencies if the positive selection is strong enough. This process is known as a *selective sweep*, and the associated increase in frequency of linked alleles is called *genetic hitchhiking*. If the clonality is not strict, then the degree of linkage between the selected locus and neighbouring loci is determined by the amount of recombination between the loci. One outcome of a selective sweep is that genetic variability will be purged at the selected locus and linked loci (Guttman, 1997).

In *E. coli* selective sweep has been noted for some loci and it does not affect the entire *E. coli* genome as would be expected under the assumption of strict clonality.

The construction of a gene tree makes possible the identification of phylogenetically significant clusters of taxa and the determination of the relatedness of these clusters to each other. This relatedness is reflected in the branching order of the gene tree. All evolutionary forces affect the structure of gene trees in some way but only recombination can cause gross changes in the inferred relatedness of taxa between loci. Recombination is the only evolutionary force capable of changing the branching order of a gene tree.

The Clonal Paradox

It is indeed paradoxical that on the one hand the populations of *E. coli* appear to be highly clonal, and on the other hand intragenic recombination is also quite frequent. How can high levels of linkage disequilibrium be reconciled with frequent recombination?

It may be that recombination acts strictly at a very local or fine genomic scale; therefore, any impact it has on *E. coli* evolution is overshadowed by the mutational process in the long run. Conceivably, mutation and recombination might make about equal contributions to the generation of new alleles and, therefore, recombination may not preclude the perpetuation of clonal structure in natural populations.

Recombination is very often viewed as a force that homogenizes genetic variation within a species; this homogenization is a principal reason why gene pools do not undergo unlimited divergence. Recombination maintains the genetic integrity of the species. If the species is composed of structured populations, recombination can also play a second role: recombination between individuals of different clones or populations increases genetic diversity within the clone or population. Hence, recombination may also be a mechanism, complementing mutation, that generates variability within clones and can potentially drive clonal divergence.

An alternative proposal may be to abandon the clonal concept. One such alternative relies on recurrent selective sweeps in a structured population (Guttman, 1997). Population structure can generate linkage disequilibrium. *E. coli* may not have much geographic population structure, but it does have good ecological population structure. *E. coli* inhabits numerous and varied niches, it is a gut commensal, a pathogen in the urinary tract, gut, blood and skin; it can transiently occur as a free-living water and soil bacterium. In these niches, its strains have further specialized to exploit particular resources. Unfortunately, this ecological population structure of *E. coli* has not been well studied.

A selective sweep is the rapid spread of an advantageous allele through a population and the subsequent purging of genetic variability at the selected and linked sites. Patterns of genetic polymorphism indicating a recent selective sweep have been noted in natural *E. coli* isolates. When a selected allele is advantageous only to specific strains living under particular ecological conditions, then the extent of a selective sweep defines an ecological subpopulation or ecotype. A selective sweep purges the standing genetic variation and, so creates much more extensive linkage disequilibrium within clones or ecotypes than in the overall species. This is exactly what has been found in the case of some pathogenic *E. coli* strains.

The Smart Bacteria: Cognition and Natural Genetic Engineering

In the past half a century it emerged that bacteria have several cognitive, computational and evolutionary capabilities. Analysis of metabolism, regulation of protein synthesis, and DNA repair revealed that bacteria continually monitor their external and internal environments and compute functional outputs based on information provided by their sensory apparatus (Shapiro, 2007). Critical studies of genetic recombination, lysogeny, antibiotic resistance and transposable elements showed the existence of many of bacterial systems for mobilizing and engineering DNA molecules. Research being conducted in many laboratories on cell-cell signaling, symbiosis and pathogenesis has established that bacteria utilize sophisticated mechanisms for intercellular communication and some are even able to dictate the basic cell biology of 'higher' plants and animals to satisfy their own needs. Indeed, even though small in size, bacteria are smart and knowledgeable.

Microorganisms play an essential role in maintaining the biosphere and performing most energetic and geochemical transformations on the planet. They need not be considered as 'lower' forms of life, simpler and less capable than eukaryotes just because their cells are smaller and apparently lack internal cellular structure.

In its initial phase, molecular biology was concerned with defining how living cells functioned at a physico-chemical level. Later, it showed us a vast world of highly complex intracellular machinery, signal transduction, regulatory networks and very sophisticated and precise control processes. Increasingly, computational models are being proposed to account for the operation of subcellular systems, the cell cycle, cellular differentiation, and the development of multicellular organisms (Bray, 1995; Gerhart and Kirschener, 1997).

There has been a strong paradigm shift in biosciences. The focus of classical molecular biology was on matter. This has shifted to information as the essential parameter for understanding how living systems work. Informatics is the key to explaining cell biology and cell activities. According to the traditional mechanistic view, the structure of biological molecules determines the actions of cells in some kind of linear way. But it has now emerged that biological molecules change their structures upon interacting with other molecules; also that these structural changes contain information not only about the external environment but also those within the cell. What a cell does is really a function of the information it has about itself and its surroundings (*i.e.* about past molecular interactions). Contemporary research aims to learn how cellular processes are controlled adaptively to ensure survival and reproduction in response to the countless molecular events that take place in each cell cycle. This *informatic approach* enables us to consider complex, non-linear, goal-oriented processes with all kinds of feedbacks and decision points (Shapiro, 2007).

Bacterial Cognition

A most important point is that bacteria are much smarter than humans at regulating complex operations. The fast growing bacterial cell represents an ultimate just-in-time production facility. An *E. coli* cell can divide once every twenty minutes; highly reliable coordination is achieved for millions of biochemical reactions and biomechanical events. *E. coli* cells replicate their DNA at almost 4000 base pairs per second, yet suffer an error frequency of less than one nucleotide misincorporation per every genome duplication (*i.e.* 2 x 4.6 million base pairs are duplicated every forty minutes; Cooper and Helmstetter, 1968). Such extraordinary precision is achieved not mechanically but by pressing into service two systems of expert error monitoring and correction: (1) exonuclease proofreading in the polymerase itself, which spots and corrects over 99.9 per cent of all mistakes as soon as they are made (Kunkel and Bebenek, 2000); and (2) the *methyl-directed mismatch repair (MMR) system*, which detects and deals with over 99 per cent of any errors that escaped the exonuclease (Modrich, 1991). In this way, this elegant proofreading system boosts the 99.999 per cent precision of the polymerase to over 99.99999999 per cent (Shapiro, 2007). At both stages of the error correction process, the distinct roles of sensory and repair components have been clarified. For the MMR system, the sophistication is so impressive that the molecules discriminate newly replicated from old DNA, and they only correct the newly synthesized strand (Radman and Wagner, 1988).

DNA replication proofreading elegantly exemplifies a judicious combination of *cognition (error detection)* and *response (error correction)* for many different bacterial processes. Bacteria constantly gather information from inside and outside the cell and function adaptively, on the basis of what they have sensed. There is no clear sharp line between cellular cognition and computation but in general, the former is applied mostly to upstream sensory operations whereas the second computation is usually applied to functional decisions based on cognitive inputs (Shapiro, 2007).

Critical molecular scrutiny reveals cognitive and computational phenomena in many of the classic bacterial systems. A good example is that of bacterial chemotaxis. The chemotaxis control circuit is a striking paradigm of biological interlocking cytoplasmic feedback loops and receptor interactions in the cell membrane; *E. coli* chemotaxis can achieve remarkable functional bandwith–the system can even guide bacterial swarming over chemical concentration gradients of more than six orders of magnitude (Bray, 2002). Bacteria are pastmasters in effectively using chemotaxis to locate nutrients, steer clear of toxic chemicals, sense pH, and interact with host organisms in symbiosis and pathogenesis. All these capabilities attest to the power of the functionality and adaptiveness of this cognitive system. Cognition and information processing also have essential roles in damage and accident repair. In the SOS response to DNA damage or replication fork failure, the RecA protein behaves like a sensory microprocessor: it monitors the accumulation of single stranded DNA and derepresses expression of SOS repair functions in response to such accumulation (Sutton *et al.*, 2000). Once the SOS system is activated, sensory and information-processing checkpoints ensure that cell division and DNA replication do not resume until genomic damage has been repaired.

Nanofabrication

This refers to certain intricate kinds of sensing that operate in such complex structures as the helical flagella that propel bacteria when they swim through fluid environments. Dedicated proteins 'usher' components of each new flagellum in an unfolded condition for transport to the external tip of the growing structure, where the components become incorporated and extend the helical filament. Once flagellar biogenesis has been completed, the components can no longer be delivered and they become transcriptional repressors that shut down synthesis of the now superfluous polypeptides (Aldridge and Hughes, 2002)–the bacterial cell makes use of the multifunctional usher/repressor proteins for sensing the completion of a morphogenetic process and for integrating it with genome expression. The operational molecules (ushers, transporters, enzymes) perform informatic roles also.

Bacteria as Natural Genetic Engineers

It is amazing how bacteria can move DNA around. Molecular genetics is deemed to have originated with the study of transformation in gram-positive bacteria, *i.e.*, their ability to incorporate information from exogenous DNA showing that DNA is the hereditary material. Hayes (1968) showed how autonomous plasmids promote DNA transfer between bacterial cells. Norton Zinder and Larry Morse then reported that bacterial viruses could transfer genetic information from one cell to another and sometimes even clone pieces of the bacterial chromosome.

When antibiotic chemotherapy began to be widely practised, it was theorized how bacteria could evolve resistance–mutations would change cell structures so that the cells lost sensitivity to antibiotic action, or so that the antibiotic could not enter the cell to act on its target. This theory was amply confirmed by laboratory experiments and yet it turned out to be inapplicable to most antibiotic resistant bacteria found in hospitals in the world (Shapiro, 2007).

Natural antibiotic resistance is generally acquired by the expression of new functions for inactivating antibiotics or for eliminating them out of the cell. The resistance mechanism can involve chemical modification of a cellular antibiotic target so that it becomes sensitive. In the majority of cases, resistance was acquired through additional functions by the bacteria–the functionalities being encoded by plasmids, phages and transposable elements (see Craig *et al.*, 2002). Bacteria benefited evolutionarily by acquiring mobile fragment of DNA in their genomes and systems for transferring DNA from cell to cell.

The advent of widespread complete genome sequencing has boosted understanding of the role of natural genetic engineering in bacterial evolution enormously. Operons that code for multiple antibiotic resistances are created by lambda-like integration systems called integrons (Hall and Collis, 1995). These integrons enlarge and shrink by the insertion and excision of single protein coding sequence cassettes. Integrons are by no means limited to antibiotic resistance determinants. A large 'super-integron' encoding pathogenicity determinants has been reported in the genome of *Vibrio cholerae* by Mazel *et al.* (1998). Several bacteria harbour integrons or sequences for the integrase protein that mediates cassette insertion and excision (Holmes *et al.*, 2003).

Bacterial genomes typically contain large DNA pieces which extend over tens of kilobases and are very different from the surrounding genomic DNA in base composition. These fragments have been critically examined in pathogenic organisms, where they contain coding sequences for virulence functions and are thus called pathogenicity islands (Dobrindt *et al.*, 2004). Their different base composition from the rest of the chromosome suggests as if these islands were imported from other species. Besides pathogenesis, islands have also been reported to encode such adaptive functions as magnetotaxis, symbiosis, exoenzyme production, xenobiotic degradation, and toxicity to insects. Pathogenicity islands and the magnetosome island seem to be products of the kinds of natural genetic engineering studied in the laboratory.

Genomic data strongly point to the possibility that bacteria use natural genetic engineering systems to gain large pieces of DNA that code for complex adaptive functions from other species. This is basically an extended version of the hypothesis about bacteria as a global superorganism, *viz.* that there is one large distributed bacterial genome, from which genomes adapted to particular niches arise by transfer and integration of DNA segments from different cells into one particular cell type (Sonea and Mathieu, 2001).

The concept of bacteria as natural genetic engineers is further supported by several instances where these small cells use DNA rearrangements to control protein synthesis or to modify the structure of the proteins they produce (Table 2.1). All these examples involve external proteins where structural changes are critical to avoid immune recognition, to change surface attachment properties, or to allow various interactions with different cell types. Importantly, bacterial DNA restructuring and protein engineering also involve two characteristics that are often thought to be limited to 'higher' organisms–repeat DNA sequences (Shapiro, 2002) and reverse transcription (Doulatov *et al.*, 2004).

The DNA pieces that move through the genome, the places they move to, and the sequences they rearrange are flexible and predictable (Shapiro, 2007).

Table 2.1: Five Recently Reported Cases of Bacterial DNA Rearrangements Involved in Regulating their Protein Synthesis and Protein Structure

Example	*Reference*
Fimbrial phase variation in *E. coli*	Blomfield (2001)
Borrelia VSG expression	Barbour *et al.* (2000)
Changes in R64 plasmid sex pilus structure	Gyohda *et al.* (2004)
Neisseria pilus and opacity protein changes	Saunders *et al.* (2000)
Bordetella reverse transcriptase	Doulatov *et al.* (2004)

Bacterial Mastery of Cell-cell Interactions

Besides natural genetic engineering there are several other phenomena that show even greater abilities of bacteria to reorganize themselves in evolution. The origin of mitochondria from alpha proteobacteria and of chloroplasts from cyanobacteria is well known (Gray, 1999). Genome sequence data also support some other major symbiotic events in the history of life as, for instance, (1) the generation of Gram-negative bacteria from symbiosis between an archaea and a Gram-positive bacterium (Gupta, 1998, 2000); and (2) the descent of the first eukaryotic cell from symbiosis between an archaea and a Gram-negative bacterium (Gupta, 2000; Rivera and Lake, 2004; Embley and Martin, 2006). At the molecular level, all these events involved a careful and correct integration of separate genomes, metabolisms, envelopes and external structures into a viable cell. Orchestrating the required molecular processes would certainly have been much more complicated a task than DNA restructuring.

Symbiosis can materialize only after a strong coordination of several processes in distinct cell types. We are now learning about how bacteria coordinate the processes for effective intercellular communication. Conceptual tools enhance our understanding about intercellular and multicellular information processing. One such tool is the ability to think in functional terms about bacterial activity at the population rather than the single cell level (see Shapiro, 1998). One chief focus of current researches is on biofilms–thin colonies spread over a surface (Parsek and Fuqua, 2004; Branda *et al.*, 2005). Multicellularity provides several advantages for pathogens not only in initial colonization but also for protection against host defenses (Costerton *et al.*, 1999; Davies, 2003). Many bio(geo)chemical transformations are being performed by consortia which link up energetically feeble, single cell processes into thermodynamically robust changes with large overall drops in free energy. Indeed, Margulis *et al.* (2000) even suggested that just this kind of metabolic consortium was the progenitor of the first eukaryotic cells.

Reconceptualization of Basic Cellular Processes

During the past six decades, our understanding of the molecular mechanisms that govern basic cellular processes, such as DNA replication, transcription, or cell division, has changed substantially. In 1968, Nobel Laureate Arthur Kornberg postulated that DNA replication involved only a few proteins (polymerase and DNA ligase). At that time, not much was known about such basic features of the replication process as primers for initiating polymerization or the differences between copying of leading and lagging strands; and nothing was known about the requirements for primases, helicases, topisomerases and clamps. Kornberg's paper reflected the simplistic and reductionist thinking prevailing in the early decades of molecular biology–thinking that was based on linear, sequential and unitary concepts of how biological systems operate. Nothing was known half a century ago about any ubiquitous multimolecular complexes or signal transduction networks. Pre-DNA ideas of genotype and phenotype prevailed in genetics–nobody had any ideas related to cognition, communication and computation.

All this has changed, necessitating reconceptualization. Two facts have inspired such rethinking: the studies of regulation initiated at the Institut Pasteur (France), and the molecular study of transposable elements. Jacque Monod's findings about bacterial metabolism contain all the basic ingredients of transcriptional regulation and cellular signal transduction: *cis*-acting signals in DNA, allostery, receptors, protein phosphorylation, protein-protein interaction, nucleoprotein complexes and second messengers. The operator was regarded as the prototype of the various genetic motifs in DNA that allow cells to perform the functions of genome compaction, replication, transmission, expression and repair (Shapiro and Sternberg, 2005). The second example is the molecular confirmation in bacteria (later also in several eukaryotes) of B. McClintock's cytogenetic discoveries in maize which revealed how inextricably multidirectional information transfer between the genome and the rest of the cell has to be. Genomes harbour sequences and code for molecules that are used by cells to restructure DNA for such adaptive purposes as resource utilization, biocide resistance, and intercellular communication. In fact, how and when DNA restructuring takes place is a complex expression of connections between signal transduction networks, natural genetic engineering functions and the genome (Shapiro, 2007).

Research in molecular biology during the past half a century warrants abandoning the mechanistic and atomistic concepts of the pre-DNA era and adopting a more organic, cognitive and computational view of cells and genomes. According to Shapiro, "There are no units, only interactive systems. Bacteria continually pick up and process information about the environment, internal conditions and other cells to decide on appropriate biochemical and biomechanical actions. Comparisons to electronic information systems are useful because they allow us to think concretely and scientifically about complex information processing". However, while doing so, we should not forget that digital electronic computing systems are much simpler than the distributed analog processors in living cells.

It is high time that we recognized that bacterial information processing is far more powerful than human technology. Even though small, the bacterial 'smarts' are extremely sophisticated not only at coordinating processes that involve millions of individual events but also at making them so highly precise and reliable. The versatility shown by bacteria in managing the biosphere's geochemical and thermodynamic transformations is amazing–these processes are certainly more complex than the largest human-engineered systems.

Horizontal Gene Transfer (HGT) and Microbial Relationships

Horizontal gene transfer (HGT, also called Lateral gene transfer, LGT) has taken place widely throughout bacterial evolution (Ochman *et al.*, 2000). Genomic islands contain clusters of horizontally transferred genes. Most genomic islands appear to have many conserved features, *e.g.*, an abrupt change in GC content as compared with that of the rest of the genome, the presence of direct repeats flanking the genomic island, the presence of an integrase gene at the junction and use of tRNA genes as the integration sites (Hacker and Kaper, 2000). Genomic islands can be detected by distribution of GC content which in turn can be estimated by counting G and C residues within a window sliding along the genome. Unfortuantely, the resolution of this method is usually not high enough to detect genomic islands; if the window is large, the GC content changes are averaged, whereas if the window size is small, the GC content changes become masked by random fluctuations.

In Rhodopseudomonas palustris, Zhang and Zhang (2004) found that some metal efflux transporters, multidrug efflux pumps and type IV secretion systems can be created by horizontal gene transfer. This explains why *R. palustris* survives in many diverse environments and acquires the necessary nutrients while resisting toxic compounds.

HGT (see Lawrence and Ochman, 1998; Rivera *et al.*, 1998; Doolittle, 1999) among prokaryotes is not random. Informational genes are more difficult to transfer than operational genes. Environmental, metabolic, and genetic differences among organisms also restrict HGT, so that prokaryotes preferentially share genes with other prokaryotes having properties in common, including genome size, genome G+C composition, carbon utilization, oxygen utilization/sensitivity, and temperature optima; this further complicates attempts to reconstruct genomic tree of life (Simonson *et al.*, 2005). A new method of phylogenetic reconstruction which is based on gene presence and absence is termed *conditioned reconstruction (CR)* and has improved the chances for reconstructing prokaryotic evolution. It can also detect past genome fusions, such as the fusion that may have created the first eukaryote. This genome fusion between a deep branching eubacterium, possibly an ancestor of a cyanobacterium and a proteobacterium, with an archaeal eocyte (crenarchaea), appears to be the result of an early symbiosis (Simonson *et al.*, 2005).

The massive HGT involving genetic crosstalk has the theoretical potential to erase much of the history of life recorded in DNA (see Doolittle, 1999).

The origin of the eukaryotes undoubtedly was a milestone in the evolution of life, because they are very different from prokaryotes in spatial organization. Eukaryotes have an extensive system of internal membranes traversing the cytoplasm and enclosing the mitochondrion chloroplast, and nucleus. This compartmentalization necessitated several unique innovations–the most dramatic being the nucleus, a specific compartment for storing and transcribing DNA, for processing DNA and RNA and possibly even for translating mRNAs (Hentze, 2001).

The prokaryotes, with their simple cellular organization, are believed to have preceded the eukaryotes. What is not clear is which prokaryotic groups branched first, because the root of the tree of life is uncertain and in flux.

The HGT Revolution

Before the advent of genome sequencing, it was widely believed that prokaryotic genomes were evolving clonally–the family tree derived from any one gene would look like the family tree from any other gene. Most researchers felt that reliable organismal trees could be calculated from sequences of individual genes, particularly rRNA genes which were easy to sequence, and it was assumed that trees calculated from rRNA would probably be the same as those calculated from any other genes. It was not known that HGT could potentially alter gene trees.

When many complete genomes became available, the pace of discovery accelerated, and recent studies of the evolution of life, based on analyses of complete genomes, uncovered the flaws in the old view of clonal evolution. Scientific opinion has now shifted and favours a significant role for HGT in prokaryotic genome evolution (Simonson *et al.*, 2005). HGT is now known to be rampant among genomes. Not all genes are equally likely to be horizontally transferred; informational genes (involved in transcription, translation, and related processes) are rarely transferred, whereas operational genes (involved in amino acid biosynthesis, and many other operational activities) are readily transferred. Biological and physical factors that may have affected HGT include intracellular structural constraints among proteins (the complexity hypothesis), interactions among organisms, and interactions with the physical environment (*e.g.* Bult *et al.*, 1996; Simonson *et al.*, 2005). The sequencing work of Bult *et al.* on the methanogen *Methanococcus janaschii* revealed that its genome consisted of certain groups of genes that were much more similar to eukaryotic genes than those from bacteria, whereas other groups of genes were much more closely related to their bacterial homologs. The *M. jannaschii* genes for translation, transcription, replication, and protein secretion turned out to be more similar to eukaryotes than to bacteria. These and other genome studies indicated that eukaryotes were a mixture of eubacterial and archaebacterial genes with an unusual distribution. The operational genes were primarily from the eubacteria, and the informational genes were from the archaea (see Figure 2.1). This finding suggested that the archaea were a chimera of eukaryotic and eubacterial genes (Koonin *et al.*, 1997).

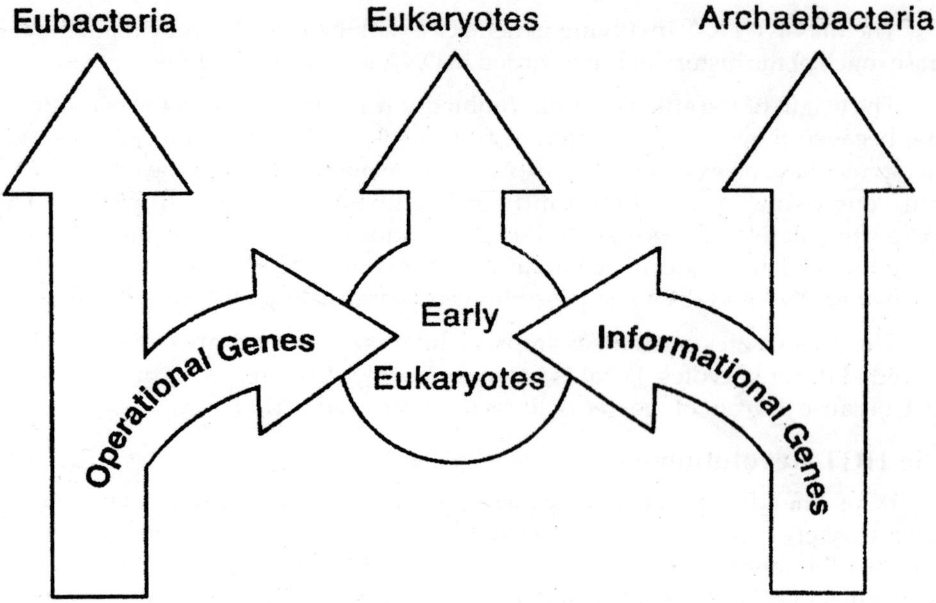

Figure 2.1: Possible Contribution of Operational Genes and Informational Genes from Eubacteria and Archaea, Respectively, to the Evolution of Eukaryotes (After Simonson *et al.*, 2005). (Copyright (2005), National Academy of Sciences, U.S.A.).

HGT is also widely prevalent in the eubacteria (Ochman *et al.*, 2005), *e.g.* in *Aquifex aeolicus*. Comparative analyses of *E. coli* open reading frames (ORFs) revealed that 675 *E. coli* ORFs have greatest similarity to *Synechocystis*, 231 to *M. jannaschii*, and 254 to the eukaryote *Saccharomyces cerevisiae* (Blattner *et al.*, 1997). Ochman *et al.* have argued that 755 of 4,288 *E. coli* ORFs have been horizontally acquired in 234 lateral transfer events, because *E. coli* diverged from *Salmonella* many million years ago. The three chief molecular mechanisms known to drive horizontal transfer are transformation, conjugation, and transduction.

Jain *et al.* (1999) showed that operational genes have been continually transferred much more frequently among prokaryotes since the last common ancestor of life or cenancestor. To explain why operational genes undergo HGT more frequently than informational genes, Jain *et al.* proposed the complexity hypothesis, according to which informational genes are less likely to undergo horizontal transfer, because their products are members of large complexes with many intricate interactions.

Operational genes, on the other hand, usually do not form part of large complexes, and so are more readily transferred. It is also emerging that gene exchange is not only occurring within species, but extensive exchanges take place within larger groups of prokaryotes consisting of multiple species as well (see Simonson *et al.*, 2005).

HGT seems to have had great impact on the evolution of life on Earth. It is perhaps the major agent involved in spreading genetic diversity among prokaryotes by moving genes across species boundaries. By rapidly introducing newly evolved genes into existing genomes, HGT circumvents the slow step of *ab initio* gene creation and so speeds up genome innovation (the acquisition of novel genes by organisms), although not necessarily gene evolution. A collection of organisms probably share genes by HGT and yet need not be in physical contact as an exchange community.

Conditioned reconstruction is a newly developed mathematical algorithm that has greatly improved the prospects of recovering the tree of life in the presence of HGT. It applies to whole-genome-based phylogenetic reconstructions (Lake and Rivera, 2004). CR analyses use the absences and presences of genes as character states and, through the use of a reference genome, they extract additional information that cannot be obtained by other types of analyses. For example, by restricting the analyses to only the genes present in a reference genome R, it is possible also to estimate the number of gene pairs that are missing in both genomes A and B. As this critical information is not available without the reference genome, it allows use of a very general class of mathematical (Markov) models to reconstruct the tree of life.

In CR, the dynamic deletions and insertions of genes that occur during genome evolution, including the insertions introduced by HGT, provide the information needed to reconstruct phylogenetic trees. CR carries some potential to reconstruct deeper branchings in the tree of life than is possible with sequence analyses; this is because whole gene characters evolve more slowly than nucleotides, amino acids, and even gene inserts.

One remarkable property of CR is that it can rigorously identify the merger of genomes–a process that till recently could not be analyzed by using gene sequence. The application of this method has provided evidence that the eukaryotic genome was actually formed by a fusion of the genomes from two disparate prokaryotes (Simonson *et al.*, 2005).

Implications of the Ring of Life

Three notable theories proposed for the origin of eukaryotes are autogenous, chimeric, and genome fusion theories. The results derived in the CR analyses argue against autogenous (*i.e.*, tree of life) theories in which eukaryotes evolved clonally from a single, possibly very ancient, prokaryote. Chimeric theories refer to the acquisition of genes by eukaryotes from multiple sources through unspecified mechanisms. Available data also argue against them. Several genome fusion theories have been proposed in which the eukaryotic genome originated from two diverse genomes (Gupta *et al.*, 1994; Martin and Muller, 1998). These are strongly supported by CR analyses. Symbiotic relationships are fairly common among organisms living together and, in rare cases, this leads to endosymbiosis, the intracellular capture of former symbionts. Given a genome fusion, and in the absence of other mechanisms that could produce fusions, it appears that an endosymbiosis may have been the probable cause (Simonson *et al.*, 2005).

The ring of life implies that prokaryotes predate eukaryotes, because two preexisting prokaryotes contributed their genomes to create the first eukaryotic genome. This appears to place the root of the ring below the eubacterial- and eocytic-eukaryotic last common ancestors (Figure 2.2).

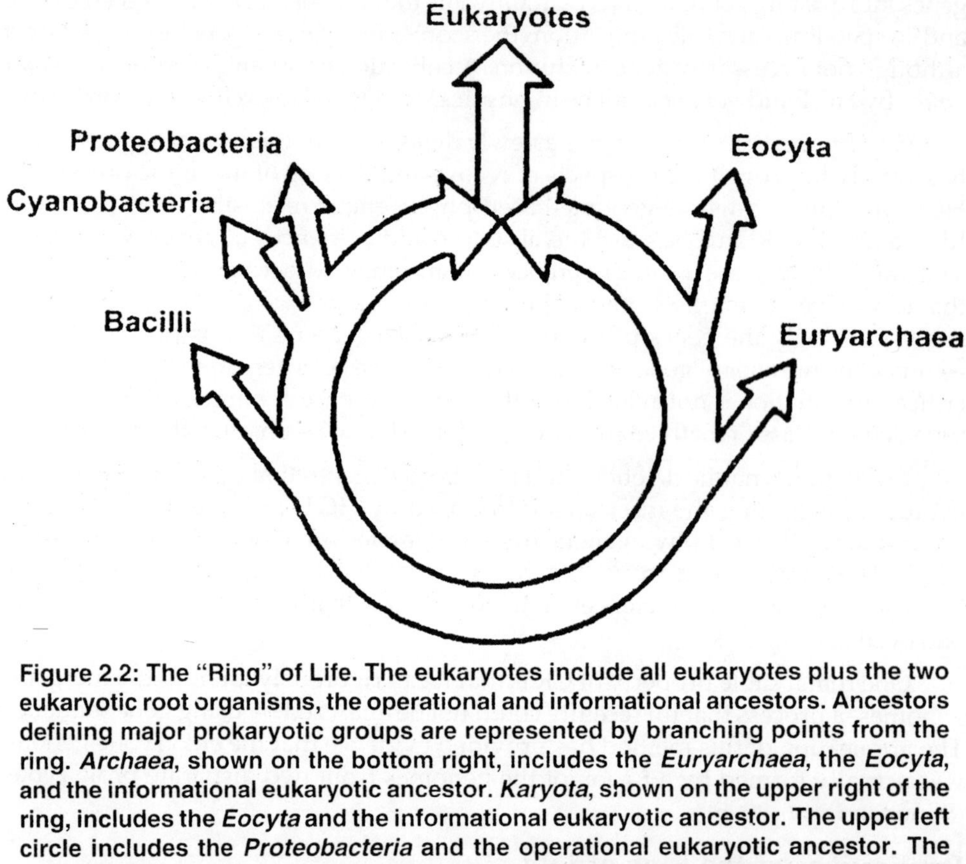

Figure 2.2: The "Ring" of Life. The eukaryotes include all eukaryotes plus the two eukaryotic root organisms, the operational and informational ancestors. Ancestors defining major prokaryotic groups are represented by branching points from the ring. *Archaea*, shown on the bottom right, includes the *Euryarchaea*, the *Eocyta*, and the informational eukaryotic ancestor. *Karyota*, shown on the upper right of the ring, includes the *Eocyta* and the informational eukaryotic ancestor. The upper left circle includes the *Proteobacteria* and the operational eukaryotic ancestor. The most basal node on the left represents the photosynthetic prokaryotes and the operational eukaryotic ancestor (after Lake, 1988; Woese *et al.*, 1990; Simonson *et al.*, 2005). (Copyright (2005), National Academy of Sciences, U.S.A.).

Horizontal Gene Transfer–Direct Visualization

Besides transformation and phage-mediated transduction, conjugation is an important mechanism for horizontal gene transfer in bacteria (Ochman *et al.*, 2000). Ever since the first evidence for sex by conjugation in *E. coli* (Lederberg and Tatum, 1946) by which prototrophic progeny was obtained by mixing two different auxotrophic parents, the phenomenon of horizontal gene transfer has been demonstrated to be involved in widespread transfer among bacterial populations of genes conferring antibiotic resistance, metabolic functions, and virulence determinants (Babic *et al.*, 2008).

Conjugational DNA is transferred by the F plasmid unidirectionally from an F^+ donor cell to an F^- recipient cell. The F plasmid has all the genes needed for conjugation and for regulation of DNA mobilization and its own unidirectional transfer. At low frequencies, this plasmid can integrate into the chromosome of the host cell, generating an Hfr (high frequency of recombination) strain. Chromosomal genes of the Hfr bacterium can be mobilized and transferred to a recipient. Sometimes, F excises from the chromosome of Hfr, creating an F′ molecule which carries chromosomal genes as well as the conjugation genes (Low, 1991). Both Hfr and F′ can serve as DNA vehicles in horizontal gene transfer between bacteria.

The contact between mating cells occurs through a tube-like structure known as the F-pilus. DNA is transferred from the donor to the recipient in single-stranded form. It is converted to duplex DNA by the synthesis of the complementary strand in the recipient cell. When the conjugational transfer stops, double-stranded donor DNA is either circularized (in the case of F′ transfer) or, in the case of Hfr transfer, incorporated into the recipient chromosome by RecA-dependent homologous recombination or degraded by RecBCD exonuclease (Lloyd and Buckman, 1995).

Many details about the mechanism and consequences of conjugation are still unknown and include the role of the F pilus in DNA transfer during conjugation, the fate of the transferred DNA, the global frequency of the horizontal gene transfer (versus the frequency of inheritance of individual genetic markers), and the pattern of inheritance of donor DNA present in the initial transconjugant cell. To address these issues, Babic *et al.* (2008) developed an experimental system that allows to distinguish the transferred donor DNA from both donor and recipient DNA, and to visualize DNA transfer and recombination by means of fluorescence microscopy, at the level of individual living cells. This enabled Babic *et al.* to quantify the ongoing transfer of DNA during conjugation and to make time-lapse movies to track the fate of the newly acquired DNA in individual cells through any number of cell divisions.

By using a fluorescent protein fusion, SegA-YFP, Babic *et al.* visualized this process in real time and in single cells of *Escherichia coli*. It was seen that the F pilus mediates DNA transfer at considerable cell-to-cell distances. Transferred DNA integrated by recombination in up to about 95 per cent of recipient cells, whereas in the remaining cells it was completely destroyed by the RecBCD helicase/nuclease. Tracking the acquired integrated DNA through successive replication rounds revealed that it occasionally split and segregated with different chromosomes, leading to the inheritance of different gene clusters within the cell lineage. The incidence of DNA splitting corresponded to about one crossover per cell generation (Babic *et al.*, 2008).

The tubular F pilus which projects from the surface of the donor cell is required for conjugation. According to Achtman *et al.* (1978) and Panicker and Minkley (1985), the pilus only facilitates contact between the mating cells and that DNA is transferred soon thereafter through a pore in the membrane. But according to Harrington and Rogerson (1990), DNA can be transferred from the donor to the recipient even when mating cells are physically separated. Babic *et al.* observed SeqA-YFP foci appearing in 43 of 753 recipients that were not in direct cell wall contact, and the maximal observed distance to the closest donor was 12 μm. No distant transfer of DNA was noted in the presence of 0.01 per cent SDS (sodium dodecyl sulphate, which depolymerizes F pili) or when the transfer-deficient *traA* donors (which lack the pili) were used. Clearly, the appearance of distant SeqA-YFP foci depends on conjugation and on the presence of F pili. These results support the idea that the F pilus, besides establishing contact between mating cells, also acts as a conduit for single-stranded DNA transfer during conjugation.

Another important finding discussed by Babic *et al.* is that direct, physical, measure of recombination generates significantly higher recombination frequencies than those measured genetically as the acquisition of specific genetic markers (ranging from 10 to 30 per cent depending on specific genetic marker, donor strain, and mating conditions). Clearly, conjugational recombination is very efficient when donors and recipients are genetically identical strains.

The technique used by Babic *et al.* enabled them not only to visualize but also quantify the DNA of any sequence as it is being transferred from one cell to another, also to watch its stable genomic acquisition via genetic recombination (horizontal gene transfer) in real time. The observations also implicated F pilus-mediated conjugation at cell-to-cell distances of up to 12 μm and enabled estimation of the frequency of intragenomic crossover events, or sister-chromatid exchanges. This experimental system should now be successfully extended to monitor horizontal gene transfer by indefinitely following the fate of DNA acquired in intra- and inter-species crosses (Babic *et al.*, 2008).

Lateral Gene Transfer and Microbial Systematics

The objective of most workers who study evolution is to learn about the natural historical process that gave rise to various taxa, to classify the taxa efficiently, and to make generalizations about them. The quantitative importance of lateral gene transfer (LGT/HGT) inferred from genomic data has been well acknowledged by microbiologists but conflicts with the conceptual foundations of the traditional phylogenetic system erected to achieve these goals (Bapteste and Boucher, 2008). For microbial evolution, Bapteste and Boucher (2008) suggested developing an alternative concept of natural groups for which they introduced a new notion–the composite evolutionary unit. They also argued that a comprehensive database containing overlapping taxonomical groups should be created for the classification of microbes in which LGT is so widespread.

Molecular Phylogenetics

An ambitious molecular phylogenetics project was proposed by Zuckerkandl and Pauling (1965). It was expected to function as a powerful time machine, enabling the identification of genetic, ultrastructural and metabolic features of ancient life forms for which no fossils could be found. Through their congruence (the agreement between phylogenies obtained by using different data sets), genes helped in reconstructing the Tree of Life (TOL). By using robust phylogenetic algorithms accurately to model the complex evolution of molecules (Falsenstein, 2004), and by sequencing large number of phylogenetic markers (Cavalier-Smith, 1981), several TOLs were created (Schwartz and Dayhoff, 1978; Ciccarelli *et al.*, 2006). Lateral inheritance (as opposed to vertical descent) has emerged as strong evolutionary force in microorganisms (Koonin *et al.*, 2001; Hanage *et al.*, 2006; Lo *et al.*, 2007). For archaea, bacteria and some unicellular eukaryotes, individual gene histories sometime markedly differ from species history, and their species trees and gene trees do not have to show much identity with one another on a broad evolutionary scale. A single microbial genome can comprise genes arising from multiple phylogenetic sources, yet it does conflict with the conceptual foundations of the phylogenetic system.

Consequently, the traditional TOL reconstruction project could not be applied to prokaryotic organisms. To address this issue, Bapteste and Boucher (2008) proposed alternative concepts to the traditional phylogenetic projects to deal with microbial evolution and systematics: (i) a redefinition of natural groups; (ii) the description of a new type of evolutionary unit originating from LGT; and (iii) the realization of an interactive taxonomical data-base (comprising overlapping groups) to make possible a more natural classification. The third of these solutions would constitute a transition possibly as significant as the change from a linear system of classification to a nested hierarchy that occurred thousands of years ago (Bapteste and Boucher, 2008).

Evolutionary Trees

Two kind of evolutionary trees have been built (i) Genome trees are based on the statistical properties of the genome, on the presence or absence of genes, on the chromosomal gene order or on average sequence similarity, as calculated in BLAST analyses. (2) Phylogenomic trees are based on vertically inherited orthologs (see Snel *et al.*, 2005). Genome trees enable comparison of the evolutionary information found in various genomes, but do not reflect the exact course of organismal evolution and should not be interpreted as phylogenies. Also, such phonetic trees are complex to interpret because some of the groupings obtained can result mainly from lateral relationships, whereas others result from vertical ones. Genome trees only reveal prevailing trends in the evolution of genome-scale gene sets (Wolf *et al.*, 2002). In contrast, phylogenomic Trees (species Trees) are reconstructed to understand the pattern of natural relationships between species (Brochier *et al.*, 2005) on the basis of strictly vertically inherited markers. For this, molecular datasets are pruned to exclude those genes which give signals. The problem is the paucity of data for which to confidently assess a strictly vertical transmission, resulting in skeletal microbial phylogenomic trees, built on a very small amount of information. This is exemplified by the latest TOL (Ciccarelli *et al.*, 2006).

Alternative Approaches to Microbial Phylogenetics and Systematics

One important definition of a natural group has been prompted by the work of Splitter (1988), Collier and Muller (1998) and Brooks (2001) and could benefit microbial phylogenetics. According to Splitter, a natural group is 'natural when it is causally efficacious, relative to some explanatory theory' that is, natural groups are real when they have a real causal impact and real consequences on the biological world. Under this definition, evolutionary units constitute natural groups. The consequences of such a perspective are profound. Firstly higher taxa (*e.g.* the Proteobacteria) might not be considered as a natural group under this definition because there is no such thing as a real causal impact of the Proteobacteria phylum (*i.e.* no single physiological feature is shared by all Proteobacteria that is not a general feature of bacterial cells). Such higher taxa are an arbitrary way of classifying the living world rather than the natural one. Second, since multiple evolutionary units of all sizes play some role in different biological processes, natural groups in a revised systematics should be quite diverse.

Composite Evolutionary Units

Microbiologists are well aware of various phylogenetically diverse, yet functionally integrated groups. Coevolving associations of multiple phylogenetically distinct microorganisms occur frequently in nature and such evolutionary units usually show emerging properties that none of their constituent parts harbours alone. One example is that of syntrophic microbial consortia, composed of multiple organisms with various physiologies, which can perform chemical reactions that would be energetically unfavourable if carried out by a single microbe. This kind of relationship has been reported between closely associated methanotrophic archaea and sulfate-reducing bacteria found in anoxic marine sediments (Nauhaus *et al.*, 2007). Here, the archaeal partner metabolizes methane and the bacteria use a resulting metabolite as an electron source. Ecological and environmental pressures influencing LGT, and consequently the genetic units composing organisms, generate new evolutionary units by the association of different genes or pathways within organisms. A good example is significant LGT between Sulfolobales and members of the Thermoplasmatales, two phylogenetically distant phyla that frequently share thermoacidophilic environments (Ruepp *et al.*, 2000). The evolution of the hyperthermophilic bacteria Thermotogales may also have been influenced by their uptake of DNA from the archaea that often share their environment (Nelson *et al.*, 1999).

Composite evolutionary units of all sizes and levels can appear in nature. These units depend on parts which might have different origins, some global biological process being responsible for their association while selection is acting on the emerging higher-level phenotype. When a biological function is selected, the composition of its lower-level structural components is sometime flexible (*i.e.* bacteria can synthesize the essential isoprenoid building block isopentenyl diphosphate through two analogous pathways, one using 1-deoxy-D-xylulose-5-phosphate as a precursor and the other using mevalonate (Boucher *et al.*, 2003). The list of genes able to fulfill a function can be extensive. Studies on LGT point to the existence of several levels of selection and the presence of many biological 'individualities' in complex interactions in the microbial world. This warrants a richer view of biodiversity and one that comprises more evolutionary units than the mere 'species' and 'genes' usually considered in traditional phylogenetics, and hence more natural groups to classify (Bapteste and Boucher, 2008).

Agrobacterium tumefaciens–The Crown Gall Bacterium

Till recently, gene transfer from bacteria to plants was thought to be limited to the bacterial genus *Agrobacterium*. But other bacterial groups also contain species capable of interkingdom genetic exchange (Gelvin, 2005; Broothaerts *et al.*, 2005). 'Horizontal' gene transfer is a major factor in evolution. In terms of ongoing processes, bacterial conjugation (gene transmission from one bacterium to another through direct physical contact) is the commonest known example. The horizontal transfer of genetic material between species from different phylogenetic kingdoms–from bacteria to plants, for instance–can also take place. It has previously been documented only for the bacterial genus *Agrobacterium*. Broothaerts *et al.* now show that several bacterial species outside this genus are also capable of interkingdom 'sex'.

Along with the related species *A. rhizogenes, A. vitis* and *A. rubi, Agrobacterium tumefaciens* incites tumours by transferring a portion of a large extrachromosomal DNA element (the tumour-inducing or Ti, plasmid) to its plant hosts. When inside the plant cell, this transferred DNA (the T-DNA) targets the nucleus, where it eventually integrates into the host genome. The expression of T-DNA-encoded oncogenes makes the plant overproduce or become overly sensitive to hormones, leading to loss of control of cell-growth and to tumour formation.

Scientists have made use of the ability of *Agrobacterium* spp. to transfer DNA to plants for genetic-engineering purposes: oncogenes may be deleted from the bacterial T-DNA and substituted with some gene of interest. Once transferred, these 'transgenes' can confer new characteristics on plants, including herbicide or pathogen resistance, altered growth or nutritional qualities, or the ability to produce drugs or edible vaccines–antigens that can direct immunity against pathogens in humans or animals that ingest the plants (Gelvin, 2005). Under laboratory conditions, *Agrobacterium* is also capable of genetically transforming fungal and animal cells.

Although *Agrobacterium* can transfer genes to plants and, when suitably modified, is the most effective vector for gene transfer in plant biotechnology, the complexity of the patent landscape has created several obstacles to the effective use of this technology for agricultural improvements worldwide (Gelvin, 2003; Roa-Rodriguez and Nottenberg, 2003). Broothaerts *et al.* (2005) have shown that *Agrobacterium* is by no means the only bacterial genus that can transfer genes to plants–several species of bacteria outside the genus can also be modified to mediate gene transfer to a number of diverse plants. These plant-associated symbiotic bacteria (*e.g.* species of *Rhizobium* and *Sinorhizobium*) can be made competent for gene transfer by acquisition of both a disarmed Ti plasmid and a suitable binary vector. This alternative to *Agrobacterium*-mediated technology for crop improvement, besides affording a versatile 'open source' platform for plant biotechnology, can potentially lead to novel uses of natural bacteria-plant interactions to achieve plant transformation.

Broothaerts *et al.* used *Rhizobium* species NGR234, *Sinorhizobium meliloti* and *Mesorhizobium loti*–representatives of two different bacterial families. They first introduced a 'disarmed' Ti-plasmid that lacks a T-DNA region into these bacteria. This Ti-plasmid contained certain virulence genes whose protein products are needed to act on a T-DNA and transfer it to plants. Then a second plasmid containing a T-DNA region was introduced into these non-*Agrobacterium* strains. This T-DNA had a gene coding for resistance to the antibiotic hygromycin, as well as *gusA*, a gene that directs the expression of the marker enzyme β-glucuronidase. To avoid any mixing up, the T-DNAs introduced into *Agrobacterium* and non-*Agrobacterium* species contained different molecular 'tags'.

These bacteria were then used to infect several different plant species, representing two major higher-plant groupings: species from the dicots (which are highly susceptible to *Agrobacterium*-mediated transformation) included tobacco and *Arabidopsis*; from the monocots (which are much less susceptible), rice was used. The production of plants expressing β-glucuronidase and showing resistance to hygromycin indicated successful transformation by all bacterial species tested. This was confirmed by DNA-blot analysis and by the sequencing of T-DNA-plant-DNA junctions recovered from plants transformed with non-*Agrobacterium* species (Gelvin, 2005).

Broothaerts *et al.* also reported that non-*Agrobacterium* spp. are capable of the full range of genetic transformation mechanisms shown by their *Agrobacterium* counterparts. Although non-*Agrobacterium* bacterial species could be engineered to genetically transform several plant species, the transformation efficiency was relatively low and ranged from less than 1 per cent to almost 40 per cent of that of *Agrobacterium*-mediated transformation.

The above findings will surely have strong implications for plant science and biotechnology.

Chapter 3
Nanobiology: Viruses as Nano-organisms

Introduction

As the vast majority of viruses have at least one dimension less than around 100 nm, they are best considered as nano-organisms. Viruses lie at the borderline of living and nonliving. Most microbiologists included them as acellular microorganisms in their textbooks. Except for a few big viruses, which may be regarded as microorganisms, the overwhelming majority are nano-organisms rather than microorganisms. Viruses differ from cellular organisms in that any given virus contains either DNA or RNA, not both (a few exceptions to this have been reported recently). This nucleic acid is covered with a protein coat. There are much fewer proteins (enzymes) in a virus than in a cell. A good example of a well-studied virus is tobacco mosaic virus (TMV). The concept of the virus as an infectious agent has been strongly influenced by studies on the mosaic disease of tobacco and its causal agent TMV. Viruses as pathogens were discovered at the end of the 19th century by Martinus Beijerinck who noted that the sap from infected tobacco plants, when passed through porcelain filters, retained its infectivity to tobacco plants; he concluded that it was a new form of infectious agent, a 'contagium vivum fluidum'. He named it a virus (In cases where bacteria caused some plant disease; the filter retained the pathogen).

The TMV was studied in detail in the 1930s and was isolated and characterized. It was well suited for such studies because of its abundance in its host plant, its stability and ease of transmission. Later studies on certain mutants provided useful genetic information.

The 1950s saw much excitement about TMV and many fundamental concepts about this and other viruses emerged. It was concluded that the virus is a nucleoprotein. In the 1950s also the nature of the viral protein and the structure of the virus were determined and it emerged that the viral RNA was the infectious agent (protein was not needed for infectivity); the sequence of the coat protein was analyzed and natural and induced mutants were examined. Viable virus could be successfully regenerated by reconstituting it from its isolated protein and nucleic acid components.

The beautiful structures and symmetries in the submicroscopic world of viruses were first detected in the 1960s but precise details of their structural geometry remained obscure. Particularly revealing electron micrographs of the negatively stained herpes simplex virus were made in 1960 in Glasgow. When devoid of their floppy envelopes, the protruding prisms of the morphological units or capsomeres appeared to be arranged in the 5-3-2 symmetry of an icosahedron. The capsomeres were hexameric in cross-section, with the exception of 12 at the apices that were pentameric.

D. Caspar and A. Klug, in 1962, proposed that the icosahedral regularity of the capsid and the less regular distribution of the capsomeres could be reconciled by the concept of quasi-symmetry. Some parallels with the families of polyhedra characterized by Buckminster Fuller, the architect of geodesic domes, were also noted and it was later found that there are remarkable similarities in the designs of geodesic domes, viruses and fullerenes and these designs combined structural economy with operational efficiency (Kemp, 1998).

There has been a striking revolution in our understanding of the viral world. Despite their profound importance for disease, not much was known about viral relationships. Viral origins have been obscured by their great diversity, complexity, and small size. Viruses evolved clever solutions not only to the problems of packing information into genomes of limited size but also penetrating and exploiting their host. Until recently, viruses could only be imaged by low resolution electron microscopy (EM) and most viruses were not amenable to crystallographic approaches. Spectacular advances in imaging techniques, changes in the manner that viruses are studied, and increasing structural evidence have contributed to a major change–viruses fall into lineages identifiable by features in common with ancient precursors (Bamford *et al.*, 2002). The best-supported lineage has "double-barrel trimer" coat proteins and icosahedral virions (Burnett, 2006). Its members infect not only Gram-positive and Gram-negative bacteria but also animals. They include very large viruses infecting insects, algae, and amoebae. Structures have been elucidated for coat proteins of viruses infecting animals, bacteria, and algae, but not for any archaeal virus. This crucial missing link has now been found in the thermophilic archaeal *Sulfolobus* turreted icosahedral virus (STIV) (Khayat *et al.*, 2005). There is impeccable structural evidence that at least one lineage originated billions of years ago, before life separated into its three domains: Bacteria, Archaea, and Eukarya.

Dramatic progress has occurred in describing viral architecture since the first crystal structure of a complete viral particle, or virion was published by Harrison *et al.* (1978). This plant virus has an eight-stranded barrel, with a "viral" jellyroll fold, for its coat protein. As most other viruses, including simple animal viruses, also have this fold, a possible relationship was suspected (Rossmann and Johnson, 1989). Larger viruses (such as adenovirus) were more difficult to study by means of crystallography. Even elucidating the structure of the coat protein alone, as recently done for the STIV major coat protein (MCP, Khayat *et al.*, 2005), was very difficult for molecules like the 967-residue adenovirus hexon whose EM images revealed that this molecule, although trimeric, has a pseudohexagonal shape and the viral hexons pack closely in the adenovirus virion.

One crucial issue in viral architecture is how multiple copies of the coat protein locate correctly for error-free assembly. The answer may possibly be that they lie in similar environments.

Abrescia *et al.* (2004) studied the organization of the coat proteins in bacteriophage PRD1 (and adenovirus). The 240 closely packed pseudohexagonal trimers (P3 or hexon) form arrays of 12 in the facets, which bend or curve to bring the proteins together at the icosahedral edge. An additional separate location between trimers 1-4 yields facets of any size. The elongated PRD1 tape-measure protein stretches from the vertex to the middle of an edge, where it forms a hook (see Burnett, 1985; Abrescia *et al.*, 2004).

Recent commendable progress in structural virology (Bamford *et al.*, 2005) would have been impossible without cryo-EM as a supplement to crystallography. In cryo-EM flash-frozen virions are protected from radiation damage and imaged without stain by using electrons to obtain views of many particles in different orientations. The first EM reconstruction of the adenovirus virion at 35-? resolution (Stewart *et al.*, 1991) revealed that the facets were rounded, rather than planar, to bring edge hexons into more intimate contact. Instead of the 27 virions used in the past, several thousands can now be used to obtain resolution <10 ?. Cryo-EM can image very large virions (>1,000 ? in diameter). It does not need large quantities or crystals, and can even reveal protein secondary structure. Impressive examples of the many virions imaged are *Paramecium bursaria* Chlorella algal virus 1 (Yan *et al.*, 2000) and the largest one known, mimivirus (5,000 ? in diameter) (Xiao *et al.*, 2005). If EM is combined with crystallography, even better images are obtained as shown by the approach to STIV– here, EM has been particularly valuable because MCP crystallizes as a monomer (Burnett, 2006).

Several critical researches on some viruses have prompted the conclusion that viruses may be descendants of ancient self-reproducing systems that enabled early cells to develop by shuttling genes between the protected environments in divisible lipid vesicles. There is evidence of a primordial role for viruses (and barrels), rather than to a later appearance.

Structure and Organization

Viruses, viroids and other similar organisms are acellular in the sense that they lack a typical cell structure. Any given virus contains either RNA or DNA but not both. Cellular organisms, on the other hand, contain both RNA and DNA in their cells.

Four basic types of virus organization are shown in Figure 3.1. Most enveloped viruses attack animals and their envelope is partly derived from the host plasma membrane.

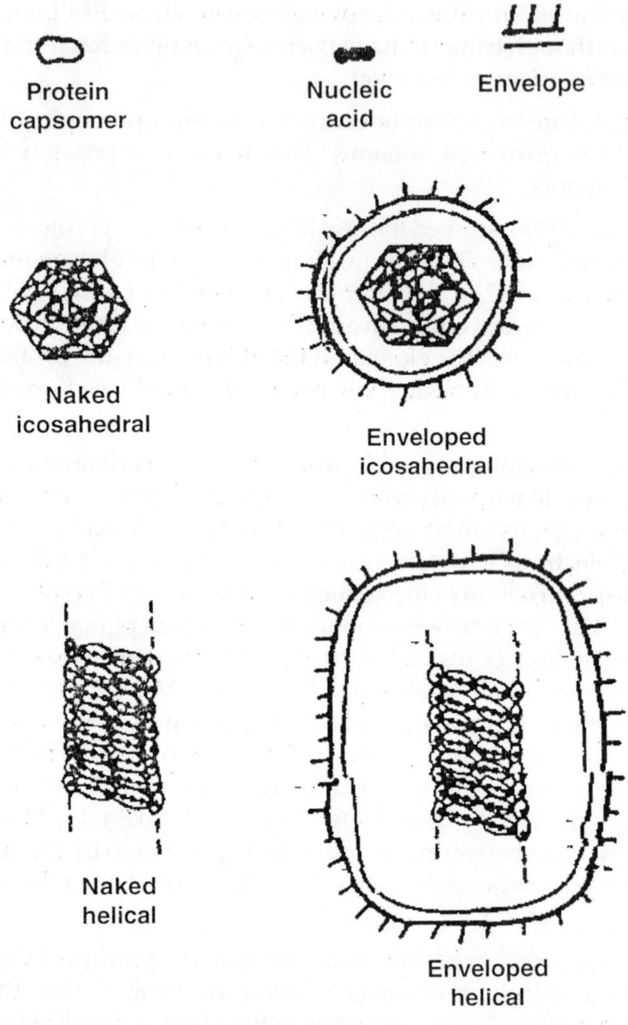

Figure 3.1: Four Basic Types of Virus Structure.

Some viruses have a highly complex structure. The best studied are the tailed bacteriophages (Figure 3.2) and, in particular, the coliphages. Whereas the head of the phage is an almost conventional icosahedron containing the viral nucleic acid, there is a very complex tail structure attached to it containing at least five different protein types making up the sheath, core, collar, end plates and fibres.

Viruses may be classified on the basis of mechanism of replication and on the relationship of the genome to messenger RNA (see Figure 3.3). Figure 3.4 illustrates some representative members of animal viruses.

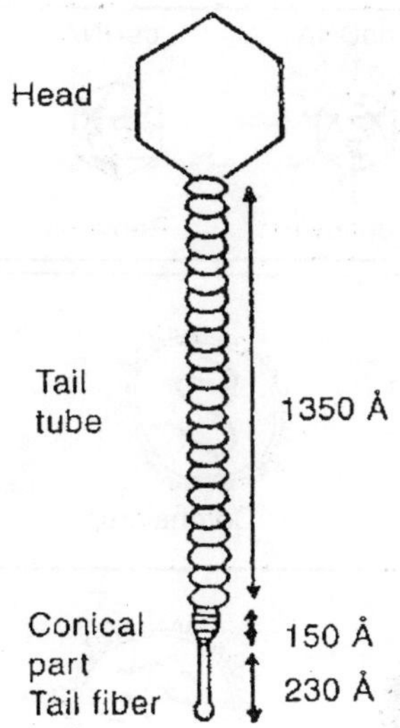

Figure 3.2: The Structure of a Lambda Phage Particle (A tailed bacteriophage).

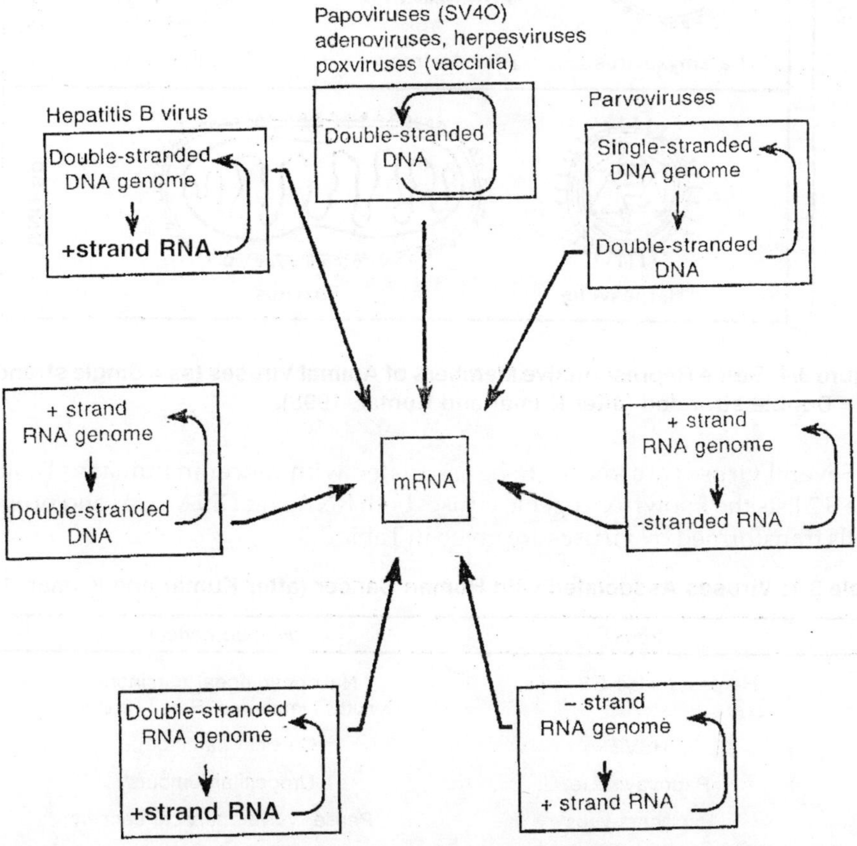

Figure 3.3: Classification of Viruses Based on the Mode of Replicatio.n and on Relationship of Genome to mRNA. Arrows within boxes show information flow during replication. Arrows to mRNA originate at template for mRNA synthesis (after Kumar and Kumar, 1998).

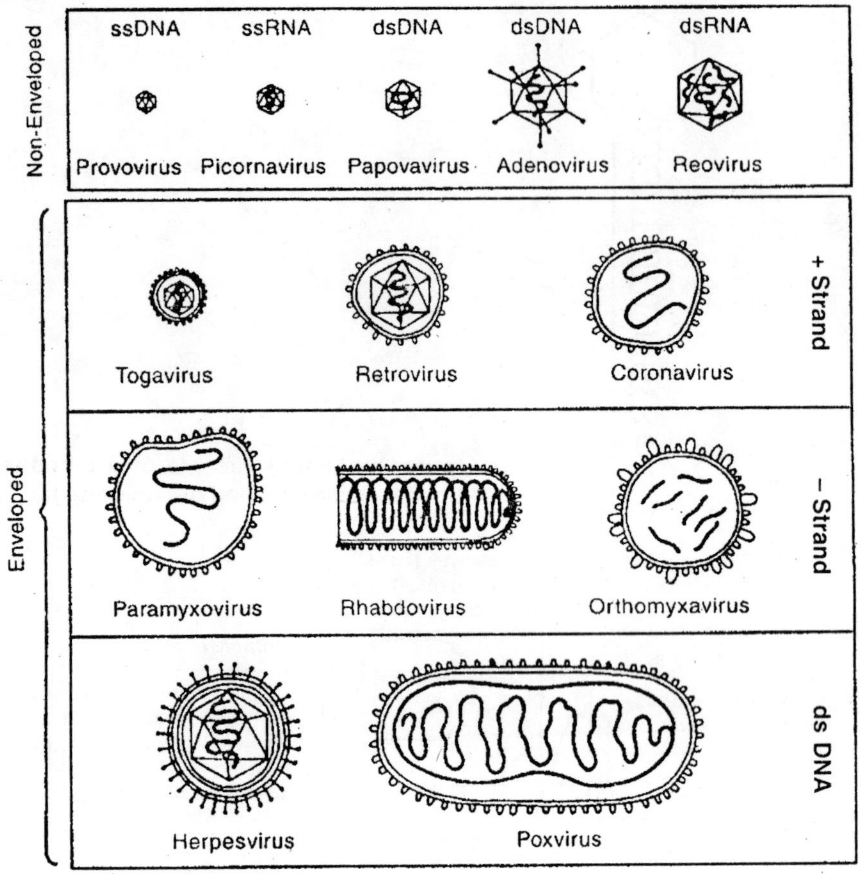

Figure 3.4: Some Representative Members of Animal Viruses (ss = Single stranded; ds = Double stranded (after Kumar and Kumar, 1998).

Several viruses are known to be associated with cancer in humans (Table 3.1). Table 3.2 lists the known oncogenic viruses both RNA and DNA virus and properties of cells transformed by viruses are given in Table 3.3.

Table 3.1: Viruses Associated with Human Cancer (after Kumar and Kumar, 1998)

Virus	*Human cancer*
Herpes viridae EB virus	Nasopharyngeal carcinoma Burkitt's lymphoma, B cell lymphoma
HSV2	Cervical carcinoma
Papovaviridae	Urogenital tumours
Papilloma virus	Penile, vulval, cervical cancers
Hepadnaviridae HBV	Primary hepatocellular carcinoma
Retroviridae HTL virus	Adult T cell leukaemia

Table 3.2: Oncogenic Viruses (after Kumar and Kumar, 1998)

A. RNA virus

- I. Retroviruses
  1. Avian leucosis viruses
  2. Murine leucosis viruses
  3. Murine mammary tumour virus
  4. Leukosis-sarcoma viruses of various animals
  5. Human T cell leukaemia virus

B. DNA Virus

- I. Papovavirus
  1. Papillomaviruses of man, rabbits and other animals.
  2. Polyomavirus
  3. Simian virus 40
  4. BK and JC viruses
- II. Poxvirus
  1. Molluscun contagiosum
  2. Yaba virus
  3. Shope fibroma
- III. Adenovirus

  Many human and nonhumantypes
- IV. Herpes virus
  1. Marek's disease virus
  2. Lucke's frog tumour virus
  3. Herpes virus saimiri
  4. Epstein–Barr virus
  5. Herpes simplex virus types 1 and 2
  6. Cytomegalovirus
- V. Hepatitis B virus

Viral Quasispecies

Viral species is a complex, self-perpetuating population of diverse, related entities that act as a whole. The standard definition of a biological species does not apply to viruses. Essentially, a virus is a genetic program that carries the simple message 'Reproduce me!' from one cell to another. Because a virus represents only one or a few of the messengers vying for the attention of its host, it has to use certain biochemical tricks to recruit the host's replication machinery for its selfish purpose. Often those ploys result in the host cell's death.

The genome of a single-strand RNA virus such as HIV, which comprises only 10,000 nucleotides, is small and simple compared with that of most cells. Yet, from a molecular standpoint, it is exceedingly complex.

Table 3.3: Properties of Cells Transformed by Viruses (Kumar and Kumar, 1998)

I. Altered cell morphology:

Fibroblasts become shorter, parallel orientation is lost

Chromosomal aberrations appear.

II. Altered cell metabolism:

Increased growth rate; increased production of organic acids and acid mucopolysaccharides.

III. Altered growth characteristics:

Loss of contact inhibition, formation of heaped-up growth (microtumours), capacity to divide indefinitely in serial culture, capacity to grow in suspension or in semisolid agar.

IV. Antigenic alterations:

Appearance of new virus specified antigens (T antigen-TSTA)

Loss of surface antigens, cells become agglutinable by lectins.

V. Capacity to induce tumours in susceptible animals.

The concept of sequence space is an invaluable tool for interpreting what is a viral species. In biology, the members of a given living species must show common traits and must be at least potentially able to produce offspring by recombining their genetic material. At the genetic level, a biological species is represented by a huge variety of differing DNA molecules.

A small number of neutral mutations trend to continuously replace the existing wild type in the population. The genome of a species would therefore drift steadily but randomly through a certain volume of sequence space. Wild-type populations will localize sharply in sequence space after completing an advantageous or neutral shift. Mutations appear blindly, irrespective of their selective value. No single neutral or advantageous mutation would occur more frequently than any disadvantageous one.

Biologically, the quasispecies is the true target of selection. All the members help to perpetuate the stable population. The fitness of the entire population is what matters, not the fitness of individual members. Chemically, the quasispecies is a multitude of distinct but related nucleic acid polymers. Its wild type is the consensus sequence that represents an average for all the mutants, weighted to reflect their individual frequency. Physically, the quasispecies is a localized distribution in sequence space that forms and dissolves cooperatively in very much the same way that molecules of water pass through phase transitions as they freeze or evaporate. Its stability is limited by the error threshold–a kind of 'melting point' for the genome information.

Population dynamics of a virus depend on the error rate of its replication process. These figures are highly simplified representations of the sequence spaces that might contain a viral population. Because the error rate directly determines the size and integrity of a quasispecies, it is the most telling characteristic of a virus. The error rate is the probability that an error will occur when one nucleotide in a sequence is being copied. It depends both on the type of nucleotide substitution taking place and on its position in the sequence.

A Hot New Virus

A new virus has been isolated from the near-boiling water of a thermal pool in Yellowstone National Park. Its host is the hyperthermophilic archaeon *Sulfolobus sulfataricus*, which thrives at temperatures above 80°C and a pH of 2. Although very few viruses of Archaea have been described, it appears that archaeal viruses may be just as diverse as the more extensively characterized viruses of Bacteria and Eukarya. The viruses that infect the archaeal halophiles are so far confined to ones that have the same virion morphology and even occasional sequence similarity with the familiar tailed bacteriophages, but the viruses of the hyperthermophiles are a strange and diverse group with highly diverse virion morphologies. Indeed the new virus looks quite unlike anything described before. It is an icosahedrally symmetric virus which differs in detail from previously-described viruses of this shape.

The most dramatic morphological feature of the virion is the protruding "turrets" that extend 13 nm above the capsid surface at the 12 five-fold symmetrical positions of the icosahedron. The turrets may have a role in attaching the virus to the cell and initiating infection. The virus has been named STIV, for *Sulfolobus* turreted icosahedral virus (Rice *et al.*, 2004). It has a small, 17,663-bp circular double-stranded DNA genome, with 36 probable protein coding genes. Only three of the predicted proteins match sequences in the public databases; all three are of unknown function, with two from other *Sulfolobus* viruses and the third from a *Sulfolobus* genome. The preponderance of "pioneer" sequences is strongly reminiscent of the situation in the tailed bacteriophages (Hendrix, 2004).

Archaeal Virus with a Capsid that Spans All Domains of Life

Of the three major domains of life (Eukarya, Bacteria, and Archaea), the least understood is Archaea and its associated viruses. Many Archaea are extremophiles, with species that grow at some of the highest temperatures and extremes of pH of all known organisms. Phylogenetic rRNA-encoding DNA analysis places many of the hyperthermophilic Archaea (species with an optimum growth ^{3}80°C) at the base of the universal tree of life, suggesting that thermophiles were among the first forms of life on earth. Very few viruses have been identified from Archaea as compared to Bacteria and Eukarya. Rice *et al.* (2004) reported the structure of a hyperthermophilic virus isolated from an archaeal host found in hot springs in Yellowstone National Park, USA. Sequencing the circular double-stranded DNA viral genome showed that it shares little similarity to other known genes in viruses or other organisms. By comparing the tertiary and quaternary structures of its coat protein with those of a bacterial and an animal virus, conformational relationships were found among all three, suggesting that some viruses may have a common ancestor that preceded the division into three domains of life >3 billion years ago.

Redefining Virology

Human cytomegalovirus (HCMV) is a giant DNA herpesvirus with more than 200 genes. Its DNA is infectious by itself, and the additional proteins that the virus brings into the host cell further enhance the efficiency of infection. Herpes simplex virus (HSV) has only 84 known genes and can serve as a guide reference for HCMV. The latter brings into the host cell all that should be necessary to enable its genes to be expressed. Gene expression results in production of lots of infectious progeny and curtailment of any inimical response on the part of the host cell. Yet it seems that this is not quite enough. HCMV particles contain not only DNA (as is to be expected of a DNA virus) but also four species of mRNA (Bresnahan and Shenk, 2000; Roizman, 2000).

These four mRNA species are transcribed from one immediate-early gene, two early genes, and one late gene. On entering a host cell, the viral capsid is transported to the nuclear pore where its DNA load is released into the nucleus. The four mRNAs are most likely localized in the viral tegument–a layer of protein between the capsid and the envelope; they remain in the cytoplasm where they are translated into proteins in the absence of newly made viral gene products. This is unexpected–the current dogma is that in cells infected with DNA viruses, all newly synthesized proteins arise from mRNA transcribed from the virus genome after infection (Roizman, 2000).

What are the early events in the infection of host cells by HCMV? After fusion of the viral envelope with the plasma membrane, the capsid (with some associated tegument proteins) is transported to the nuclear pore where viral DNA is released into the nucleus. There, the DNA forms a circle and is transcribed by the cellular transcription machinery. Other tegument proteins remain in the cytoplasm or are independently transported to the nucleus. Viral mRNAs carried into the host cell with the capsid are translated in the cytoplasm. At least one of the proteins encoded by the viral mRNAs is associated with the endoplasmic reticulum-Golgi network.

It appears that the four mRNAs may provide a structural framework for the packaging of tegument proteins in virus progeny.

The tegument is a structure unique to herpes viruses. It consists of an amorphous layer of protein located between the capsid containing the DNA and the envelope, the membrane that forms the outer layer of the virion. Herpes viruses acquire the tegument just before or during envelopement and, in some of them, the tegument is either exchanged or is modified after addition of the envelope. Of the 84 known viral proteins of HSV, at least 12 are components of the tegument. HCMV, with more than 200 open reading frames, may have twice this number of tegument proteins. After entry of the virus into the host cell by fusion of the viral envelope with the plasma membrane, the capsid (with some associated tegument proteins) is transported to the nuclear pore where the DNA is released. Other tegument proteins either remain behind in the cytoplasm or are transported to the nucleus independently of the capsid. Many HSV tegument proteins are dispensable for viral replication in cultured cells but turn out to be crucial for viral pathogenesis in experimental animal systems (Roizman, 2000).

Tegument proteins create an intracellular environment within the host cell so that the invading pathogen can take over the cell machinery for production of virion progeny.

The work of Bresnahan-Shenk has struck a death blow to the classical definition of a virus, *viz.* any given virus has either DNA or RNA but not both. The original definition was designed to differentiate between obligate intracellular parasites containing both types of nucleic acid, and viruses that hitherto were known to contain either RNA or DNA; HCMV is the first DNA virus to be found with DNA and RNA co-existing before transcription of the viral genome. Although the amount and nature of the mRNAs in HCMV readily differentiate this virus from intracellular parasites, the definition of viruses has lost its key discriminator (Roizman, 2000).

Bacteriophages

Usually, whenever a cell is infected with a virus it alerts the host immune system by expressing detectable markers on its surface. These telltale markers are recognized by the infected cell, preventing viral replication and stopping infection. Yuan *et al.* (2006) studied cells infected with herpes simplex virus 1 (HSV-1) and reported that the virus reduced the surface expression of CD1d molecules–these are proteins that bind viral lipids and present them to natural killer T cells during antiviral defense. This was done not by reducing synthesis levels nor by promoting endocytosis from the cell surface, but by preventing the recycling of internalized CD1d to the lysosomal membrane. Reducing the levels of CD1d at the cell surface lowers the ability of the infected cells to stimulate natural killer cells and also helps HSV-1 to evade the immune surveillance machinery, especially during latent infections.

Bacteriophages are viral parasites of bacteria. Virion particles can exist independently outside the host but all bacteriophages are obligate intracellular parasites and must enter a host bacterium to replicate. Bacteriophage infection is initiated when the virion interacts with host cell surface receptor molecules (Hayes, 1968). Many bacteriophages are highly specific for their receptors and do not interact with receptors whose structure is even slightly different. It is this specificity that forms the basis of numerous phage typing methods for the identification of bacterial species or subspecies.

In contrast, some bacteriophages do productively infect a range of bacterial species. Of these broad-host-range phages, P1 and Mu are the best studied. Bacteriophage P1 is a generalized transducing virus capable of plaque formation on several enteric species besides *Escherichia coli*. Bacteriophage Mu produces progeny virions capable of adsorption to and plaque production on different bacterial species due to the differential orientation of the invertible viral G segment region (van de Putte *et al.*, 1980). Bacteriophages that can interact with a wide range of host species are important for the control of the composition and genetic diversity of microbial communities as well as the processes of transductional gene exchange and the transfer of antibiotic resistance genes through those communities (Jensen *et al.*, 1998). The prevalence of broad-host-range phages relates to the origin of virus particles which constitute such a large percentage of the dissolved organic carbon in marine ecosystems and which are present in large numbers in other ecosystems as well (Hennes and Simon, 1995).

To address the question of the distribution of broad-host-range bacteriophages, Jensen *et al.* examined several independent phage collections. One was obtained from sewage treatment plants with *Sphaerotilus natans* as a host, while the others were obtained from a freshwater pond and sewage with *Pseudomonas aeruginosa* and *E. coli* as hosts. The experiments demonstrated that broad-host-range bacteriophages can be readily isolated from aquatic environments and that some of these bacteriophages are capable of generalized transduction. The frequency with which broad-host-range bacteriophages are isolated is increased by the use of isolation protocols that include two bacterial host species for enrichments rather than one.

Their results suggest that a multiple-host enrichment protocol may be more effective for the isolation of broad-host-range bacteriophages by avoiding the selection bias inherent in single-host methods. At least two of the broadhost-range bacteriophages mediated generalized transduction. Possibly, broadhost-range bacteriophages play a key role in phage ecology and gene transfer in nature.

Genomic Diversity in Bacteriophages

The genomes of many closely related organisms, particularly prokaryotic pathogens, have been or are being sequenced. But a clearer picture of the diversity of microorganisms and the mechanisms behind such diversity can emerge only when we move towards non-pathogenic organisms, and also when we realize that much of microbial genomic diversity actually occurs in the genomes of the viruses which infect them.

Pedulla *et al.* (2003) reported the sequencing of ten genomes of mycobacteriophages (in sum nearly 1 Mbp). They compared them to each other and to four previously sequenced mycobacteriophage genomes. Mycobacteriophages infect mycobacteria, some of which cause tuberculosis and leprosy, although many mycobacteria grow harmlessly in soil. A total of 1659 open reading frames were detected among the 14 mycobacteriophage genomes, out of which approximately one half had no matches in data bases. An additional 37.5 per cent of the identified ORFs could only be found in mycobacteriophage, meaning that nearly 88 per cent of the total ORFs identified by Pedulla *et al.* are unique to the mycobacteriophage gene pool. Their results also suggest that the unexplored genetic diversity locked up in bacteriophage is quite high. Indeed bacteriophages may well represent the largest unexplored reservoir of sequence information in the biosphere (see Papke and Doolittle, 2003; Pedulla *et al.*, 2003). This view is supported by observations in the marine and thermal environments where genetic diversity of uncultured marine bacterial viruses (isolated as particles) from two different samples demonstrated that 65-75 per cent of the cloned DNA had no significant matches in GenBank (see Breitbart *et al.*, 2002).

Lateral gene transfer must have played a role in the evolution of many phages. Besides the mosaic genomic structure (see Papke and Doolittle, 2003), Pedulla *et al.* reported that up to 13 per cent of the total ORFs they defined are homologous to ORFs from non-mycobacteriophage species, such as bacteria. Further, the occurrence of lateral gene transfer among the mycobacteriophage populations is even more frequent than that apparently occurring between mycobacteriophage and their hosts or other bacteriophages (see Papke and Doolittle, 2003).

Approximately 90 per cent of the identifiable homologues are shared by one or more of the sequenced mycobacteriophage genomes, although none is common to all of them.

Much of the gene sharing and exchange process studied by Pedulla *et al.*, however, affects unrelated genes which themselves have little sequence similarity. In fact, the mass sequencing of so many mycobacteriophage genomes strongly supports an interesting theory proposed by Hendrix (2000) regarding the evolution of mycobacteriophages and perhaps all tailed bacteriophages. Hendrix proposed that illegitimate recombination occurs arbitrarily throughout the phage genome, and that the nonrandom appearance of recombination events is the consequence of a natural selection force that eliminates 'offspring' that have DNA inserted in the middle of useful relevant genes (Papke and Doolittle, 2003). Two important lessons that emerge from the work of Pedulla *et al.* are: first that phages provide a vast reservoir of exchangeable genes which overlaps that of their hosts, and may drive its evolution. Second, collections of phage genomes possibly can provide valuable pointers to creating theoretical models which view evolution's pattern as a web rather than a tree, and which attempt to measure the relative importance of lateral versus vertical inheritance.

Lysogenic conversion by bacteriophages appears to be advantageous, efficient and, in contrast to conjugative transfer of plasmids, does not require intimate contact between bacteria. Bacteriophages can carry large blocks of DNA and can survive harsh conditions that eliminate bacterial populations. Therefore, DNA important to a population can be preserved until a host for lysogenic conversion is reintroduced into an environmental niche. Bacteriophages also spread DNA directly to an entire population of bacteria; this eliminates the need for clonal expansion of a specific population (Miao and Miller, 1999).

Viruses as Vectors for Gene Transfer

As gene therapy is a most promising approach to curing cancers and genetic diseases (Friedmann, 1996), viruses such as adenoviruses, adeno-associated viruses (AAV), retroviruses and lentiviruses have been adapted as gene transfer vectors by incorporating the gene of interest into the viral genome. Gene transfer techniques based on viral vectors have the advantage of being efficiently applied *ex vivo* and *in vivo;* they show higher transduction efficiencies than most nonviral transfection methods. But one drawback of current viral-vector technology is that the gene is transferred not only to the desired target cells but also to other cells susceptible to infection. The latest generation lentiviral and adenoviral vectors have been responsible for some inadvertent transduction of antigen-presenting cells, which increases the risk of inducing neutralizing antibodies against the transgene product (Chuah *et al.*, 2003). As viral vectors are not specific, the risk of inadvertent germline gene transfer after somatic gene therapy is increased, generating safety and ethical concerns (Marshal, 2001). *In situ* gene delivery by administration of vector locally into the target tissue sometime reduces gene transfer into distal nontarget tissues but does not prevent it. Further, some viral vectors can trigger a strong inflammatory immune response, or are associated with an increased risk of insertional leukemiagenesis

(Buckley, 2002). Another limitation is that mass production of recombinant viruses for clinical use is time-consuming and labour intensive, besides posing liability considerations for manufacturers. These concerns prompted Yamada *et al.* (2003) to develop a gene transfer method that does not involve viral genomes and is easily amenable to mass production, while showing high transfection efficiency and high cell and tissue specificity.

Hepatitis B virus (HBV) envelope L particles are hollow nanoparticles with a peptide involved in liver-specific infection by hepatitis B virus in humans. Yamada *et al.* (2003) demonstrated the use of L particles for the efficient and specific transfer of a gene or drug into human hepatocytes both in culture and in a mouse xenograft model (see Smith, 2002). In this model, intravenous injection of L particles carrying the gene for green fluorescent protein (GFP) or a fluorescent dye produced observable fluorescence only in human hepatocellular carcinomas but not in other human carcinomas or in mouse tissues. When the gene encoding human clotting factor IX was transferred into the xenograft model using L particles, factor IX was produced at levels adequate for the treatment of hemophilia B. The L particles are easily produced in yeast, do not contain potentially hazardous virions and are highly specific to human hepatic cells. These features are advantageous in overcoming some of the limitations and risks of viral technologies, avoiding, for instance, inadvertent gene transfer into undersired cell types or into germline cells after somatic gene therapy (Marshal, 2001). In view of the production criteria and particle stability, L particles appear to be more promising than conventional liposomes for drug delivery. The L particles can accommodate drugs as well as genes. These advantages should greatly facilitate targeted delivery of genes and drugs to the human liver (Yamada *et al.*, 2003).

RNA Virus Vectors

Genetically engineered animal viruses have allowed the construction of vectors to direct the expression of heterologous proteins in different systems. Transfected DNA molecules of SV40 (approximately 5,000 bp in length) allowed the first rescue of defined viral mutants (see Palese, 1998). Subsequently, the molecular engineering of herpes simplex viruses (genome approximately 150 kbp) and of vaccinia viruses (genome approximately 190 kbp) were major milestones in virology. Techniques were then developed to specifically alter the genomes of adenoviruses, adeno-associated viruses (NAS, 1996), and many other DNA-containing viruses. It has become possible to rescue human cytomegalovirus and Epstein-Barr virus by contransfecting into cells or cosmids containing overlapping fragments of the respective viral genomes. The genomes of DNA viruses can be harnessed to generate mutants and recombinant viruses expressing foreign proteins. Similar success has occurred in retroviruses by the study of many novel constructs of RNA tumour viruses. Retrovirus-based vectors can express foreign genes in animals over periods of weeks and months (NAS, 1996).

There has been considerable progress in developing suitable genetic engineering techniques for nonretroviral RNA viruses. RNA viruses or their components are used to express foreign genes. Infectious Sindbis virus has been made from a cDNA clone. This virus is an alphavirus containing a 12-kb, single-stranded, positive-sense, capped and polyadenylated RNA genome. When its genomic (naked) RNA is introduced into cells, infectious virus forms. Two approaches have been developed to express foreign proteins from Sindbis virus constructs: (i) A self-replicating, self-limiting replicon has been made by replacing the genes for the virion structural proteins with that of a reporter gene. (ii) A chimeric virus could be made by introducing a subgenomic promoter which drives the expression of the heterologous protein (see Palese, 1998). These replicons/vectors amplify to high levels, thereby killing the transfected/infected cells. Palese (1998) described an ingenious way to circumvent the cytopathic properties of these vector systems, so that the killing is avoided. This ingenious feat was accomplished by using certain specially-made noncytopathic replicons of Sindbis virus by Agapov *et al.* and proved for the first time that a virus/vector expression system which was thought to be lytic can be modified so that it becomes noncytopathic and still effectively and stably expresses a foreign protein.

Bipartite noncytopathic vectors have been developed which express levels as high as 30 µg of a foreign protein per 10^6 cells. The advantages of RNA virus-based vector systems are the following: (i) the use of these rapidly selectable noncytopathic RNA vectors for cell culture studies. After efficient RNA transfection by electroporation, puromycin selection is imposed, and a population of cells expressing the gene of interest can be raised quickly. This enables use of the system to trans-complement viral functions; (ii) these replicon-based vectors often express heterologous proteins to high levels for prolonged periods, so serving as excellent vaccine delivery systems; (iii) they do not contain a complete complement of all viral genes and therefore no infectious particles are produced which could lead to a more generalized infection spreading to other tissues. These vectors therefore are quite safe; (iv) because the vector component derived from the virus is small and does not express structural proteins, the host immune response to the vector tends to be limited. This has relevance to evolving better gene therapy enterprise. (v) This RNA virus vector (or for that matter other RNA virus-based systems) are unlikely to cause cell transformation. Because these RNA viruses lack a DNA phase, unwanted integration of foreign sequences into chromosomal DNA, is avoided (see Palese, 1998).

Long-term expression of foreign genes is often desirable for certain medical applications but other purposes are better approached using self-limiting RNA vectors. For instance, alphavirus expression vectors, which are cytolytic (and hence suicidal), may enhance the efficacy of DNA vaccines. Administration to mice of plasmids that transcribe self-amplifying alpha-virus RNA replicons has proven effective in inducing an enhanced immune response against diverse infectious agents.

Other positive-sense RNA viruses (*e.g.* poliovirus) also have been engineered to express foreign genes/epitopes. These vectors are rather limited in their ability to express large foreign sequences, but they share advantages with other RNA virus vectors in that they do not seem to (down)modulate the immune system as do many large DNA viruses, including poxyviruses and herpes viruses. Thus, they induce highly protective immune responses which strengthen mucosal defenses.

The large family of negative strand RNA viruses also is now amenable to genetic engineering. These viruses with segmented or nonsegmented genomes do not in general modulate the immune system. In fact, some members are extraordinarily good inducers of cellular (CTL) and humoral immune responses. The fact that they lack a DNA phase makes them safer for use in humans. Finally as negative strand RNA viruses do not show measurable rates of homologous RNA recombination, they are quite stable and safe. Expression of foreign proteins/antigens is now known for genetically engineered influenza viruses, rabies virus, measles, respiratory syncytial, SV5, and Sendai viruses. Attempts are underway to express antigens of viruses or bacteria for which we do not have vaccines available. The next step would be for viral vectors to express multiple foreign proteins for use against multiple disease agents.

Nuclear Import and Export in Influenza Virus Infection

Several enveloped viruses, *e.g.*, influenza virus, herpes viruses, hepatitis B-virus and RNA tumour viruses, replicate in the nucleus, but the virus particles themselves mature at the plasma membrane or at cytoplasmic membranes. There is a spatial separation between genome replication and viral maturation. Therefore these viruses have not only to get their genome into the nucleus to initiate infection, but also need to ensure that, after replication, the genome is transported to cytosolic budding sites. Each of these virus families use different strategies to solve the nuclear trafficking problem.

Infection with influenza virus involves a complex series of nuclear import and export events. Early in infection, incoming viral ribonucleoproteins (vRNPs) are imported into the nucleus. Later, viral transcripts are exported from the nucleus, newly synthesized structural proteins are transported back into the nucleus and, finally, newly assembled vRNPs are exported. All these import and export steps, and, in particular, the bidirectional traffic of vRNPs rely on the transport machinery of the cell, but are regulated both by viral and cellular factors. The viral M1 protein serves as the master organizer in determining the directionality of vRNP transport (Whittaker *et al.*, 1996).

The influenza virus genome is split into eight, separate, single-stranded RNA segments, which are packaged individually into helical vRNPs. The vRNPs, in turn, are packaged into mature virions. In addition to the RNA segment, the vRNPs contain the viral nucleoprotein (NP; 56 kDa) that forms the vRNP central core. The RNA in each vRNP wraps around the outside of a chain of NP molecules, one molecule of NP being bound per 25 nucleotides of RNA. The whole virion contains some 700 copies of NP. Each vRNP also has a single, trimeric polymerase complex located at one end of the vRNP and associated with the 5' end of a terminal, double-stranded panhandle structure. The complete vRNPs are 30-110 nm in length, depending on the size of the RNA, and their diameters of 10-20 nm approach the 23-nm limit for passage through the nuclear pore complex (Whittaker *et al.*, 1996). Except for the M1 matrix protein all the other structural proteins of the virus are integral to the viral envelope.

The following four crucial steps are involved in the penetration and uncoating of influenza virus:

(A) The virus particle binds to cell surface sialic acid.

(B) The particle is internalized by receptor-mediated endocytosis into endosomes. The low-pH environment of the endosome activates the M2 channels. Protons enter the interior of the virus particle via the M2 channel and induce a change in the nucleocapsid, priming it for uncoating. Haemagglutinin (HA) undergoes an acid-triggered conformational change that results in fusion of the viral and endosomal membranes.

(C) The nucleocapsid is released into the cytoplasm and the viral ribonucleoproteins (vRNPs) dissociate from each other and from the M1 protein.

(D) The vRNPs are imported rapidly and efficiently into the nucleus through the nuclear pores. NPC, nuclear pore complex; vRNP, viral ribonucleoprotein (Whittaker *et al.*, 1996).

There is considerable evidence to support the view that M1 is required for vRNP export (see Whittaker *et al.*, 1996).

DNA Shuffling by Viral Vectors

Several factors have hampered the current gene transfer methods for the development of truly effective human gene therapy, foremost being the generally inefficient and nonspecific delivery of gene delivery vectors *in vivo*. An *ex vivo* model for retrovirus-mediated gene delivery has made encouraging progress toward a definitive genetic treatment of children with an immunological disorder. Direct *in vivo* delivery of clotting Factor IX with an adeno associated virus vector has also shown great promise (see Miyanohara and Friedman, 2000). But the broad application of genetic transfer methods to the correction of human disease requires the design of vectors that can survive the rigors of production and the onslaughts of host defenses and that can home in specifically and efficiently to the intended cells in the body to deliver their therapeutic genes. Virologists have developed methods to modify viral coat proteins to improve the gene delivery properties of virus vectors, but quite often the understanding of virus structure and function is not good enough to produce stable, highly concentrated preparations of tissue-targeted vectors. Powell *et al.* (2000) proposed a potent approach for solving some of the inadequacies of current protocols letting nature do the work of designing virus envelope proteins that improve the properties of vectors through a process of in vitro genetic recombination called DNA shuffling. In this, a variety of parental retroviral envelope genes are 'shuffled' by DNase digestion and PCR-based synthesis. The mixture of recombinant envelope genes is introduced into packaging cell lines to produce a library of virus particles with different envelope chimeric proteins on their surface Specific virus particles with envelopes conferring the desired phenotype are then selected by infection of appropriate target cells or by physical selection, as in the case of ultracentrifuge selection of virus particles containing envelope proteins resistant to stress forces (Miyanohara and Friedman, 2000).

The method of "DNA shuffling" to produce mix-and-match combinations of retroviral envelope fragments is derived from several different parent MLV virus strains that infect only murine cells (ecotropic viruses). The collection of chimeric recombinant envelope molecules was introduced into an MLV retroviral backbone and then used to transfect packaging cells to produce a library of infectious viruses containing many different recombinant envelope genes, thereby producing envelope proteins with new physical properties.

The described shuffling method has produced retroviruses with at least two new phenotypes: a new tropism and an enhanced stability to ultracentrifugation. However, there are many more altered phenotypes lurking in the library, waiting to be selected for. The impressive lesson from these studies is that combinatorial libraries and selection of chimeric molecules can be more useful than rational vector design for the rapid identification of molecules with desired new properties. There is little likelihood that these recombinants could have been designed nearly as easily through the more traditional approach of first defining the structure-function relationships of the envelope or other viral proteins and using that knowledge to design recombinant molecules from basic principles. Unquestionably, nature seems to be far better at virus design than are most virologists, and the methods of in vitro recombination and selection of the desired phenotype after in vitro evolution will be extremely useful for isolating viruses with such altered properties to illuminate aspects of virus structure and function, such as attachment to cell surface molecules and the nature of virus-receptor interaction, cell entry, intra-cellular trafficking, and nuclear transport. And of course, these kinds of in vitro recombination methods should not be limited to retroviruses.

Infectious Viral RNAs (Reverse Genetics)

It is more difficult to manipulate an RNA genome than a DNA genome. The first RNA virus to be genetically modified was Qβ phage. Although the RNA molecules were chemically modified during RNA replication *in vitro* using cumbersome procedures, the potential power of reverse genetics as a tool for studying RNA viruses was quite clear. The advent of recombinant DNA technology in the 1970s enabled virologists to convert viral RNA genomes into complementary DNA copies and then replicate them as plasmid inserts in bacterial hosts for easier genetic manipulation. Amazingly, the plasmid containing the complete cDNA of the Qβ phage RNA proved fully infectious when introduced into bacterial hosts and was capable of completing the full viral replication cycle. Presumably, transcription of the viral RNA was randomly initiated, and the RNA was processed mysteriously to the correct viral sequence. Later, this technique was applied to some other viruses and viroids. The work of Almazán *et al.* (2000) represents an engineering feat. Because coronavirus contains the longest viral RNA genome by far (and is probably one of the longest stable RNAs in nature), this approach may show the way for the reverse genetic studies for all RNA viruses.

Certain cloning methods have made possible the construction of infectious cDNA or RNA for most RNA viruses, but one remaining virus that has so far resisted the onslaught of cloning attempts is coronavirus. Coronaviruses include many economically and medically important viruses, *e.g.*, porcine TGEV, mouse hepatitis virus (MHV), avian infectious bronchitis virus (IBV), and human coronaviruses, the last of these being responsible for many common colds and, possibly, gastroenteritis and neurological illnesses such as multiple sclerosis. These viruses contain a positive-sense RNA genome of 27-32 kb, which is more than twice the size of the largest genomic RNA of the conventional RNA viruses. The viral RNA is replicated by an RNA-dependent RNA polymerase entirely in the cytoplasm, independently of the nucleus, although nuclear factors are probably involved in viral RNA synthesis (Lai, 2000). More than two-thirds of the viral RNA sequences are devoted to making gene products involved in viral RNA synthesis. The enormous size of the coronaviral RNA posed a daunting obstacle to the construction of the infectious cDNA or RNA for these viruses. Almazán *et al.* (2000) have now solved these problems.

They resurrected the old trick of using cDNA transfection to drive the production of viral RNA *in situ*, but with one notable difference. In previous cDNA transfection studies, the viral cDNA was expressed by random initiation of transcription, probably from within the plasmid sequences; however, Alamzan *et al.* placed the viral cDNA under a specific promoter (cytomegalovirus immediate-early promoter), and the ends of viral RNA were carefully engineered to match their natural sequences. In any case the cDNA transfection approach was intuitively thought to be artificial because the viral RNAs in question (*e.g.*, coronavirus) naturally replicate only in the cytoplasm; thus, the production of viral RNA in the nucleus would introduce additional roadblocks to the replication of viral RNA in the cytoplasm. Past examples of the successful use of the cDNA transfection approach for RNA viruses mostly involved viruses that normally replicate in the nucleus, such as hepatitis delta virus, viroids, and influenza virus. For cytoplasmic viruses (*e.g.*, most of the negative-strand RNA viruses), an alternative cDNA transfection approach was developed in which viral cDNA (under a T7 promoter) transfection is coupled with the expression of T7 polymerase through a recombinant vaccinia virus, which replicates in the cytoplasm. Such an approach ensures that viral RNA is transcribed directly in the cytoplasm, where it replicates.

The successful cloning of an infectious TGEV cDNA constitutes a major breakthrough for coronavirus research. Here, the spike protein alone is sufficient to determine the pathogenicity of the virus (Lai, 2000).

Large RNA Virus Genome as Artificial Bacterial Chromosome

Almazan *et al.* (2000) showed that the application of two strategies, cloning of the cDNAs into a bacterial artificial chromosome and nuclear expression of RNAs that are typically produced within the cytoplasm, is useful for the engineering of large RNA molecules. A cDNA encoding an infectious coronavirus RNA genome has been cloned as a bacterial artificial chromosome. The rescued coronavirus conserved all of the genetic markers introduced throughout the sequence and showed a standard mRNA pattern and the antigenic characteristic expected for the synthetic virus. The cDNA was transcribed within the nucleus, and the RNA translocated to the cytoplasm. Interestingly, the recovered virus had essentially the same sequence as the original one, and no splicing was observed. The cDNA was derived from an attenuated isolate that replicates exclusively in the respiratory tract of swine. During the engineering of the infectious cDNA, the spike gene of the virus was replaced by the spike gene of an enteric isolate. The synthetic virus replicated abundantly in the enteric tract and was fully virulent, suggesting that the tropism and virulence of the recovered coronavirus can be modified. This work opens up the possibility of employing this infectious cDNA as a vector for vaccine development in human, porcine, canine and feline species susceptible to group 1 coronaviruses (Almazan *et al.*, 2000).

Coronaviruses are single-stranded, positive-sense RNA viruses and have the largest genomes known among RNA viruses. The construction of cDNA clones encoding large-size RNA molecules of biological interest, like coronavirus genomes, has been hampered by the instability of those cDNAs in bacteria.

Reconstitution of Retroviral Fusion and Uncoating in a Cell-free System

To better understand the molecular events underlying retroviral uncoating, occurring after membrane fusion and leading to the formation of an active reverse transcription complex, Narayan and Young (2004) developed a cell-free system that recapitulates the early steps of retroviral replication by using avian sarcoma and leucosis virus as a model retrovirus. The substrates used in this system were viral particles that are trapped before completing membrane fusion. These virions were induced to fuse out of endosomes and the viral cores were released into solution where they can be manipulated biochemically. This system revealed that membrane fusion is not sufficient to stimulate the formation of a reverse transcription complex. Instead, ATP hydrolysis and cellular factors >5 kDa in size are required. Furthermore, later steps of avian sarcoma and leucosis virus reverse transcription were stimulated by nuclear factors. This cell-free system should allow for the definition of retroviral uncoating mechanisms and facilitate identification and characterization of the cellular factors involved.

Antiviral Therapy–Bacteriophages as Antibiotics?

We have entered the golden phase of information production via high-throughput technologies and computational advances. The prospects for systems biology in studies of virus-host interactions have become very exciting with recent advances in noninvasive technologies to image such fundamental processes as transcriptional regulation, signal-transduction cascades, protein-protein interactions and cell trafficking) in living cells (Silver and Way, 2007) and animal models. Integration of ultrastructural and live-animal imaging technologies with functional genomics and computational analysis should enable characterization of the features of membrane-associated RNA replication complexes in several viruses and generate a 'systems-based model' that can boost our understanding of the general principles and mechanisms of positive-strand RNA virus replication as well as of how the viruses' evade the immune system.

In antiviral therapy, combination therapies are currently the norm. A systems biology approach that allows formulation of multicomponent drugs to achieve optimal efficacy with reduced mechanism-based toxicity will significantly impact antiviral drug discovery and development. In vaccine development, one promising approach is based on the development of novel peptide microarrays that use antigen peptides as fixed probes and serum antibodies as targets to identify antibody reactivity patterns (and potentially valuable clinical biomarkers) involved in virus infections.

Antibacterial activity against *Vibrio cholerae* in the waters of the Ganga was recorded by Ernest Hankin in 1896. Frederick Twort in 1915 and Felix d'Herelle in 1917 showed that some viruses infect bacteria. By 1920 bacteriophages had been isolated from several bacterial species and could be used as antibacterial agents. Phage therapy products were sold in the United States in 1930s. Eli Lilly marketed seven different phage preparations to fight *Staphylococcus*, *Streptococcus* and *E. coli*. In the former Soviet Union also active phage preparations were licensed for use in humans.

Antibiotics were introduced for curing all bacterial infections only in the 1940s, whereafter bacteriophage therapy became unpopular in most of the world, except in the Soviet Union and Eastern Europe. The widespread development of antibiotic resistance in pathogenic bacteria has renewed the need for alternative strategies to control microbial infections. There are thousands of different bacteriophages, each of which may infect one or several types of bacteria. While some phages infect several related species of bacteria, they do not infect antigenically unrelated bacteria. Phages are common in all natural environments and are directly related to the number of bacteria present.

When a bacteriophage contacts a suitable host bacterium, its tail fibres bind to the precise receptor molecules that distinguish that particular bacterium as a suitable host. The phage then injects its nucleic acid into the cell.

The two main groups of bacteriophages–lytic and lysogenic–have different strategies for getting their host to replicate them. Lytic phages instruct the machinery in the host cell to make more phages. Fully viable progeny bacteriophages burst and kill the bacteria. The released bacteriophages attack new bacteria. This process continues until all the bacteria are eliminated.

Lysogenic phages attach their strands of genetic material to the bacterial DNA. The phage DNA is replicated along with the bacteria, generation by generation. These phages can also cut their piece of DNA free from the host's DNA at any time, and instruct the host cell to produce a large number of phages; this bursts the bacterial cell and releases newly formed bacteriophages.

Only the lytic phages are suitable for developing therapeutic phage preparations. Lysogenic phages are not suitable as they do not destroy bacteria immediately. They can also transfer virulence genes and those mediating resistance to antibiotics to other bacteria.

Bacteriophage therapy has now been rediscovered as a potent weapon against antibiotic-resistant bacteria. Active bacteriophages have been selected against pathogenic bacteria. Therapeutic preparations of specific phages have been successfully used in burns, sepsis, gastroenterological infections, paediatric infections, also in surgery against primary and nosocomial infections.

Since the 1980s, phages against *Klebsiella* and *Acinetobacter* have been isolated and developed into therapeutic preparations. A recent intestiphage preparation includes 23 different phages active against a range of enteric bacteria. An enteric-coated pill has been developed, using phage strains that survive the drying process. The Eliana Institute in Tbilisi, Georgia, has focused on nosocomial infections, where multi-drug-resistant organisms pose serious problems. Clinical studies have confirmed the high effectiveness of bacteriophage therapy in combating those bacterial infections which do not respond to treatment with the available antibiotics (see Sandeep, 2006).

In Poland and Europe bacteriophages have been successful in treating purulent meningitis in infants caused by *Klebsiella pneumoniae*, subphrenic abscess, suppurative bacterial infections, chronic suppurative infections of the skin caused by *Pseudomonas*, *Staphylococcus*, *Klebsiella*, *Proteus* and *Escherichia*, postoperative septic Staphylococcal infections, and septic arthritis of the knee and osteomyelitis.

One merit of phages is that they cannot infect the cells of organisms more complex than bacteria because the surface properties of these cells are not susceptible to the bacteriophage invasion. Bacteriophages can be isolated from wherever their specific host bacterium grows–in water, skin, air, food, body surface, throat, intestine, faeces, sewage, soil, ocean depths and hot springs.

Advantages

1. Each bacterial species has at least one complementary bacteriophage that specifically infects it. Phage therapy is possible in all bacterial infections.
2. If a suitable bacteriophage is introduced onto an infected wound, it multiplies as long as there are bacteria to infect and destroy. When all the bacteria have been destroyed, the action of the phage ceases.
3. Being highly specific, phages will not harm other beneficial bacteria present in the intestine and other parts of the body and will not affect the microbial ecosystem in the body. There is no chance of super infection with other bacteria. The microbial imbalance caused by treatment with many antibiotics often leads to serious secondary infections involving relatively resistant bacteria, often extending hospitalization time, expense and mortality. This will not occur with specific bacteriophage therapy.
4. For people allergic to antibiotics, phage therapy could be a useful alternative. Allergic reaction to phage therapy has never been reported, possibly because phages are widespread in all habitats.
5. Phage therapies can be administered to patients in the form of pills, injections, enemas, nasal sprays or ointments.
6. No major side effects of phage therapy have been described.
7. Because bacteriophages grow exponentially, a single dose is usually enough to treat an infection.
8. An important feature of phage therapy is that bacteriophages do not infect human or animal cells.
9. Bacteria can develop resistance against both antibiotics and phages. Unlike antibiotics, bacteriophages can mutate in step with evolving bacteria. If

bacteria become resistant to bacteriophages, other bacteriophage species can attack those new resistant strains. Mutations that enable bacteria to resist antibiotics do not enable bacteria to resist bacteriophages. Development of a new antibiotic is very expensive and time-consuming but developing a new phage can potentially be accomplished in days, at much less expense.

Drawbacks

1. In view of the high specificity of phages, the disease-causing bacterium has to be correctly identified before the administration of phage therapy.
2. The gastric acidity needs to be neutralized prior to oral phage administration.
3. Bacteriophage with a lytic lifecycle within a well-defined *in vitro* environment does not ensure that the bacteriophage will always remain lytic under normal physiological conditions found in a body. It may change to lysogenic cycle in some circumstances.
4. Bacteriophages are viruses and, so, in general, can swap genes with each other and with other organisms with which they come into contact. So there is some chance of spread of antibiotic resistance in bacteria.
5. Doctors feel scared to give live bacteriophage to the patients.

Viruses and the Cellular DNA Damage Response

Cellular DNA is constantly being attacked by both endogenous and exogenous sources. Cells face a continuous challenge of how to maintain their genome in a healthy state. Each cell has to repair over 10,000 DNA lesions every day (Lindahl, 1993). Cells have the needed machinery to monitor damage and ensure the fidelity of replication (Bakkenist and Kastan, 2004). When cells sense damage, they initiate signaling pathways that activate checkpoints to prevent replication of damaged DNA. Which DNA repair proteins are pressed into service depends on the type of lesion encountered as well as the phase of the cell cycle (Lilley *et al.*, 2007).

To tackle UV-induced damage, nucleotide excision repair (NER) proteins unwind, nick and re-synthesize the damaged DNA. When the damage produces an oxidized or alkylated nucleotide or eliminates a nucleotide, the base excision repair pathway performs the repair. When an incorrect base has been inserted, mismatch repair pathways actively correct the mistake. Different kinds of DNA lesions are so treated as to yield either single or double-stranded breaks. While double-stranded breaks (DSBs) activate the ataxia-telangiectasia mutated (ATM) kinase and its substrates, single-stranded DNA activates checkpoint pathways controlled by the ATM and Rad3-related (ATR) kinase (Sancar *et al.*, 2004). ATR responses are a must for the cell to deal with replication stress by facilitating replication fork arrest when necessary. DSBs constitute the most severe damage faced by the cell as there is no complementary sequence to ensure faithful repair. DSBs can be induced by endogenous challenges such as immunoglobulin recombination or exogenous insults such as ionizing radiation.

Two main pathways for repairing DSBs in eukaryotic cells are: homologous recombination (HR) and non-homologous end joining (NHEJ). The HR requires members of the Rad52 epistasis group, whereas NHEJ is a more error-prone repair pathway that induces religation of DNA ends. The primary actors in the NHEJ pathway are DNA ligase IV, the X-ray repair cross-complementing group 4 (XRCC4) and DNA-dependent protein kinase (DNA-PK), which consists of a catalytic subunit (DNA-PKcs) and its associated proteins Ku70 and Ku86. Some proteins, such as the Mre11 complex are involved in both NHEJ and HR (Sancar *et al.*, 2004).

Many viruses can interact with the DNA damage sensing and repair machinery (Table 3.4). These viruses have learnt how to remove, circumvent or exploit various aspects of the DNA damage response of the host cell. Some strategies include activation of repair proteins or targeting of specific cellular factors for degradation or mislocalization.

Table 3.4: Selected Examples of Viruses that Interact with the Cellular DNA Damage Machinery (after Lilley *et al.*, 2007)

Virus	*Family*	*Genome Structure*	*Signalling Pathways*	
			Activated	*Inhibited*
Herpes simplex viruses 1, 2 (HSV1, 2)	Herpesviridae	Ds, linear	ATM	NHEJ/ ATR
Epstein-Barr virus (EBV)	Herpesviridae	Ds, linear	ATM	–
Human cytomegalovirus (HCMV)	Herpesviridae	Ds, linear	ATM	ATM, ATR
Adenovirus	Adeoviridae	Ds, linear	–	ATM, ATR, NHEJ
Simian virus 40 (SV40)	Polyomavirinae	Ds, circular	ATM	–
Polyomavirus	Polyomavirinae	Ds, circular	ATM	–
Adeno-associated virus (AAV)	Parvoviridae	Ss, linear	ATM	–
Hepatitis B virus	Hepadnaviridae	Partially ds, circular	–	NER
Retroviruses	Retroviridae	Ss, linear	–	–
Human immunodeficiency virus 1 (HIV-1)	Retroviridae	Ss, linear	ATM, ATR	–
Mimivirus	Mimiviridae	Ds, linear	N/A	N/A
Omega and lamda bacteriophages	Bacteriophages	Ds, linear	N/A	N/A

ds: Double-stranded DNA; ss: Single-stranded DNA.

(Except for Retroviruses and HIV-1, the nucleic acid is DNA. In Retroviruses and HIV-1, it is RNA).

During infection, the aim of viruses is to hijack the cell whereas that of the host is to defend itself. Two common traditional defenses are the interferon response and apoptosis. This antiviral arsenal also includes the cellular DNA damage response machinery. Critical study of interactions between viruses and cellular DNA repair proteins has not only revealed new complexities of the virus-host interaction but has also supported the idea that viruses can throw light on key regulators of cellular pathways through the proteins they target (Lilley *et al.*, 2007).

The cellular DNA repair machinery can recognize viral genetic material as 'damage' (Weitzman *et al.*, 2004), but some viruses can inhibit or circumvent the host-cell response. For other viruses, the cellular response seems to be beneficial, and viruses have evolved ways to hijack cellular DNA repair proteins to facilitate their own replication.

Several DNA viruses press into service cellular signaling cascades characteristic of a DNA damage response. These viruses include herpes simplex virus (HSV)-1, HSV-2, Epstein-Barr virus (EBV), polyomavirus and simian virus 40 (SV40). For HSV-1, polyomavirus and SV40, the DNA damage response is beneficial because viral replication is less efficient when damage signaling is abrogated ((Dahl *et al.*, 2005; Lilley *et al.*, 2005). A distinctive aspect of the signaling induced by SV40 infection is that the viral T-antigen serves as a substrate for the activated ATM kinase. As ATM could also possibly be required to phosphorylate the T-antigen of related viruses, this may be a general strategy adopted by polyomaviruses (Dahl *et al.*, 2005).

Exploitation of the cellular damage response is by no means restricted to DNA viruses. The interaction of retroviruses with the cellular DNA repair machinery has been well studied by Skalka and Katz (2005). DNA-PKcs is required for efficient transduction by retroviral vectors. Subsequently, other PI3-like kinases also were reported to be activated following HIV-1 infection (Lau *et al.*, 2005) and possibly have a role in provirus integration. Using caffeine or a small molecule inhibitor to block the activity of ATM inhibited HIV transduction and replication (Daniel *et al.*, 2005).

The report that the DNA damage response benefits the replication of some viruses has important clinical implications (see Lilley *et al.*, 2007). A small molecule inhibitor of ATM was efficacious in a conceptual study with HIV-1; it even demonstrated antiviral activity against a drug-resistant strain. Whether other pathological viruses will be similarly responsive to DNA repair-based therapies is, however, not known.

Not much is known about the viral triggers that induce DNA damage signaling. Some possibilities are incoming or replicating nucleic acid and virally encoded proteins. In the case of HIV-1, the requirement of a functional integrase for the activation of targets downstream of ATM suggested that proviral integration might be the trigger (Lau *et al.*, 2005). Specific HIV-1 proteins are also known to trigger cellular damage signaling (Zimmerman *et al.*, 2006).

Viral Inhibition of the Cellular DNA Damage Response

Some viruses need to overcome another antiviral obstacle, the DNA damage response of the host cell. Just as viruses have learnt how to block the interferon response and prevent apoptosis, some effective counter-strategies against components of the cellular DNA damage response have also evolved.

Common inactivation strategies adopted by viruses are usually based on either degradation or mislocalization of repair proteins central to the cellular defense. Adenovirus employs both strategies to prevent the host cell from 'repairing' the viral

DNA–repair that would result in covalently linked viral genomes (Stracker *et al.*, 2002). Infection by mutant adenoviruses which cannot inactivate DNA repair proteins induces damage responses involving activation of both ATM and ATR signaling cascades, and recruitment of many cellular DNA repair proteins to viral replication centers (Stracker *et al.*, 2002; Carson *et al.*, 2003).

Some viruses encode specific proteins that target DNA damage proteins. The human polyomavirus JCV inhibits NHEJ by encoding a late protein that binds to and mislocalizes Ku70 (Darbinyan *et al.*, 2004).

Some viruses first exploit and then inactivate specific cellular DNA repair pathways. SV40 infection activates ATM, and this is required for efficient viral replication. However, Wu *et al.* (2004) reported that SV40 T-antigen targets Nbs1 to create an environment in which the viral genome can be maximally amplified. During SV40 infection, T-antigen recruits a cullin-containing ubiquitin ligase to degrade the Mre11 complex.

Recent researches tend to suggest that classifying cellular DNA damage responses as either beneficial or harmful for a particular virus is premature–each virus tends to interact with DNA repair pathways on multiple levels. Only future research can reveal how these viruses spatially and/or temporally regulate or manage this kind of dual response.

Consequences for the Cell

Being intracellular parasites, viruses depend on host cell functions for their replication. They must ensure that the health of the cell is maintained long enough to maximize production of viral progeny. Taking control of the cell cycle is crucial for viral replication. To ensure an abundant supply of nucleotides and other essential replication factors, many viruses cross normal checkpoint control and force the cell into S phase. In normal cells, retinoblastoma protein (Rb) binds to the transcription factor E2F to prevent unscheduled S-phase entry. Viruses therefore target Rb to promote S-phase (O'Shea, 2005). When a cell suffers DNA damage, it initiates checkpoints to allow time for repair before the damaged DNA can be replicated. When viral infection triggers a DNA damage response, the effects on checkpoint control are possibly synergistic. A virally induced damage response selectively activates checkpoints, which supports maintenance of an S-phase environment. This is illustrated by polyomavirus, in which T-antigen inactivation of Rb promotes S-phase entry. This interaction also induces an ATM-dependent DNA damage response that leads to activation of checkpoints, extension of S-phase and increased production of progeny virus (Dahl *et al.*, 2005).

Apoptosis

Viruses have developed several mechanisms to prevent apoptosis. These include restricting the function of the pro-apoptotic protein p53, even though p53 is often activated during infection. Some viruses that activate and then limit p53 function are adenovirus, SV40, EBV and KSHV. HCMV and HSV-1 infections also lead to the stabilization and phosphorylation of p53, but not much is known about how these viruses prevent the pro-apoptotic function.

Another concern between the DNA damage response and apoptosis is poly (ADP-ribose) polymerase 1 (PARP1). PARP1 perfoms p53-independent roles in modulating histones in response to repairable DNA damage, as well as roles in promoting apoptosis when the damage is irreparable. It also binds to single-stranded breaks in DNA where it mediates repair. It could possibly also have an antiviral role as, by binding directly to EBV and KSHV genome, its activity negatively correlates with viral replication in KSHV-infected cells (Deng *et al.*, 2005). Work is warranted to understand ways in which viruses target PARP1.

Transformation and Genomic Instability

Viral proteins can transform cells. Typically, viral gene products are maintained post-transformation but in some cases, viral oncogene cannot be detected after the cell has been transformed. This kind of 'hit and run transformation' is mediated by the adenovirus proteins E4orf3 or E4orf6 in conjunction with E1A (Nevels *et al.*, 2001).

Bartkova *et al.* (2005) and Gorgoulis *et al.* (2005) reported that aberrant, pre-cancerous cell division can trigger the cellular response to DNA damage. Induction of the damage response arrests proliferation or directs the cell towards apoptosis, so limiting the malignant progression. The DNA damage response was not seen in mature tumors, suggesting that an inactivation of the response might be involved in the switch between early lesions and cancer. Viral inactivation of the DNA damage response might conceivably precipitate this switch and could be a new mechanism whereby these viruses contribute to cellular transformation (Lilley *et al.*, 2007).

Munger *et al.* (2004) discussed multiple links between papillomaviruses and oncogenic potential and which might involve cellular DNA repair pathways. Some viruses such as HCMV cause chromosomal breaks and aberrations, which leads to genomic instability (Fortunato *et al.*, 2000). Viral integration may also have long-term consequences for the host cell. Retroviruses preferentially integrate into actively transcribing genes and affect transcription patterns (Lewinski and Bushman, 2005). Further, these viral genomes contain unusual DNA structures such as inverted repeats or palindromic sequences that could cause some genome instability.

Consequences for the Virus

What happens to the virus depends on the cellular environment it meets upon infection. Many infections do not result in productive viral replication and this often depends on the DNA damage response environment of the host.

Viruses that establish latency could possibly differentially manipulate the DNA repair machinery in favour of lytic replication or latency. Viruses which integrate into the host genome can be susceptible to the DNA damage environment of the cell, especially if they make use of cellular DNA repair proteins to complete the process. Studies in severe combined immune deficiency (SCID) mice have indicated some role for DNA-PK in inhibiting rAAV integration (Song *et al.*, 2004).

DNA repair proteins are able also to process the genomes of non-integrating viruses. The potential for viruses to capture cellular DNA repair genes or to evolve repair functions has attracted interest. The giant Mimivirus is unusual in coding for several DNA repair proteins, *e.g.* Rad2, Rad50, MutS and UV-damage endonuclease homologues (Raoult *et al.*, 2004). The lambda bacteriophage Omega contains Ku homologues which cooperate with a DNA ligase from the host cell to facilitate NHEJ, resulting in circularization of the incoming phage genomes (Pitcher *et al.*, 2006). These viruses have probably captured DNA repair genes from their hosts to help their own replication. This might also be true for mammalian viruses (Brown *et al.*, 1997).

Giant Viruses

These are nucleocytoplasmic large DNA viruses (NCLDVs) which infect algae (phycodnaviruses) and amoebae (*Mimivirus*). Filee *et al.* (2006) described an unusual abundance in these giant viruses of islands of bacterial-type genes, including seemingly intact prokaryotic mobile genetic elements. This led them to suggest that NCLDV genomes undergo successive accretions of bacterial genes. The viruses probably acquire bacterial genes within their bacteria-feeding eukaryotic hosts, and such acquisition seems to be driven by the strong coupling of recombination and replication in NCLDVs.

These NCLDV viruses make up a diverse group that infects several different eukaryotic hosts including algae (phycodnaviruses), protists (*Mimivirus*) and metazoans (poxviruses, African swine fever virus, iridoviruses). According to Iyer *et al.* (2006), they either replicate exclusively in the cytoplasm or begin their cycle in the host nucleus before passing into the cytoplasm. In keeping with this lifestyle, they have most of the genes needed for DNA metabolism, replication and transcription in addition to those involved in virion assembly.

Acanthamoeba polyphaga mimivirus is a very large double-stranded DNA virus (genome size, 1.2 megabase pairs). By examining images of infected amoebae with electron tomography and cryo-scanning electron microscopy, Zuberman *et al.* (2008) inferred how the genome is released from and packaged into the icosahedral viral capsid. Whereas other DNA viruses seem to use a single icosahedral vertex both for loading DNA during viral biogenesis and for releasing it upon entering the host cell, mimivirus uses two distinct portals. (a) When feeding its genome into newly assembled viral capsids, a passageway at the center of an icosahedral face is used; (b) when releasing its DNA, the mimivirus capsid undergoes a large conformational opening of five icosahedral faces around a single vertex. This so-called stargate functions as a membrane-lined sleeve through which the whole viral genome can escape promptly after infection. These entry and exit strategies may also be used by other large DNA-containing viruses, especially those that, like mimivirus, have an internal membrane and code for proteins related to the DNA-packaging ATPases that are involved in bacterial DNA segregation, another process during which a large amount of DNA passes through a membrane portal.

Giant viruses are considered to be monophyletic based on a common set of 30 homologous genes. The genomes of some of these giant viruses have been sequenced. The sequences range between 330 kb (phycodnaviruses) and 1.2 Mb (*Mimivirus*), suggesting that their genomic repertoire is larger than that of other DNA viruses (30-200 kb) and equivalent to that of several parasitic bacteria (Koonin, 2005). Most giant virus genes are lineage-specific and their genome size is correlated with the number of paralogous genes (Suhre, 2005) indicating frequent successive gene duplications during giant virus genome evolution.

Almost all giant virus lineages contain genes with bacterial and eukaryotic homologs (van Etten, 2003; Iyer *et al.*, 2006). These genes may point to an ancient origin predating the divergence of the three kingdoms of life in the universal tree of life. Each giant virus lineage would then be expected to have retained a diverse assemblage of genes resembling bacterial or eukaryotic genes (Raoult *et al.*, 2004). Another possibility is that giant viruses are regressive or highly derived cells that suffered a simplification process. These hypotheses are at variance with the traditional view of evolution in which virus and phage are considered to have evolved chiefly by accretion of genes from disparate sources (Suhre, 2005). In this context, analysis of the *Mimivirus* genome showed that most genes homologous to eukaryotic genes are only distantly related to those of their host (an amoeba) (Claverie *et al.*, 2006), contradicting the idea of numerous lateral gene transfers with its amoebal host. According to Filee *et al.* (2006), NCLDVs are able to acquire foreign DNA sequences, and that some may indeed have acquired significant numbers of bacterial genes.

The location of the bacterial-like genes is not random. They are typically localized in specific genomic regions toward the ends of the genome, possibly as islands. In *Mimivirus* they are positioned within the first and last 250 kb of the genome. On the other hand, NCLDV core genes and genes with eukaryotic affinities are located toward the middle of the genome. In phycodnaviruses, even though bacterial-like genes are more dispersed, they still show this tendency. As in most viruses, there is a great abundance of orphan genes but these are distributed fairly evenly over the entire genome.

Suhre (2005) pointed out that the genomes of *Mimivirus* and phycodnaviruses are terminally redundant (*i.e.* carry duplicated regions). As previously suggested for vaccinia viruses by Yao and Evans (2001), insertion of foreign genes into a resident gene in the terminally redundant region leaves the other resident copy (located at the other end) intact. This probably favours viral derivatives in which inserted DNA occurs within one or the other end (Filee *et al.*, 2006). Filee *et al.* also detected several mobile genetic elements previously known to be specific to prokaryotes. These include apparently intact insertion sequences (ISs), which are probably chief agents of lateral gene transfer in prokaryotes (Frost *et al.*, 2005), homing endonucleases and inteins. In fact, giant viruses also have fairly large numbers of prokaryotic-like homing endonucleases and inteins reinforcing the idea that these viruses are actually chimeras of genes from disparate sources. Inteins are defined segments of proteins that excise and rejoin the flanking peptides. They occur in the DNA polymerase gene of *Mimivirus* (see Ogata *et al.* 2005) and in the phycodnavirus NY2A ribonucleotide reductase gene. At least two-thirds of known inteins occur in prokaryotes; those found in the eukarya are restricted to yeasts.

Filee *et al.* suggested that most bacterial-like genes detected in giant viruses had been recently and independently acquired in *Mimivirus* and phycodnavirus lineages and possibly also indicate that the resident bacterial DNA has been acquired several times during successive events. Acquisition of bacterial genes and mobile genetic elements seems to be a continuous process. These findings go against the idea that they could be inherited from a common ancestor. On the basis of their work on bacteriophage T4 and poxviruses, Filee *et al.* proposed that it is the extraordinarily tight connection between replication and recombination that leads to exceptionally high recombination frequencies facilitating promiscuous incorporation of foreign genes into the viral genome. The presence of repeated sequences such as ISs would then favour this process by supplying the homologies to drive intermolecular recombination. Their presence might also lead to further rearrangements in the host viral genome during subsequent viral replication cycles.

Filee *et al.* also suggested that either the host (or its symbiont) of a giant virus provides both the needed source of bacterial DNA (from grazing of bacteria) and the cell compartment that enables this DNA to come across or find actively replicating viruses. These systems therefore would provide an extensive interface for one-way genetic exchange between prokaryotes and eukaryotes. In this context, it may be recalled that bacterial gene acquisition by viruses was first observed in the form of transducing bacteriophages.

Aggresomes and Virioplasm

The replication of many viruses is associated with specific intracellular compartments called virus factories or virioplasm. These are thought to provide a physical scaffold to concentrate viral components and thereby increase the efficiency of replication. The creation of virus replication sites results in rearrangement of cellular membranes and reorganization of the cytoskeleton. Similar rearrangements can occur in cells in response to protein aggregation, where aggresomes and autophagosomes are produced to facilitate protein degradation. Some viruses appear to induce aggresomes and autophagosomes to generate sites of replication (Wileman, 2006).

Autophagy is a cellular response to starvation and as a quality control system that can remove damaged organelles and long-lived proteins from the cytoplasm. It is involved in several developmental pathways and disease processes and may provide defense against pathogens (Shintani and Klionsky, 2004; Talloozy *et al.*, 2002). Autophagy begins with the sequestration of an area of the cytoplasm within a crescent-shaped isolation membrane. These membranes mature into large double-membraned vesicles (diameter 500 to 1000 nm) termed autophagosomes, which eventually fuse with endosomes and lysosomes.

Autophagy has an important role in the removal of protein aggregates from cells. Protein aggregates, *e.g.*, are associated with neurodegenerative conditions such as Huntington's disease–they are first delivered to the microtubule organizing center (MTOC) by dynein-dependent retrograde transport along microtubules. When the degradative capacity of proteasomes is exceeded, protein aggregates accumulate in perinuclear inclusions called aggresomes. Aggresomes are surrounded by vimentin filaments and recruit chaperones, proteasomes, and mitochondria, suggesting a site specialized for protein folding and degradation. Many protein aggregates that cannot be refolded or degraded by proteasomes are eventually removed from aggresomes by autophagy; this delivers lysosomes for degradation.

Virus Replication in Factories

The factories generated by large cytoplasmic DNA viruses such as vaccinia virus and irido viruses contain viral DNA and structural proteins concentrated within inclusions resembling aggresomes. Like aggresomes, factories are located very near to the MTOC and recruit vimentin, cellular chaperones, ubiquitin, and mitochondria.

The close structural similarity between aggresomes and virus factories points to the possibility that aggresomes provide defense against infection that confines viruses within inclusions at the MTOC in preparation for degradation by proteasomes and/ or autophagy. The dual role of dynein (in the formation of virus factories and in the delivery of protein aggregates to aggresomes) raises the possibility that cells view viruses as protein aggregates and transport them to the MTOC. This view is supported by the fact that many viral core particles are of similar size (60 to 100 nm) to the aggregates being transported to aggresomes. For many viruses, transport to the MTOC for storage and eventual degradation probably protects cells from infection. For the large cytoplasmic DNA viruses, however, recognition by elements of the aggresome pathway may provide a site for replication (Wileman, 2006).

At least some viruses replicate on (or within) structures used by cells to remove protein aggregates. This could be important for virus replication–virus yields fall if the structures are not made or if viruses are denied access to them.

Genetic Exchange Among Terrestrial Bacteriophages

According to Hendrix (2002), microbes are the most numerous entities in the biosphere, with bacteriophages constituting the majority of these organisms. Although mechanisms of genetic exchange between individual microbes have been well-investigated, the importance of these processes in structuring virus populations is not clear. The rate of genetic exchange in virus populations has been estimated to be relatively low (see Lin *et al.*, 2004; Lindstrom *et al.*, 2004; McCullers *et al.*, 2004; Silander *et al.*, 2005), but only a few medically or agriculturally important viruses have been studied. Bacteriophages happen to be the most numerous entities in the biosphere and yet, not much is known about their genetic structure. Phages often control the population sizes of their hosts and hence influence large-scale ecological and biogeochemical processes attributed to bacteria (Fuhrman, 1999). Sex seems to be generally favoured as it promotes linkage equilibrium which, besides *de novo* mutations, provides the genetic variability so important for natural selection (Otto and Lenormand, 2002). This points to the importance of estimating the rate of genetic exchange in phages and to know whether sexual processes act to structure biological populations that comprise a major portion of our global ecosystem. Silander *et al.* (2005) presented a biogeography study involving 25 previously undescribed bacteriophages from the Cystoviridae clade, a group characterized by a dsRNA genome divided into three segments. Bacteriophages from this clade were first isolated in 1971 from bean straw infested with *Pseudomonas syringae* pathovar *phaseolicola* (Pp) (Vidaver *et al.*, 1973). These viruses are characterized by a tripartite dsRNA genome, and the three RNA segments per phage particle are referred to as L, M, and S (large, medium, and small, respectively). More phages from this clade were isolated later from the leaves of several agricultural species. Intrasegment recombination in

Cystoviruses occurs infrequently at a rate of ?10^{-7} per segment per generation (Onodera *et al.*, 2001), but exchange of segments between viruses occurs readily when host cells are multiply infected (Mindich *et al.*, 1976). These characteristics are biologically like those of segmented RNA viruses of medical and agricultural importance (*e.g.* Influenza-A). Reassortment between very distantly related phages in the Cystoviridae clade has also been shown *in vitro* (Qiao *et al.*, 2000).

When multiple Cystoviruses infect a single host cell, they undergo (*i*) rare intrasegment recombination events, and (*ii*) frequent genetic reassortment between segments. Analyzing linkage disequilibrium (LD) within segments, Silander *et al.* did not find intrasegment recombination in wild populations, consistent with (i). An extensive analysis of LD between segments supports frequent reassortment, on a time scale similar to the genomic mutation rate. The absence of LD within these phages is consistent with expectations for a completely sexual population, despite the fact that some segments have >50 per cent nucleotide divergence at 4-fold degenerate sites. This extraordinary rate of genetic exchange between highly unrelated individuals is unprecedented in any taxa (Silander *et al.*, 2005).

Genetic Exchange and the Viral Species Concept

Reliable methods for demarcating species boundaries in virus taxonomy are not available because of meager understanding of what constitutes a species in naturally occurring viral populations. It may be informative to apply the biological species concept–viruses being capable of genetic exchange (McCullers *et al.*, 2004), this method can separate species according to barriers to genetic mixis. If viral species can be accurately delineated, it follows that effective population sizes may also be estimated. Not much is known about the population size of most natural viral populations, although this may have a crucial role in processes such as disease emergence. Woolhouse *et al.* (2001) suggested that pathogens with higher mutation rates produce more genetic variants and are, therefore, more likely to be generalists. This may also be true for larger population sizes, because a more appropriate measure of genetic variation is the population mutation rate. This means that determination of viral species through the biological species concept and subsequent measurements of population size might provide a valuable clue to the likelihood of emergent disease. Conceivably, in many viral groups, the rate of genetic mixis may be a continuous trait, unlike the situation in most eukaryotic taxa, in which genetic exchange is largely either present or absent, implying that, instead of distinct genetic clusters such as those seen in eukaryotic and most prokaryotic taxa (see Cohan, 2002), loosely cohesive genetic clouds are observed (Silander *et al.*, 2005).

According to Marilyn Roossinck of the Samuel Roberts Noble Foundation in Ardmore, Oklahoma, a virus within a fungus is vital if that fungus and the grass it lives in are to survive high soil temperatures. They studied *Dichanthelium lanuginosum*, which grows in soils heated by geothermal activity in America's Yellowstone National Park. Plants that lack the fungus *Curvularia protuberata* fail to survive in soils above 38°C.

A virus was detected inside the fungus. When plants were infected by a fungus hat had been completely 'cured' of the virus, they shriveled and died at high temperature. When the fungus was restored, the systems' heat tolerance also revived. The virus-fungus combination also protected some tomato plants against high soil temperatures.

Tethering HIV-1 Virus

Mammals employ several different mechanisms to avoid the spread of intracellular pathogens. Even if an infectious agent crosses the first lines of defence, the infected cells press into service some restriction factors which block its further dissemination. One such restriction factor prevents HIV-1 from leaving infected cells and is counteracted by an accessory viral protein called Vpu. Neil *et al.* (2008) characterized this and obtained clues about how it functions.

To escape the infected cell, HIV-1 emerges through the cell surface, is wrapped in an envelope derived from the host cell membrane, and pinches off. But even if the virus reaches this stage of release, Vpu-deficient HIV-1 particles usually remain trapped at the cell surface.

The effect of Vpu on virus release is enigmatic because it appears to be unrelated to the protein's other function of reducing cellular levels of CD4, the HIV-1 receptor on T cells of the immune system (Göttlinger, 2008). Vpu's effect on release also does not require other HIV-1 components, and its importance for virus release can vary greatly among cell lines.

Varthakavi *et al.* (2003) and Neil *et al.* (2006) showed that Vpu counteracts a human antiviral factor which suppresses HIV-1 release by tethering mature viral particles to the cell surface after they have completely pinched off. Retention of HIV-1 particles at the surface of infected cells is also induced by human interferon-α, a protein that 'jump-starts' a cell's antiviral defences (Poli *et al.*, 1989). Neil *et al.* (2007) reported that Vpu counteracts the effect of interferon-α on HIV-1 release. They have now identified an interferon-α-induced human protein that fulfils all the criteria for the tethering factor antagonized by Vpu. The protein, named tetherin, is expressed only by those cells which require Vpu for HIV-1 release. Decreasing tetherin levels in cells which normally produce this protein trigger the release of Vpu-deficient viruses, tetherin expression in cells that typically lack this protein selectively inhibits the release of Vpu-deficient HIV-1. Neil *et al.* (2008) suggested that tetherin is highly potent, with only minute quantities sufficing to effectively inhibit HIV-1 lacking Vpu.

Both ends of tetherin are inserted in the cell membrane through its unusual pair of membrane anchors (Kupzig *et al.*, 2003). The central portion of the protein which faces the outside of the cell interacts with the same region of another tetherin molecule. According to Neil *et al.* (2008), tetherin molecules that end up in the viral envelope may keep the virus in check by interacting with tetherins associated with the cell surface. One unanswered question is: how does Vpu counteract the effects of tetherin? No reduction in tetherin levels in the presence of Vpu was detected. An earlier study (Bartee *et al.*, 2006), however, reported that levels of the protein now identified as tetherin are decreased by an unrelated human virus (Kaposi's sarcoma-associated herpesvirus); this finding also points to the broad antiviral activity of the protein. The viral protein involved in this case, K5, is a ubiquitin ligase, which adds the molecular tag ubiquitin to proteins, thereby marking them for degradation. K5 is structurally quite similar to some human ubiquitin ligases, at least one of which can strongly bring down the cellular levels of tetherin (Göttlinger, 2008). In fact Vpu also decreases the normal cellular levels of tetherin. These observations point to the possibility that Vpu uses a cellular ubiquitin ligase to dispose of tetherin, just as it does for CD4. But this does not necessarily rule out other possibilities, *e.g.*, tetherin relocalization by Vpu, rather than its degradation.

Only HIV-1 and some of its related viruses produce Vpu. This prompts the issue of how other related viruses deal with tetherin. HIV-2 (the less virulent human AIDS virus) is a protein that chiefly facilitates viral entry; it substitutes for Vpu, prompting virus release (Bour *et al.*, 1996). This warrants a study of whether this protein also antagonizes tetherin.

As the amino-acid sequence of tetherin differs among mammals, some HIV-1-related animal viruses probably cannot overcome human tetherin; this prevents them from becoming human viruses. Conversely, there is need for investigating whether tetherin contributes to the inability of HIV-1 to efficiently escape from most rodent cells (Mariani *et al.*, 2000). Lack of knowledge about this aspect has thwarted efforts to develop small-animal models of HIV-1 infection. Even in human cells, Vpu might not always be able to overcome the strong effect of tetherin, as the release of infectious Vpu-positive HIV-1 is usually inhibited with high doses of interferon-α. All this underscores the urgent need to determine how tetherin works, and how Vpu fends it off. A clear understanding of these issues may potentially enable strategies to limit the spread of HIV-1 and other viruses that target humans (Göttlinger, 2008).

HIV-AIDS Drugs

Over 20 million people infected with the *Human Immunodeficiency Virus* (HIV) live in developing countries. They have no access to drugs available on the market, *Zidovudine* (AZT) is too expensive for most patients.

Antiretroviral treatments lengthen the life span of people having HIV / AIDS. A combination of three antiretroviral drugs is today the most effective treatment available to suppress the replication of HIV. The triple therapy is expensive (US$ 12,000 to 18,000 per patient annually) and a rigorous and difficult regime with a daily intake of about 15 tablets is needed to prevent the drug resistance. Undesirable side effects occur.

The triple therapy is the standard form of care in the industrialized countries. Antiretrovirals such as *Zidovudine* as a monotherapy no longer constitute an effective treatment of HIV / AIDS, and yet this drug is the best way to minimize mother-to-child transmission of HIV (Hakansta, 1998).

AZT became available on the market in 1987. It was originally developed for cancer treatment. AZT is out of reach for most infected people in developing countries.

A modified version of AZT called *Zidovir-100* has been developed in India by Cipla. It is much cheaper than the Glaxo drug AZT. Recently, the United Nations Organization and the World Bank have established the *Joint United Nations Program on HIV/AIDS* (UNAIDS). The main focus of UNAIDS is to promote the development, evaluation and availability of appropriate HIV vaccines for worldwide use, especially in developing countries. UNAIDS has already made arrangements with .Giaxo-Wellcome to buy AZT at subsidized prices.

Epistasis in RNA Viruses

Many examples of epistatic gene interactions in plants and animals have been known for a long time. Recently, analogous interactions have been reported in some viruses (Sanjuan *et al.*, 2004a; Bonhoeffer *et al.*, 2004). Epistatic gene interactions in RNA viruses have some relevance to viral recombination. They are important in evolutionary biology whenever multilocus genetics matter (Wolf *et al.*, 2000). Epistatic interactions also play a conspicuous role in speciation and reproductive isolation (Gavrilets, 2004; Michalakis and Roze, 2004). They are central to the evolution of genetic recombination.

Sanjuan *et al.* (2004, 2004a) studied vesicular stomatitis virus (VSV), which does not show recombination. They examined several single-nucleotide mutations generated by directed mutagenesis. The individual fitness effects (deleterious or beneficial) of these mutations had already been established. The authors then studied the fitness effects of pairs of mutations that are either deleterious or beneficial when single. It emerged that singly beneficial mutations when paired exhibit negative epistasis, whereas singly deleterious mutations when paired exhibit positive epistasis (see Michalakis and Roze, 2004). But in both cases, the mutation pairs act antagonistically–their combined effect is less than that expected from their individual effects. Bonhoeffer *et al.* (2004) analyzed positive epistasis in another RNA virus, *viz.*, human HIV-1. This virus, unlike VSV, recombines frequently. If recombination in HIV-1 had been selected for because of gene interactions, epistasis between pairs of genes would be somewhat negative. Surprisingly, Bonhoeffer *et al.* found that gene interactions in HIV-1 show positive epistasis. This work involved several mutants of HIV-1 and their analysis by two different methods. Both the methods revealed positive epistasis in HIV-1.

According to Michalakis and Roze, epistasis between loci, like dominance between different alleles at a given locus, may be considered as either the outcome of evolution or the unavoidable result of the manner organisms function. For instance, according to the metabolic control theory of deleterious mutations affecting enzyme activities, epistasis should be either synergistic or antagonistic depending on whether the concerned mutations influence the same enzyme (always synergistic) or not. When the mutations affect different enzymes, the nature of epistasis depends on whether selection maximizes metabolic flux (antagonistic) or the concentration of an intermediate in the pathway (partly synergistic), or to optimize either the flux or the quantity of metabolic product (mostly synergistic) (Szathmary, 1993).

Some studies are being planned to decide how and whether a particular kind of epistasis is dependent on the environment. One implication of the recent researches is that the pattern of epistasis in RNA viruses may not be compatible with current genetic theories of sexual reproduction and recombination–these theories assume that mutations affecting fitness show negative epistasis. While VSV exhibits no recombination, HIV-1–recombines frequently according to Bonhoeffer *et al.* It may do so to repair single-strand breaks in the RNA genome. This, and the fact that both VSV and HIV-1 can suffer very high mutation rates, warrant caution against extrapolating results obtained with these simple organisms, even though their contribution to our understanding of evolution is clear.

Interferon

Half a century ago virologists tried hard to understand why an inactivated virus made it difficult for a normal virus to infect cells–a process called interference. Isaacs and Lindenmann (1957) found a likely answer to this question–the inactivated virus triggered the infected cells to produce a protein that suppressed replication of the live virus. The protein involved called interferon, has since turned out to be a useful therapy for hepatitis C and many cancers.

Isaacs and Lindenmann adapted a standard experimental model for infection studies which uses the chorioallantoic membrane–the membrane surrounding the growing embryo–in fertilized chicken eggs. The quantity of virus generated by infected membranes after each experiment was measured by a bioassay which relies on the ability of viruses to cause clumping of red blood cells. The authors used the influenza A virus. Rather than infecting the membrane by injecting the virus into the cavity between the membrane lining the shell and the embryo, they peeled away the membrane and cut it into 6-8 fragments. Each piece was placed in a test tube containing nutrient medium. This yielded more experimental values per egg and also greater experimental flexibility.

When first developed, interferon was greatly hyped. In reality it never became a miracle cure for cancer or viral infections. However, in view of its likely applications; it was patented in 1957.

Viral Vectors for Protein Production in Insects

Members of the baculovirus (BV) family of insect pathogens are reliable vehicles for carrying genes into insect cells. Recombinant BV vectors are being developed for gene expression and protein production systems in both insect cell cultures and living larvae. Therapeutic agents, vaccines, reagents, and pesticides are some of the products that can be made by this unusual method (Knight, 1991). High production levels of biologically active products can be achieved in this way. Under the control of a strong BV–derived promoter sequence, productivity in insect systems can exceed that in mammalian systems by thousands of times (Miyajima *et al.*, 1987). The method is quite fast as well as versatile because insect expression systems will produce almost any kind of protein. Over 200 genes cloned from viruses, bacteria, fungi, plants, and animals have been expressed in insect cell lines. Insect cells process these proteins correctly, performing functionally important posttranslational modifications such as glycosylation in the same way as mammalian cells. They can successfully produce functionally authentic proteins. As BVs do not affect vertebrates or plants, the potential for unintended transfers of genes between species, including into or out of humans, is quite low (Luckow, 1991; Knight, 1991).

BVs are rod–shaped viruses that contain a circular double–stranded DNA genome 100 to 150 kilobases long. The most commonly studied BV vector is that from *Autographa californica* nuclear polyhedrosis virus (Ac–NPV). Foreign genes placed under the control of its polyhedrin promoter usually can be expressed at high levels. For this vector, cultures of cells from *Spodoptera frugiperda*, the fall armyworm, are the standard host system.

Polyhedrin, the major structural protein of the virus, occludes BV particles into polyhedra. This enables the virus to survive in the wild. In the natural life cycle, insects feeding on contaminated plants ingest the polyhedra, which uncoat and replicate when they are inside cells that line the insect's digestive tract. A second form of the virus circulates through the insect's blood, causing a systemic infection. In the laboratory, intact insects can be infected early and insect cell lines can be infected with recombinant BV vectors.

Scale-up of production may favour larval culture over insect cell culture. Aseptic cell culture systems for large-scale fermentation are expensive and larval systems are cheaper and easier. Also, larval expression is stable. For pharmaceutical products, cell cultures are preferred as they are cleaner.

Silkworms can act as small-scale pharmaceutical factories. The silkworm *Bombyx mori* can be mass cultured easily. The availability of suitable BV vectors for inserting foreign genes into silkworm may help produce proteins more precious than silk. These plump insects–which weigh up to 5 grams each–are easy to infect by injecting BV vectors. After a few days, the foreign gene product may be secreted into the insect haemolymph (the fluid of the open circulatory system). Bleeding the larva yields 0.5 to 1.0 milliliter of product–containing fluid, and products can be easily purified from it by chromatography.

Silkworms are also susceptible to another BV, *B. mori* nuclear polyhedrosis virus (BmNPV). The gene and its promoter for the major coat protein of BmNPV have been isolated. Work is underway on improving expression rates of both pharmaceutical proteins and pest control agents in *B. mori* larval cultures through modified transfer vectors (see Maeda *et al.*, 1985; McCutchen *et al.*, 1991).

Experimental use of silkworm larvae to produce the lymphokine interleukin-3 (IL–3) is quite promising. One larva can give 1 mg of IL–3 in 4 days after injection.

Applications

BV is a good biological pest control agent. The implication of its oral transmission route is that it can be sprayed onto crops for protection. Unmodified, BV is slow acting, meaning a greater percentage of crops is lost before it takes effect. Hence, BV does not yet compare with conventional chemical insecticides.

To enhance the effectiveness of BV, the toxin and related factors within the BV genome have been studied. BV normally contains a gene called *egt* that blocks molting in the insect host, giving it more time to eat. Deleting the *egt* gene from BV significantly reduces insect feeding damage. Adding genes for exogenous toxins also enhances BV effectiveness. Splicing of the gene that encodes a paralytic insect toxin produced by predatory mites into BV can kill infected insects in as little as 2 days (Knight, 1991). Or, the gene for an insect–selective scorpion toxin can be added to a BmNPV vector. *B. mori* larvae infected with this recombinant virus stop feeding 40 hours after infection, instead of the usual 96. For crop pests, the same scorpion toxin has been added to an AcNPV vector, decreasing the average time to kill *Heliothis virescens* larvae from 125 hours to 88 hours (see Knight, 1991).

Chapter 4
The Archaea and Extremophilic Microorganisms

The discovery of the Archaea was a milestone that challenged the paradigm that life splits neatly into two distinct groups, one prokaryotic and the other eukaryotic. There are in fact three primary lineages (or domains) of evolutionary descent: Bacteria, Eukarya, and Archaea (see Friend, 2007).

The Archaea constitute one of the two groups of prokaryotes, the other being bacteria. Archaea are distinguished from bacteria by numerous genetic and physiological differences. Those members of the archaea that have been cultivated in the lab produce methane or are tolerant of unusually high temperatures or salinities. Uncultivated archaea, such as those now known to be abundant in the deep sea, may have other distinctive functionalities. Some archaea fix nitrogen. Evolutionary studies place the archaea, bacteria and eukaryotes as three fundamentally different domains of life.

It was the methanogens that provided the first evidence that there was more to life than bacteria and eukaryotes. Karl Stetter has been a hunter extraordinaire of microbes that live in boiling, sulfide-rich waters. Norman Pace combined molecular phylogenetics with strategies to extract DNA sequences directly from natural communities of microbes. He explored the microbial world without requiring that the organisms of interest be grown in cultures, ushering in a revolution in microbial ecology.

Friend himself visited Costa Rican rainforests to collect termites that house a microbial community in their gut; this community may be the world's bioreactor *par excellence* for degrading cellulose. Many different biotechnological applications stem from these and other exotic Archaea, including the potentials to minimize our reliance on fossil fuels and to remediate some rather nasty chemical dumps.

While riding a submersible to see the microbial assault under way on the *Titanic's* hull, Friend explains how the great diversity of microbes in the world's ocean is being catalogued and how genomic methods have revealed a previously unrecognized type of photosynthesis in the seas.

The Third Domain rightly hails the discovery of the Archaea as one of the 20th century's most dramatic developments in biology.

In contrast to eukaryotes and bacteria, archaea have largely been known as hardy inhabitants of such extreme environments as hot springs or acid mines. However, recent development and availability of suitable methods for detecting even small numbers of archaea in unpurified samples have proved that these prokaryotes do participate in biogeochemical cycles that affect all spheres of life. Recent research on archaea-based oxidation of ammonia in North Sea waters and in northern European soils has involved measurements of the amounts of the gene encoding ammonia monooxygenase (the first enzyme in the nitrification pathway) and correlating these data with the presence of Crenarchaeota-specific lipids. Quantitation of ammonia monooxygenase genes in the upper 1000 m of the North Atlantic and across pristine and fertilized soils showed that the archaeal version was generally much more abundant than the bacterial enzyme. Incubation of the marine sample and estimates of the rates of Crenarchaea growth and production of nitrite yielded an oxidation flux of about 3 fmol of NH_3 per cell per day. This was then extrapolated to a global inorganic carbon fixation rate of 4×10^{13} mol of C per year (see Wuchter *et al., PNAS* (USA) 103: 12317, 2006).

Hyperthermophilic Archaea

According to Finlay (2002), dispersal limitations do not affect unicellular organisms because of their small size, high abundance, and metabolic versality. Several environmental surveys based on both 16S ribosomal RNA (rRNA) signature sequences and phenotypic characters supported Finlay's suggestions. These surveys identified apparently identical microorganisms in similar environments as far apart as the polar oceans (Massana *et al.*, 2000). Intensive studies based on use of multilocus analysis showed that pathogenic bacterial species and *Bacillus* spore formers have global panmictic distributions (Roberts and Cohan, 1995). However, such a ubiquitous distribution seems unlikely for extremophiles, whose growth requires harsh habitats that are usually discontinuous and distant (Staley and Gosink, 1999). This is exemplified by *Sulfolobus* species, which are archaea inhabiting solfataric geothermal springs and grow optimally at about 80°C and pH 3. Abrupt temperature changes cause cell cycle arrest and chromosomal DNA degradation in liquid *Sulfolobus* cultures, and no resistant spore state has been recorded for any *Sulfolobus* species. It is indeed surprising that the same *Sulfolobus* species has been isolated from geothermal hot springs throughout the Northern Hemisphere (Rice *et al.*, 2001). This prompts the puzzle of how organisms with such specific growth requirements survive dispersal across the long and inhospitable distances that separate geothermal regions (see Stetter *et al.*, 1993). To tackle this issue, Whitaker *et al.* examined the population structure in *Sulfolobus* and tested whether barriers to dispersal or ecological selection are chiefly responsible for its development. According to Whitaker *et al.* (2003), barriers to dispersal between populations allow them to diverge through local adaptation or random genetic drift. High-resolution multilocus sequence analysis has revealed that, on a global scale, populations of hyperthermophilic microorganisms are isolated from one another by geographic barriers and have diverged over the course of their recent evolutionary history. The identification of a biogeographic pattern in the archaeon *Sulfolobus* challenges the current model of microbial biodiversity in which unrestricted dispersal constrains the development of global species richness.

In general, biogeographical patterns abound in multicellular eukaryotic organisms, but are unusual for microorganisms. Several authors reported geographic patterns in bacterial distributions but their data did not explicitly show the importance of geographic barriers (Staley and Gosink, 1999; Oda *et al.*, 2002). One other test of microbial geographic structure is based on differences in the genomic fingerprint of repetitive elements (Cho and Tiedje, 2000) but data cannot be interpreted in an evolutionary context as they depend on the untested assumption of homology between similar-sized bands. The patterns of genetic divergence reported by Whitaker *et al.*, in contrast, are based on nucleotide substitutions, which are well characterized by classical evolutionary theory.

The data of Whitaker *et al.* reveal that gene flow among *Sulfolobus* populations is limited. Either the specialized growth requirements of these hyperthermophilic acidophiles prevent dispersal, or immigrants are not able to persist in established endemic populations, or both. Genetically isolated populations carry a high potential for local adaptation to specific environmental conditions.

Thermoacidophilic Methanotrophy

Aerobic methane-oxidizing bacteria (methanotrophs) play a crucial role in the global carbon cycle by converting methane to biomass and carbon dioxide. These bacteria have been isolated from many different environments such as wetlands, agricultural soils and marine sediments, as well as from more extreme environments, such as hot springs and alkaline soda lakes. Until recently, however, nothing was known about whether they survived or thrived in thermoacidic environment having pH values of approximately 1 and temperatures higher than 50°C (Semrau *et al.*, 2008).

Previous studies showed that methanotrophs grow best at moderate pH (5-8) and temperature ranges (20-35°C), although psychrophilic (growth <15°C), thermophilic (growth at 57-59°C), alkaliphilic (growth at pH >9.0), and moderately acidophilic (growth at pH 5-5.5) methanotrophs were also isolated (see Dedysh *et al.*, 1998; Kaluzhnaya *et al.*, 2001; Trotsenko and Khmelinina, 2002). All known methanotrophs group within either the γ- or α-*Proteobacteria* (see McDonald *et al.*, 2008). The metabolic pathway by which these cells oxidize methane to carbon dioxide is quite similar in all known methanotrophs: carbon is assimilated into biomass at the level of formaldehyde by either the serine or ribulose-monophosphate pathways.

Dunfield *et al.* (2007), Pol *et al.* (2007) and Islam *et al.* (2008) have described three thermoacidophilic methanotrophs that are not only phylogenetically different from what have been considered as typical strains of methanotrophs but which might also use different mechanisms for methane oxidation and carbon assimilation.

These recent reports on the isolation of thermoacidophilic methanotrophs from widespread geographical locations point to the possibility that the ability of bacteria to oxidize methane could be much more diverse than believed hitherto in terms of ecology, phylogeny and physiology. It is also possible that some methanotrophs may also be found on methane present in other environments. Indeed it is conceivable that there could be methanotrophs of other phyla as yet undiscovered (Semrau *et al.*, 2008).

Cellular organisms are classified into three superkingdoms or domains namely Archaea, Bacteria, and Eukarya (Woese *et al.*, 1990). The Archaea are further divided into two main lineages, the Crenarchaeota and the Eurayarchaeota, each of which comprises several distinct classes of organisms that grow in a variety of environments. Although some aspects of archaeal biology are unique, certain other traits resemble those in eukaryotes. These traits include the machineries involved in information storage, maintenance, and processing (Lindas *et al.*, 2008).

Cell Division

The cell division patterns in bacterial and eukaryotic cells are fairly uniform but there is considerable diversity in archaeal cell division. In the cell cycle of each lineage (Archaea, Bacteria, and Eucarya), DNA replication ensures faithful duplication of the chromosome, after which a copy of the genetic material from the mother cell segregates into 2 daughter cells. It is through this process that organisms produce progeny for continuation of their lineage.

While archaeal cell cycle progression has been well elucidated, as also their regulatory and mechanistic aspects of the replication process (Bernander, 2007), not much is known about the genome segregation machinery in this domain. In species belonging to the Euryarchaeota and in bacteria, cell division is mediated by FtsZ protein filaments which form a constricting ring structure (Margolin, 2005). In eukaryotes, division involves a contractile actin-myosin ring or, in plant cells, by septum formation at a site initially marked by actin and microtubules. In contrast to bacteria, euryarchaea, and eukaryotes, no cell division components have been identified in the phylum Crenarchaeota (Lundgren and Bernander, 2005). Lindas *et al.* (2008) showed that the crenarchaeotes use a different set of proteins during septation, the process that finally leads to formation of 2 daughter cells. Bult *et al.* (1996) had reported that the archaeal DNA replication machinery is simpler than that found in eukaryotic cells. In fact, some members of archaea, *e.g.*, *Sulfolobus acidocaldarius*, used as models to decipher the mechanisms of archaeal DNA replication, have also revealed valuable insights about the more complex eukaryotic DNA replication, especially in the studies on the replicative DNA helicase minichromosome maintenance (MCM) (Barry and Bell, 2006).

Lindas *et al.* synchronized the growth of a high proportion of *S. acidocaldarius* cells, so that the majority is at the same stage of the cell cycle. The cell cycle of this crenarchaeon had been elucidated by Hjort and Bernaderin (2001). Lindas *et al.* looked for the genes that were highly expressed concomitant to chromosome segregation and cell division. Among the candidate genes, 3 occurred in a cluster which was found to be conserved in other members of the genus *Sulfolobus*, also in other members of the crenarchaeal family. The genes in the order of transcription were named as *cdvA*, *cdvB*, and *cdvC*.

According to Lindas *et al.*, a three-gene operon, *cdv*, in *Sulfolobus acidocaladarius*, forms part of a unique cell division machinery. The operon is induced at the onset of genome segregation and division, and the Cdv proteins then polymerize between segregating nucleoids and persist throughout cell division, forming a successively smaller structure during constriction. The cdv operon is strikingly down-regulated after UV irradiation, pointing to division inhibition in response to DNA damage, reminiscent of eukaryotic checkpoint systems. The *cdv* genes show a complementary phylogenetic range relative to FtsZ-based archaeal division systems such that, in most archaeal lineages, either one or the other system is present. Two of the Cdv proteins, CdvB and CdvC, display homology to components of the eukaryotic ESCRT-III sorting complex involved in budding of luminal vesicles and HIV-1 virion release, suggesting mechanistic similarities and a common evolutionary origin (Lindas *et al.*, 2008).

The crenarchaeotes have generally been thought to be unique, and only comprised of extreme thermophiles that grow at temperatures >80°C. But we now know that mesophilic crenarchaeotes abound in nature in agricultural lands,playing yet unknown roles (Simon *et al.*, 2005). In the oceans they are estimated to constitute 30 per cent of all picoplankton and probably play important roles in ammonia oxidation and carbon fixation (Cann, 2008). These enigmatic archaea are extremely difficult to culture–only one, *Nitrosopumilis maritimus*, has been successfully cultured in the laboratory. The genomes of *N. maritimus* and that of the uncultured mesophilic crenarchaeote, *Cenarchaeum symbiosum* harbour both *ftsZ* and the *cdv* genes, suggesting that these organisms use a "hybrid" of the 2 cell division systems found in the Euryarchaeota and the Crenarchaeota. On the other hand, the DNA replication machinery of the mesophilic crenarchaeotes appears to be a mixture of those from the 2 subdomains of archaea, with the euryarchaeal ones being predominant.

DNA Replication

With the recent availability of genome sequences of the Archaea, a primary domain of life (see Woese and Fox, 1977), it has emerged that the archaeal DNA replication machinery is very similar to that in eukaryotes but is evolutionarily quite distinct from that in bacteria. The archaeal machinery is believed to be a simplified, probably ancestral, form of that in eukaryotes. The organizational simplicity of the archaeal DNA replication process together with the additional biochemical advantages allowed by the study of hyperthermophilic archaea, has generated much interest in the archaeal machinery as a model of the eukaryotic machinery (Barry and Bell, 2006).

Replication Origins

Jacob *et al.* (1963) proposed the replicon hypothesis according to which a *trans*-acting initiator protein binds to a *cis*-acting replicator DNA sequence to initiate DNA replication in bacteria. Bacterial chromosomes have a single replication origin, *oriC*, which consists of an A-T-rich region of DNA containing multiple copies of the DnaA box; the box is bound by the initiator protein DnaA. In several species, the *DnaA* gene is located adjacent to the origin; this means that the two can be coregulated (Marians, 1992).

Unlike those of bacteria, chromosomes of eukaryotes contain several replication origins. *Saccharomyces cerevisiae* (budding yeast) has clearly defined replication origins that are autonomously replicating sequences and contain conserved sequence elements, similar to the situation in bacteria; they are bound by the origin recognition complex (ORC). This complex is composed of six separate polypeptides, Orc1-6, some of which contain AAA+ (*A*TPases *a*ssociated with various cellular *a*ctivities) ATPase domains. Orc1 is also closely related to another replication factor Cdc6 (Cdc18 in *Schizosaccharomyces pombe*), which probably suggests that Orc1 and Cdc6 probably descended from a common ancestor. Instead of depending on sequence-specific DNA recognition by ORC, in higher eukaryotes, origins seem to be defined by facilitated recruitment of ORC by various other DNA binding proteins.

It was earlier supposed that because the chromosome structure in archaea resembles that in bacteria, archaeal chromosomes probably also contain one origin of replication. In *Pyrococcus abyssi* the single origin was indeed reported (Matsunaga *et al.*, 2001). The origin binding proteins in archaea turned out to be homologues of the related eukaryotic Orc1 and Cdc6 proteins. The origin in *P. abyssi* is located adjacent to the gene for Orc1/Cdc6, like that of DnaA in many bacteria. Robinson *et al.* (2004) identified and mapped two origins of replication for *Sulfolobus solfataricus* which has three ORc1/Cdc6 genes, encoding Cdc6-1, Cdc6-2, and Cdc6-3. Critical research revealed that at least some archaea have multiple origins of replication (Barry and Bell, 2006).

Critical mapping of the three replication origins in *S. solfataricus* led to the identification of origin recognition boxes (ORBs), which are inverted repeat sequence elements bound by Cdc6-1, at oriC1. According to Robinson *et al.* (2004), like the case for bacteria and *S. cerevisiae*, archaeal origins may possibly be defined by specific sequence elements.

Most archaeal genomes contain from one (*Pyrococcus* species) to nine (*Halobacterium*) Orc/Cdc6 genes. These have been classified into three major groups, and all species that have more than one Orc1/Cdc6 protein have at least one from the Cdc6-1 and Cdc6-2 subgroups.

Replicative Helicase

In bacteria, DnaB, an AAA+ protein which functions as a homohexamer acts as the replicative helicase. In eukaryotes, the replicative helicase is probably MCM (*m*inichromosome *m*aintenance *c*omplex). The MCM proteins were originally identified in yeast (Maine *et al.*, 1984). MCMs also occur in other eukaryotes and some members of the MCM family may be required for replication elongation and meiotic recombination (Blanton *et al.*, 2005).

Archaeal genomes have at least one MCM homologue. Unlike eukaryotes, however, many archaea contain only one MCM gene. Like the eukaryotic MCM proteins and bacterial DnaB, archaeal MCM is an AAA+ protein. *In vivo*, MCM interacts functionally with Cdc6 and, via GINS (*go*, *i*chi, *n*ii, *s*an [five, one, two, three in Japanese]) complex was first identified in yeast and *Xenopus*, with the primase (Marinsek *et al.*, 2006). It localizes to replication origins in *P. abyssi*. However, nothing is known about whether MCM is essential for replication. In view of the conservation of MCM from archaea to eukaryotes, it is probable that it serves as a replicative helicase.

Primase

As DNA polymerases cannot initiate synthesis *de novo*, they require a DNA or RNA primer which they can elongate. This is synthesized by a primase, which in eukaryotes and bacteria is a DNA-dependent RNA polymerase. Whereas bacterial primase is DnaG (a monomer), eukaryotic primase is a dimer consisting of a small catalytic (PriS) and a large noncatalytic (PriL) subunit. This associates with Polα and the B subunit to form the Polα/primase complex. The primase synthesizes RNA primers of 8 to 12 nucleotides (nt) long which are then elongated to around 30 nt by Polα to produce a DNA-RNA hybrid before being handed to the replicative polymerase (Frick and Richardson, 2001).

Archaea do contain homologues of eukaryotic PriS and PriL, but lack Polα and the B subunit. The small subunits of primase from both *S. solfataricus* and *Pyrococcus* species are able to synthesize both RNA and DNA primers *in vitro*. *Pyrococcus furiosis* PriS preferentially synthesizes long (up to 6 kb) DNA oligonucleotides. But the addition of PriL increases the RNA polymerase activity, decreases the DNA polymerase activity, and decreases the average product length; this points to PriL playing a regulatory role (Liu *et al.*, 2001).

Okazaki fragments have been isolated from archaea.

The primase enzyme synthesizes a primer which is extended by a DNA polymerase. In common with bacteria and eukaryotes, archaea have multiple DNA polymerases. There is a clear evolutionary divide in the distribution of polymerase families within the archaea. Archaea in the euryarchaeal phylum contain polymerases belonging to two distinct families, *i.e.*, the ubiquitous family B polymerases and a family that thus far appears unique to the euryarchaea, called family D.

Besides the family D polymerases, euryarchaea possess DNA polymerases belonging to family B, suggesting that the two classes of polymerases probably have distinct roles in the cell.

The crenarchaea do not possess a family D polymerase but usually different crenarchaeal family B polymerases play distinct roles on the leading and lagging strands (Barry and Bell, 2006).

Role of Chromatin

Most studies performed on the archaeal DNA replication machinery have used naked DNA templates. However, within the cell, DNA is compacted by association with a range of small basic proteins which potentially modulate access to the DNA template. Archaeal cells use a surprising variety of different proteins to mediate genome compaction (Sandman and Reeve, 2005). Many euryarchaea have homologues of eukaryotic histones, and these form structures analogous to the eukaryotic H3/H4 tetrasome. However, very little is known about the role of archaeal histones *in vivo*. Surprisingly, except for some mesophilic marine organisms, histones are not found in the crenarchaea (Cubonova *et al.*, 2005). The most studied crenarchaeal chromatin proteins come from the genus *Sulfolobus*.

Although much work has been done on the form and function of archaeal DNA replication and chromatin proteins, much still remains to be discovered about how these proteins interact in the context of the macromolecular assemblies found at replication origins and during progression of the DNA replication fork. Also, the regulation of these proteins during the course of the cell cycle in archaeal species has not been investigated. For *Sulfolobus*, in particular, where multiple DNA replication origins are used, it should be interesting to know whether there is some mechanism to allow coordinated regulation of origin activity, and if so, how this is mediated (Barry and Bell, 2006).

Archaeal Metabolism in Deep Ocean

Certain archaea play a critical role in the marine carbon and nitrogen cycles and their abundance in the deep sea is linked to their likely metabolic profile (Schleper, 2008).

The deep ocean harbours a rich wealth of microbial life. Understanding the roles played by deep-sea microbial communities is essential for understanding global biogeochemical cycling, which in turn is so important for all other forms of life. The Crenarchaeota (Archaea) are abundant in the deep sea. Könneke *et al.* (2005) and Wuchter *et al.* (2006) have suggested that some Crenarchaeota play a central role in the ocean's nitrogen cycle by mediating the first step in the conversion of ammonia to nitrate. These were assumed to be autotrophs as they fix inorganic carbon for growth, rather than acquiring it from organic sources; this view also implicated them in the carbon cycle.

Now Agogue *et al.* (2008) have proposed that neither of these ideas is true for all Crenarchaeota. It appears that the Crenarchaeota living in the deeper oceanic waters of the subtropical and tropical North Atlantic Ocean lack genes for the main enzyme involved in ammonia oxidation; also, that deep-sea Crenarchaeota do depend on organic sources of carbon.

Variants of genes that encode ammonia monooxygenase–the enzyme that oxidizes ammonia–have been detected (Francis *et al.*, 2005) in archaeal genomes (Hatzenpichler *et al.*, 2008). Considering that there are many more archaea than ammonia-oxidizing bacteria in many terrestrial and marine environments, archaea seem to be major contributors to the nitrogen cycle (Schleper *et al.*, 2005).

The isolation and culturing of ammonia-oxidizing archaea proved that certain Crenarchaeota can generate energy through ammonia oxidation in an oxygen-dependent process, and thereby fix carbon from inorganic sources, just like their bacterial counterparts. (This enzyme, AMO, converts ammonia into hydroxylamine; and the latter is converted by the same microbes into nitrite ions). But it was not clear whether the high abundance of the archaeal *amo*A gene (which encodes ammonia monooxygenase) in the environment reflects a high archaeal contribution to ammonia oxidation. Another unresolved issue was whether all Crenarchaeota in moderate environments contribute to ammonia oxidation.

Agogue *et al.* (2008) did not find any evidence for ammonia oxidizers in the deep sea of equatorial regions; whereas the concentrations of *amo*A genes found at all depths in polar regions were high, they were 1,000 times lower than the concentration of total Crenarchaeota found in the bathypelagic waters (1,000-4,000 metres) of subtropical regions. It was therefore concluded that most deep-sea Crenarchaeota in subtropical waters do not oxidize ammonia. It may be remembered that the concentration of ammonia in subtropical deep waters is much lower than in the equivalent polar regions, where potential ammonia oxidizers occur in high numbers (Schleper, 2008).

Agogue *et al.* also studied the ability of microorganisms taken from different locations to fix inorganic carbon. The decreases in *amo*A abundance reflected the observed patterns of carbon fixation: as the ability of marine microorganisms to fix carbon declines, so does the number of *amo*A genes in the archaeal population. This in turn suggests that deep-sea Crenarchaeota are not autotrophs. According to Schleper, the data of Agogue *et al.* do not conclusively prove that deep-sea Crenarchaeota are not autotrophic, but their conclusions are supported by evidence from several previous studies (see, *e.g.*, Ingalls *et al.*, 2006) which showed that at least some Crenarchaeota do take up organic carbon for growth.

What was not clear from the earlier studies was whether heterotrophic growth is an alternative option for otherwise autotrophic, ammonia-oxidizing archaea, or whether distinct populations of Crenarchaeota exist that are strictly either heterotrophic or autotrophic. Agogue *et al.*'s findings point to the existence of specific heterotrophic Crenarchaeota, lacking the *amo*A gene, that live preferably in deep ocean waters.

Conceivably the success of archaea in many different environments might well be based on the organisms' ability to switch from an autotrophic to a heterotrophic life-style, depending on environmental conditions. Most soil Crenarchaeota are potential ammonia oxidizers. Yet it is difficult to imagine that their large populations in terrestrial habitats are merely based on this form of metabolism, especially in habitats where ammonia and/or oxygen concentrations are low. Application of robust techniques for cultivating microorganisms in the laboratory could be indispensable for throwing light on the physiology and ecological impact of the ubiquitous but enigmatic Crenarchaeota.

Archaea in the Marine Nitrogen Cycle

Archaea, which constitute one of the three domains of life (the other two being eukaryotes and bacteria) have generally been supposed to chiefly consist of organisms living in such extreme environments as sulfidic hot springs, salt brines, and anoxic environments. Modern molecular biological techniques have revealed that archaea are widespread and grow in several types of environments such as oceans, lakes, and soils (Könneke *et al.*, 2005). Surprisingly, these nonextremophilic archaea are not closely related to cultured relatives, and we know very little about their physiology and role in biogeochemical cycling.

Planktonic archaea in the ocean comprise both Crenarchaeota and Euryarchaeota, with the former being the most abundant (Fuhrman *et al.*, 1992). Marine Crenarchaeota contribute the most abundant single group of prokaryotes in the ocean, and their physiology and role in marine biogeochemical cycles are also virtually unknown. Marine Crenarchaeota are more abundant in deep neritic waters and in the meso- and bathypelagic zones of the ocean (Herndl *et al.*, 2005). They seem to account for ?20 per cent of all prokaryotic cells in the global ocean (Karner *et al.*, 2001; Schleper *et al.*, 2005). In situ labeling and microautoradiography experiments have suggested that marine Crenarchaeota use dissolved inorganic carbon as carbon source but can also take up amino acids, reminiscent of a heterotrophic lifestyle. Könneke *et al.* (2005) isolated a crenarchaeote, *Candidatus "Nitrosopumilus maritimus"*, from a sea aquarium. This species is autotrophic and can oxidize ammonium to nitrite. Observation of positive correlations between the abundance of Crenarchaeota and nitrite in Arabian Sea, other seas, and with particulate organic nitrogen in Arctic water point to the possible involvement of Crenarchaeota in the oceanic nitrogen cycle, probably as nitrifiers; but any relevance of crenarchaeotal nitrification for the marine nitrogen cycle is obscure. Traditionally it has been assumed that marine nitrification is chiefly carried out by bacteria belonging to the β-and γ-proteobacteria (Ward, 2002).

Wuchter *et al.* (2006) enriched a crenarchaeote from North Sea water and showed that its abundance (not that of bacteria) correlates with ammonium oxidation to nitrite. A time series study in the North Sea revealed that the abundance of the gene encoding for the archaeal ammonia monooxygenase alfa subunit (*amoA*) is correlated with a decline in ammonium concentrations and with the abundance of Crenarchaeota. Remarkably, the archaeal *amoA* abundance was 1-2 orders of magnitude higher than those of bacterial nitrifiers which are usually believed to mediate the oxidation of ammonium to nitrite in marine environments. Analysis of Atlantic waters of the upper 1,000 m, where most of the ammonium regeneration and oxidation take place, showed that crenarchaeotal *amoA* copy numbers are also 1-3 orders of magnitude higher than those of bacterial *amoA*. These data point to an important role for archaea in oceanic nitrification (Wuchter *et al.*, 2006).

The observation that marine Crenarchaeota can mediate chemolithoautotrophic nitrification seems to explain their distribution in the open ocean. These prokaryotes occur over a large depth range. While their absolute cell numbers are highest in the photic zone, they decline with depth and yet dominate the prokaryotic community below the photic zone. According to Arrigo (2005), nitrate depth profiles from the ocean reveal low concentrations in the upper ocean and levels varying from ?20 to 40 μM in deeper waters of the Atlantic and Pacific Ocean. These profiles probably result from four different processes operating within the water column: (a) the uptake of inorganic nitrogen in the upper ocean waters by primary producers, (b) ammonium regeneration from decomposition of descending particulate organic nitrogen, (c) oxidation of ammonium to nitrite by the *Nitrosomonas/Nitrosospira* and *Nitrosococcus* groups, and (d) oxidation of nitrite to nitrate by such nitrite oxidizers as *Nitrobacter*. Whereas no molecular ecological study has shown large numbers of known nitrifying bacteria in marine waters (see Ward, 2002), marine Crenarchaeota do constitute ?20-30 per cent of the total prokaryotic community (Karner *et al.*, 2001).

Thermoacidophilic Archaeon from Deep-sea Hydrothermal Vents

Deep-sea hydrothermal vents serve as biological oases at the sea floor and are supported by the thermal and chemical flux from the Earth's interior (Reysenbach *et al.*, 2006). When hot, acidic and reduced hydrothermal fluids mix with cold, alkaline and oxygenated sea water, minerals precipitate to form porous sulphide-sulphate deposits. It is these structures which act as microhabitats for various prokaryotes that exploit the geochemical and physical gradients in this dynamic ecosystem (Reysenbach and Shock, 2002). Although fluid pH in the actively venting sulphide structures is generally low (pH <4.5), no extreme thermoacidophile has been isolated from vent deposits. Culture-dependent and culture-independent approaches have revealed a high diversity of bacteria and archaea associated with deep-sea vent deposits (see, for example, Nakagawa *et al.*, 2006). In these deposits, the bacteria are dominated by many novel branches of the e-Proteobacteria. However, although several archaea have been isolated from these deposits, few isolates are found in environmental 16S rRNA gene clone libraries and most environmental clones have no representatives in culture. One lineage that is particularly widespread at deep-sea vents is often associated with actively venting sulphide deposits. The 'deep-sea hydrothermal vent euryarchaeotic' lineage DHVE2 is often found in clone libraries and has been reported as the most dominant clone type in some archaeal clone libraries (Reysenbach and Shock, 2002). Culture-independent surveys undertaken by Takai and Horikoshi (1999) and Nercessian *et al.* (2003) based on ribosomal RNA genes from deep-sea hydrothermal deposits have identified a widespread euryachaeotal lineage DHVE2 (deep-sea hydrothermal vent euryarchaeotic 2). Although the seemingly deep-sea endemic DHVE2 is ubiquitous in distribution, attempts to grow these archaea in culture failed so that we know little about their metabolism. Reysenbach *et al.* (2006) have now reported the isolation and cultivation of a member of the DHVE2 group, which is an obligate thermoacidophilic sulphur- or iron-reducing heterotroph capable of growing from pH 3.3 to 5.8 and between 55 and 75°C. Reysenbach *et al.* (2006) named this new taxon as *Aciduliprofundum boonei*. It was also shown that this isolate constitutes up to 15 per cent of the archaeal population, proving that thermoacidophiles are important players in the sulphur and iron cycling at deep-sea vents.

Acidophilic Microorganisms

Certain microorganisms grow best at a pH of 3 or less and are termed 'acidophiles'. For growth at low pH, acidophiles need to maintain a pH gradient of several pH units across the cellular membrane while producing ATP by the influx of protons through the F_0F_1 ATPase.

There exist both natural and man-made acidic environments such as sulfidic mine areas and marine volcanic vents in the biosphere. Acidophiles are widely represented in the bacterial and archaeal domains (Johnson and Hallberg, 2003). They play central roles in several biogeochemical cycles including the iron and sulfur cycles. In acid mine drainage (AMD), the release of metal-rich, acidic effluents causes much environmental damage *e.g.*, drinking water contamination. Acidophiles attract biotechnological application for metal extraction from ores (Rohwerder *et al.*, 2003; Clark *et al.*, 2006). This biotechnological process is sustainable and produces only small and manageable pollutant outputs (Golyshina and Timmis, 2005). Acidophiles are a potential source of acid-stable enzymes with applications as lubricants and catalysts (van den Burg, 2003).

Many acidophiles grow in inaccessible, isolated and harsh environments such as AMD sites and geothermal vents. These environments create an inaccessible physical barrier which can reduce the colonization potential of microorganisms that grow at or around neutral pH (see Baker-Austin and Dopson, 2007).

pH Homeostasis

Like neutralophiles, acidophiles require a circumneutral intracellular pH (Table 4.1). Where acidophiles differ from neutralophiles is that they tolerate pH gradients [pH gradient (ΔpH) = $pH_{in}-pH_{out}$] several orders of magnitude greater than neutralophiles. The DpH across the cytoplasmic membrane is intrinsically connected to cellular bioenergetics because it is the chief contributor to the proton motive force (PMF) in acidophiles. Proton motive force (PMF) is a measure of the energized state of the cell membrane as a result of a charge separation between the cytoplasm and external milleu created by the membrane potential ($\Delta\Psi$) and pH gradient (ΔpH) across the membrane. In acidophiles, the PMF is primarily made up of the DpH. For example, an acidophile with a near neutral cytoplasm (pH 6) separated from an acidic environment (pH 2) will experience a net force (*i.e.* the PMF) across the cell membrane that can drive energy dependent processes (see Baker-Austin and Dopson, 2007). The influx of protons through the F_0F_1 ATPase to produce ATP boosts cellular protonation and, if not controlled, rapidly dissipates the ΔpH. The functions of proteins and nucleic acids are adversely affected by protonation, and interference caused by free intracellular protons affects such processes as DNA transcription, protein synthesis and enzyme activities (Madshus, 1988).

Recent advances in the biochemical analysis of acidophiles coupled to sequencing of several genomes have shed new insights into acidophile pH homeostatic mechanisms. Acidophiles seem to share distinctive structural and functional characteristics including a reversed membrane potential, highly impermeable cell membranes and a predominance of secondary transporters. Also, once protons enter the cytoplasm, methods are required to alleviate effects of a lowered internal pH. Baker-Austin and Dopson reviewed some recent insights regarding how acidophiles are able to survive and grow in these extreme conditions.

Table 4.1: Metabolic Characteristics and Genome Details of Selected Acidophilic Microorganisms (After Johnson, 1998; Bond and Banfield, 2001; Baker-Austin and Dopson, 2007)

Microorganism	*Metabolism*[a]	*pH Optimum*	*pH_{in}*	*Genome (Mbp)*
Acidophilic archaea				
Thermoplasma acidophilum	H	1.4	6.4	1.56
Ferroplasma acidarmanus	IO/H	1.2	5.6	1.97
Picrophilus torridus	H	1.1	4.6	1.55
Sulfolobus acidocaldarius	H	1.8	6.5	2.23
S. metallicus	IO/SO/A	3.0	NA	NA
S. tokodaii	SO/H	2.5	NA	2.69
Acidianus brierleyi	IO/SO/A/H	1.5	NA	1.80
Acidophilic bacteria				
At. ferrooxidans	IO/SO/A	1.8	6.5	2.9
Acidiphillum acidophilum	SO/A/H	1.8	6.0	NA
A. cryptum	H	3	NA	3.9

N.B.: [a]A: Autotrophic; H: Heterotrophic; IO: Iron oxidizer; SO: Sulfur oxidizer.

NA: Not available.

Acidophiles pump out excess protons out of the cell with a view to maintaining pH homeostasis. The ΔpH in *B. acidocaldarius* and *T. acidophilum* is created by active proton pumping (Michels and Bakker, 1985). All the available sequenced acidophile genomes have putative proton systems. H^+ ATPases, antiporters and symporters). Several candidate proton efflux proteins have been successfully identified in the sequenced genomes of *Acidithiobacillus ferooxidans*, *At. thiooxidans* and *At. caldus* (see Baker-Austin and Dopson, 2007). The several proton-driven secondary transporters seem to represent an adaptation of these organisms to survive in an extremely acidic environment (Golyshina and Timmis, 2005). By using mainly secondary transporters, the PMF is easily harnessed for metabolic purposes.

Acidophiles make use of cytoplasmic buffering to maintain their intracellular pH. If protons somehow penetrate the acidophile cell membrane, various intracellular mechanisms help to deal with the resulting biological damage. First, the buffering capacity of the cytoplasm to sequester or release protons may be used as a pH homeostasis mechanism. All microbial cells contain several cytoplasmic buffer molecules that have basic amino acids (*e.g.* lysine, histidine and arginine) and can sequester protons. Castanie-Cornet *et al.* (1999) reported the involvement of decarboxylation of glutamate and arginine in *E. coli* in cell buffering by consuming protons, which are then transported out of the cell. Other buffering molecules found in the cell are phosphoric acid (H_3PO_4) which (pKa of 7.2), and at near-neutral pH the addition or removal of protons has only negligible effect on the pH of this molecule. According to Zychlinsky and Matin (1983), amino acid side chains are mainly responsible for acidophile cytoplasmic buffering. Krulwich *et al.* (1985) compared the buffering capacity of bacilli with various pH optima and reported that *B. acidocaldarius* does not have any higher buffering capacity than other bacilli tested. So, although acidophiles can maintain pH homeostasis by buffering, their buffering capacity is not necessarily greater than that of neutralophiles (Baker-Austin and Dopson, 2007).

It is well known that low pH can damage DNA and proteins. However, such damage can be repaired by chaperones. The representation of several DNA and protein repair genes in several extreme acidophile genome sequences might possibly be related to problems associated with pH homeostasis, whereby biomolecules damaged by low pH require fast and efficient repair. The *P. torridus* genome contains many genes determining DNA repair proteins (Golyshina and Timmis, 2005). And chaperones involved in protein refolding were found to be highly expressed in an environmental AMD biofilm community containing *Leptospirillum* group II (*L. ferriphilum*); they made up 11 per cent of the total expressed proteome (Ram *et al.*, 2005). Chaperones have also been reported to be highly expressed in '*F. acidarmanus*' cultured during aerobic and anaerobic growth. Indeed such high prevalence of chaperones in many acidophiles and conditions suggests that damage to DNA and proteins poses a formidable challenge for survival under acidic conditions.

In recent years the genome sequencing of several acidophilic microorganisms has been completed, and this has supported previous data which suggested that acidophile pH homeostasis requires a complex cell wall structure, that the reversed DY is generated by cation transport (while not undermining the importance of potassium ions), and that proton transporters are pressed into service to remove protons once the protons have entered the cytoplasm. Further, genome sequencing has also pointed to a role in low pH growth adaptation for processes such as organic acid degradation, DNA and protein repair systems, and possibly through the maintenance of a small genome size. Using these genome sequences in a functional context through the application of high throughput transcriptomic and proteomic tools to scrutinize acid stress could further enhance our understanding of the pH homeostasis mechanisms involved.

Secondary Metabolites of Extremophiles

Fibrinolytic enzymes play an important role in thrombolytic disorders such as myocardial infarction and cerebral stroke. Plasminogen-dependent type of fibrinolytic enzymes are exemplified by streptokinase, urokinase and t-PA, but these require large therapeutic dose, have limited fibrin specificity and short half-life, and suffer from re-occlusion and bleeding complications. An ideal thrombolytic agent is one that dissolves thrombus quickly and completely. On the basis of these criteria several thrombolytic agents have been developed. One such fibrin specific fibrinolytic enzyme has been named 'actinokinase'. It is produced by a thermophilic *Streptomyces* sp. (see ARI, 2008).

The enzyme production could be scaled up in a Bio-Torn 2.5 L capacity fermenter, under controlled conditions. Whereas only 500 IU of this enzyme dissolved a clot in 20 min, standard urokinase (500 IU) took 40 min. to dissolve the clot, and streptokinase 500 IU took 80 min. (see ARI, 2008, p. 27).

Production of Polyhydroxyalkanoates (PHA) using Extremophilic Bacteria

A moderately halotolerant and alkalitolerant strain of *Halomonas* isolated from Lonar Lake (Maharashtra) has been reported by the Agharkar Research Institute, Pune to produce PHA in a medium containing 1 per cent maltose and 0.1 per cent yeast extract with 4 per cent inoculum of 7 x 10^7 cells/ml and 15-21 hr of incubation as optimum conditions at flask level.

Experiments for different incubation periods (24hr, 36hr and 48hr) conducted in 10 litre fermenters showed that 24hr is the minimum period required for PHA production by *Halomonas* sp. The polymer so obtained was characterized through ^1H NMR analysis and optical microscopy. *Halomonas* produced the PHA in 12 L SS fermenter having working volume of 10 L, the fermenter being operated at 30°C with 100 rpm agitation and 0.5 vvm aeration. After 24hr of incubation, culture broth was centrifuged and the pellet was lyophilized. PHA was extracted in a rotary film evaporator by directly mixing the dry cell mass with chloroform at 60°C for 6hr. PHA (55 per cent) was obtained on dry cell weight basis. This improved procedure was found to be time saving as compared to the conventional process using Soxhlet extraction, which requires 56hr for extraction of PHA.

Upper Temperature Limit for Life

The upper temperature limit for life is crucial for delimiting when and where life probably evolved on a hot, early Earth; the depth to which life exists in the Earth's subsurface; and the potential for life in hot, extraterrestrial environments. Geological and microbiological evidences suggest that electron transport to Fe(III) was probably the first form of microbial respiration as life evolved (Lovley, 2000). There is geological evidence that microorganisms that use Fe(III) as an electron acceptor are important components of the deep, hot biosphere (Gold, 1992). The accumulation of Fe(III) in hot sediments around marine hydrothermal vents probably made FE(III) reduction an important process in modern hydrothermal environments.

To learn more about the physiological properties of Fe(III)-reducing microorganisms growing at high temperatures, Kashefi and Lovley (2003) cultured microorganisms on Fe(III) from a water sample from an active, "black smoker", hydrothermal (300°C) vent called Finn, located in the Mothra hydrothermal vent field (at 47°55.46'N and 129°06.51'W) in the Northeast Pacific Ocean. An organism (strain 121, so named because of its growth at 121°C) was isolated at 100°C in an anaerobic medium under N_2-CO_2 (80:20) in sealed culture tubes containing formate (10 mM) as the electron donor and Fe(III) oxide (100 mmol/l) as the electron acceptor. Cell growth and Fe(III) reduction were monitored. Fe(III) became reduced to Fe(II) with formation of the magnetic mineral magnetite. No Fe(III) reduction occurred in the absence of the organism.

Strain 121 is coccoid, about 1.0 mm in diameter, with lophotrichous flagellation. The cell envelope consists of a cytoplasmic membrane, a periplasmic space, and a single outer surface layer, typical of Archaea. Analysis of its 16S ribosomal DNA (rDNA) sequence of strain 121 indicated that it is a member of the Archaea, closely related to *Pyrodictium occultum* (96.0 per cent similar) and *Pyrobaculum aerophilum* (95.3 per cent similar) (Kashefi and Lovley, 2003).

Growth at 121°C is remarkable in the light of sterilization at 121°C being typical in pressurized autoclaves to maintain water in a liquid state, as long standing popular procedure shown to kill all previously described microorganisms and heat-resistant spores. Only 1 per cent of the cells of *Pyrolobus fumarii*, which can grow at temperatures up to 113°C, were intact after autoclaving for 1 hour, and there was no evidence that the remaining cells were viable. Autoclaving did not kill strain 121, and it doubled in cell numbers after 24 hours at 121°C. Cultures incubated for up to 2 hours at 130°C still grew when transferred to fresh medium at 103°C. At temperatures below 85°C, the cells remained viable, but did not divide (Kashefi and Lovley, 2003).

The use of Fe(III) as an electron acceptor was crucial for the isolation of strain 121, as it does not use other known electron acceptors.

Life in Hydrothermal Vents

The major mechanism by which heat is lost from the interior of the Earth is the eruption of volcanic rocks at the midocean ridges. About one-third of the heat is removed from the spreading centers by convective circulation of seawater. Such a great heat loss requires that the entire volume of the oceans circulates through the midocean ridges in approximately 10 million years. Seawater interaction with volcanic rocks at near 400°C results in substantial chemical flux and contributes to buffering the composition of some elements in seawater. Cations from seawater (Mg^{+2}, Ca^{+2}, and Na^{+}) form hydroxyl-bearing alteration minerals in the volcanic rocks, releasing hydrogen ions to solution. The hot, acidic altered-seawater releases metals (Fe, Mn, Zn, and Cu) and reduced sulfur (H_2S) from the volcanic rock; these are transported by hydrothermal solutions to the seafloor and form metallic mineral deposits (Zierenberg *et al.*, 2000). As compared to elsewhere in the deep oceans, seaflood hydrothermal vents support ecosystems with enormous biomass and productivity.

Oxidation/reduction (redox) reactions support chemosynthesis. The atmosphere and hydrosphere are relatively oxidizing with an abundance of potential electron acceptors (O_2, $SO_4^{=}$, and NO_3^{-}). In contrast, the basaltic rocks that form the oceanic crust are relatively reduced because of the abundance of ferrous iron. High-temperature fluid/rock interaction forms reduced gases (H_2S, H_2, and CH_4) that dissolve in hydrothermal fluid.

Sulfur is particularly important in the subseafloor redox cycle. Microbial reoxidation of H_2S near the seafloor releases the stored energy and drives biochemical reactions. Seafloor hydrothermal systems undergo cyclic variations in temperature, vent fluid composition, and biological activity.

Although there is a potential abundance of chemical energy, deep-sea hydrothermal communities have had to adapt to extreme conditions to exploit this resource. The hyperthermophiles are defined as microorganisms able to grow at 90°C and above. Several different types are known and have been found both within the walls of black smoker chimneys and where the hydrothermal vent fluids mix with the surrounding seawater. All but two of the hyperthermophilic genera are classified by ribosomal RNA analyses as Archaea.

Evolution gives no clue as to how life can thrive near and above 100°C (see Wiegel and Adams, 1998). Most microbes, and all eukaryotic cells, are unable to survive at temperatures much above 50°C, because of the general instability of biological molecules. The three-dimensional structures of most enzymes and proteins are lost at temperatures much above 70°C, and the double-helical structure of DNA also becomes unstable. A wide variety of ubiquitous metabolites are rapidly hydrolyzed at temperatures above 90°C. How do hyperthermophilic cells circumvent these problems?

Although some examples are known of modified pathways and unusual enzymes in hyperthermophiles, their biochemistry broadly resembles that of the mesophilic world. Yet, most enzymes from hyperthermophiles are extremely stable at high temperatures, showing optimal catalytic activity above 100°C with virtually no activity at ambient temperature. They contain the same 20 amino acids as those from conventional organisms. Sequence comparisons of analogous proteins from hyperthermophilic and conventional organisms are essentially identical, so the enormous amount of sequence information now becoming available may not be of any use in elucidating stabilizing mechanisms. Three-dimensional comparisons of structures reveal no gross structural differences between hyperthermophilic proteins and their mesophilic counterparts, and both forms are stabilized by the same noncovalent interactions (see Zierenberg *et al.*, 2000). Extended protein stability at 100°C appears to be the result of some subtle, synergistic cooperative intramolecular interactions.

In DNA, denaturation, depurination, and strand cleavage are greatly minimized by the relatively high salt (Mg, K, and PO_4) concentrations typically (although not universally) found in hyperthermophiles. Many of these organisms also have histone-like proteins bound to their DNA and some contain a unique enzyme termed DNA reverse gyrase, both of which may confer thermal protection. Yet how simple organic metabolites are stabilized remains a mystery.

Many mesophilic microbes in the hydrothermal vent environment use H_2S as an energy source. H_2S is usually toxic to aerobic life forms, yet dense populations of organisms flourish in various sulfide-enriched environments.

Hydrothermal vent communities can live in sulfide-rich habitats because of having detoxification mechanisms that involve microbial symbionts. Detoxification of sulfide through binding to blood-borne components is known in several chemosynthetic clams and other invertebrates, particularly in the tube worm *Riftia pachyptila*. The abundant respiratory hemoglobin present in the plume of *Riftia* can effectively bind both oxygen and sulfide simultaneously. The blood transports the respiratory hemoglobin with the tightly bound sulfide to the internal symbiotic bacteria, providing an electron donor for bacterial chemoautotrophy, while also protecting animal tissue by sequestering the toxin as a bound form. Colonies of *Riftia* are attached to the rocks where hydrothermal fluid (12-15°C) issues out onto the sea floor. At the base of their tubes, hydrothermal fluid is enriched in H_2S and CO_2, but lacks oxygen. *Riftia*'s unusual microhabitat is the interface between the hydrothermal fluids and the ambient oxygenic bottom water where essential metabolites are taken up and transported to internal bacterial for metabolism. The steep thermal and chemical gradients provide access to the reduced compounds needed to fuel growth and the oxygen needed to burn the fuel.

Vesicomyid clams living in hydrothermal vents harbour endosymbiont-containing gills. Vesicomyid blood transports oxygen bound to hemoglobin and contains an extracellular component with a high sulfide binding affinity. Here sulfide and oxygen acquisition are spatially separated. The foot of the calm is extended down into crevices that vent sulfide-rich water, enabling sulfide uptake and transport through the circulatory system to gills that are bathed in oxygen-rich seawater circulated from above.

Although various sulfide detoxification strategies have evolved, sulfide binding is a particularly effective detoxification mechanism for animals containing sulfide-oxidizing symbionts. Other marine invertebrates that have no sulfide binding protein appear to depend on sulfide oxidation for detoxification. Both strategies allow large communities of animals to flourish in habitats that otherwise would be quite inhospitable to life (Zierenberg *et al.*, 2000).

Plant/Fungal Symbiosis and Thermotolerance

According to Petrini (1986), all plants studied in natural ecosystems are symbiotic with fungi, which obtain nutrients while either positively, negatively, or neutrally affecting host fitness (Boucher, 1985). All known species of the fungus *Curvularia* are exclusively associated with plants (Redman *et al.*, 2002). Plant adaptation to selective pressures is regulated by the plant genome (Smallwood *et al.*, 1999). Assessing the effect of the *Curvularia* endophyte on the plant thermotolerance revealed that plants containing the endophyte are more thermotolerant than endophyte-free plants (Redman *et al.*, 2002).

In the absence of thermal stress, endophyte-colonized (symbiotic) and endophyte-free (nonsymbiotic) plants showed no measurable growth or developmental differences. Upon heating root zones with thermal tape, nonsymbiotic plants shriveled and turned chlorotic at 50°C. In contrast, symbiotic plants tolerated constant 50°C soil temperature for 3 days and intermittent soil temperatures as high as 65°C for 10 days. All nonsymbiotic plants died during the 65°C heat regime, whereas symbiotic plants survived. The endophyte could be successfully reisolated from surface sterilized roots and leaves of all surviving plants, showing that both the fungus and the host were protected from thermal stress.

The field performance and survival of symbiotic plants in soils above 40°C was also much better than those of non symbiotic plants. The beneficial effect of fungal symbiosis increased with soil temperatures, demonstrating that *Curvularia* sp. provided thermal protection for *D. lanuginosum*.

Besides thermotolerance, the mutualism in this system may involve other benefits (*e.g.*, nutrient acquisition by the fungus). Several possible symbiotic mechanisms could confer thermotolerance. In planta, the endophyte produces cell wall melanin which may dissipate heat along the hyphae and/or complex with oxygen radicals generated during heat stress (Davidson *et al.*, 1996). Alternatively, the endophyte could act as a "biological trigger" and allow symbiotic plants to activate stress-response systems more rapidly and strongly than nonsymbiotic plants (Redman *et al.*, 2002).

Chapter 5
Microbial Morphology and Physiology

Bacterial Cell Walls and Surface Structures

Some of the most important structures of microbial cells are their surfaces. The surfaces are in immediate contact with the external environment. The cell walls admit nutrients and release wastes, and yet need to resist internal turgor pressure and environmental insults so as to maintain cellular shape. Walls facilitate adherence of cells to other surfaces and provide space for specialized structures such as flagella, pili, spines, capsules, S-layers, and exopolymeric substances (EPS) (Beveridge, 2006).

Exploring the native architecture of bacterial surfaces has traditionally relied on electron microscopy, which uses high vacuum and high voltages. This imposes the limitation that only dry specimens can be imaged, yet living cells and their structures strongly depend on water and its ability to interact with and configure molecular ingredients. Advances in cryo-transmission electron microscopy (cryoTEM) are overcoming this problem by allowing researchers to examine native, hydrated structures in bacteria and their associated biofilms.

Early preparations made by using TEM (transmission electron microscopy) 4-5 decades ago, revealed a centrally condensed, nonenveloped chromosome surrounded by randomly distributed ribosomes and highlighting concept of an anucleate (*i.e.*, prokaryotic) cell. These images also revealed a basic difference between gram-positive and gram-negative bacteria, which have more complex walls consisting of an outer membrane, a thin peptidoglycan layer, and a periplasmic space filled with periplasm. Gram-positive walls are sometimes 20-fold thicker than the peptidoglycan in gram-neative bacteria and usually contain secondary polymers, such as teichoic and teichuronic acids, attached to a peptidoglycan network.

Removal of water shrinks most structures and reshapes macromolecules. Cryo-techniques preserve higher-order structures, especially in the enveloping layers. Freeze-substitution showed that the periplasmic space of bacteria such as *Escherichia coli* and *Pseudomonas aeruginosa* is filled with a gel and that most of the lipopolysaccharide is found on the outer face of the outer membrane.

While freeze-substitution techniques have been commonly employed for electron microscopy, some have used neutron scattering, modelling, and chromatographic analysis to develop a better picture of the thickness, ordering, and complexity of the bacterial cell wall. Whatever may be it organization, its polymeric network should be quite strong and elastic. One group of bacterial cell wall modellers claim that such polymers are organized horizontally, while another group favour a vertical scaffold. Exciting atomic force microscopy (AFM) images of *Staphylococcus aureus* show freshly synthesized walls to consist of concentric rings, thereby supporting the horizontal arrangement model (see Völlmer and Höltje, 2004).

Recent images of frozen, hydrated sections of bacteria have enhanced knowledge of bacterial cell structure, such as the nucleoid and ordering of DNA in *Deinococcus radiodurans* (Eltsov and Dubichet, 2005). Researchers working on bacterial envelopes use this approach to reexamine both Gram-negative bacteria such as *E. coli* K-12 and *P. aeruginosa* PAO1 as well as Gram-positive *B. subtilis* 168 and *S. aureus* D_2H (see Matias *et al.*, 2003).

To produce such images, cells are immersed in a cryoprotectant such as dextran or sucrose before being rapidly frozen so that the surrounding water vitrifies. Molecular motion stops so fast that native structures do not deteriorate and cells are embedded in noncrystalline (amorphous) ice. The cells are physically, not chemically, fixed, and the vitrified specimens are sliced into 50-nm frozen hydrated sections, which can be viewed in a cryo-TEM microscope at approximately–140°C (see Dmitriev *et al.*, 2005).

Unlike other types of TEM that depend on heavy metal staining agents, no staining is required for frozen hydrated sections–staining destroys their vitrified state. Here, bacterial imaging depends on the inherent density of the proteins, lipids, carbohydrates, and nucleic acids within specimens being denser than the surrounding layer of frozen water. Because images of these bacteria reflect how macromolecules distribute within the cell, higher magnifications bring out the polymeric organization of the cell wall. In Gram-negative bacteria, the asymmetry of lipids in the outer membrane becomes obvious. The outer face of the bilayer containing lipopolysaccharides (LPS) has more phosphorus and mass per unit volume, and so manifests as a darker line because of its greater contrast than does the inner face consisting of phospholipids (Beveridge, 2006).

Unlike freeze substitutions, the O-side chains of the LPS on such "smooth" gram-negative species as *P. aeruginosa* PA01 are not seen in frozen sections–their mass is too low to be differentiated from the surrounding ice. Yet, the plasma and outer membranes, the periplasm, and the peptidoglycan layer are seen; their general arrangement resembles that in conventional and freeze-substitutions even though the dimensions are different.

The 2.5-nm thickness of the cell walls of many Gram-positive species can accommodate about 25 stacked layers of peptidoglycan.

The walls of *B. subtilis* can be readily differentiated from the surrounding ice. When images are examined through a densitometer, the mass goes from high on the inner face to low on the outer face; this supports the idea that cell wall turnover proceeds from inside to outside.

Visualizing structural details in bacterial biofilms is difficult because biofilms have extremely soft matter (EPS) interspersed with bacteria, materials that they have shed such as flagella, pili, and membrane vesicles (MVs), as well as some harder materials such as biominerals. Biofilms often are conspicuous enough to be seen without magnification, and confocal microscopy with fluorescent probes is particularly valuable for distinguishing molecular networks within biofilms. Ratiometric dyes facilitate probing chemical conditions such as pH within biofilms.

However, because light microscopes cannot discern molecular arrangements within biofilms, higher-resolution microscopy is essential. AFM only visualizes the topography of samples, and its cantilevers deform highly hydrated materials such as EPS. Some investigators use scanning electron microscopy (SEM), especially variable pressure SEMs, but this technique does not reveal fine ordering within biofilms. In TEM, chemical fixatives do not fully penetrate biofilms and organic solvents collapse the EPS. Cryo TEM also suffers from some drawbacks. Because EPS is 90–95 per cent water, its density is too low to be distinguished from external water. A frozen hydrated section of a biofilm resembles a frozen hydrated section of planktonic cells. Both show bacteria interspersed with low-density material (Beveridge, 2006).

Currently, thin sections of freeze-substituted biofilms seem to be the best cryo TEM method for revealing finely ordered structures. Specialized equipment can vitrify samples to a depth of 10-50 μm; this enables one to see a biofilm from top to bottom. For Gram-negative cells, high magnifications differentiate the O-side chains of LPS from the surrounding EPS matrix. Even within this matrix, differences in polymeric arrangements may be detected, from densely packed to loosely packed fibers. The most serious drawback of freeze-substitution is that only thin (50-nm) slices of a biofilm can be directly imaged. Tomography of thicker sections using 200-kV or higher-voltage microscopes may potentially offer better three-dimensional pictures of biofilms (Beveridge, 2006).

CryoTEM is strongly impacting molecular biology by providing 3-D structures of specific proteins that are cloned into bacteria. The technique is especially useful for analyzing noncrystallizable proteins. However, as particle size and complexity increase, 3-D analysis becomes more and more difficult (see, *e.g.*, Hunter and Beveridge, 2005). In the coming years, cryoTEM of viruses may well yield striking results revealing protein capsid arrangements, nucleic acid-packing, and receptor sites.

Larger objects such as bacteria are too big for single-particle analysis and too small for uncomplicated cryo-handling (Beveridge, 2006).

The bacterial cell wall is synthesized by the activity of glycosyltransferase (GT) and transpeptidase (TP) enzymes, Penicillin and related antibiotics act on TP enzymes, but bacteria usually become resistant to these antibiotics. GT enzymes are a prospective target for new drugs because they are essential, membrane bound, and thus accessible. A penicillin-binding protein 2 is a bifunctional enzyme from *Staphylococcus aureus* that contains both a GT and a TP domain. Structures with and without the inhibitor moenomycin bound in the GT domain throw light into the mechanism of cell-wall biosynthesis and could provide a starting point for structure-based design of antibacterials.

Visualizing Bacterial Cell Walls and Biofilms

- ☆ Advances in cryo-transmission electron microscopy (cryoTEM) enable investigators to examine native, hydrated structures in bacteria and their associated biofilms.
- ☆ Freeze-substitution reveals that the periplasmic space of Gram-negative bacteria is filled with a gel and confirms that lipopolysaccharide resides on the outer face of the outer membrane, while Gram-positive cells appear to build new wall in layers from the inside outward, with older material removed from the outer edge.
- ☆ Bacteria are difficult to image with cryoTEM. Doing so depends on the inherent density of the proteins, lipids, carbohydrates, and nucleic acids within specimens being denser than the surrounding layers of frozen vitrified water.
- ☆ CryoTEM facilitates visualizing structures within bacterial biofilms.

DNA Versus Membrane in *Bacillus*

In bacteria, chromosomal DNA mostly floats in the cytoplasm without ever acrossing a membrane barrier. But when cells conjugate, or are about to divide, the DNA has to traverse a membrane to reach its desired location. Burton *et al.* (2007) noted that during spore formation in the bacterium *Bacillus subtilis*, the plasma membrane completely encloses the DNA in a process that involves a channel-forming family of proteins called SpoIIIE/FtsK. These proteins are also remarkable DNA-transfer engines that can pump DNA between cells (Wu and Errington, 2008).

When rod-shaped bacteria (*e.g. B. subtilis*) divide, their plasma membrane constricts in a ring which then closes abruptly to form two separate membranes. Of course, to complete its division, the dividing cell has also to synthesize new cell-wall material.

Most bacteria (including *B. subtilis*) have single, circular chromosome. Chromosome segregation that normally precedes membrane fusion, is a complex and multistep process in these microbes and involves separating the replicated circular sister chromosomes and moving them into the two resulting cells. If the replicated chromosomes do not separate, DNA strands might become trapped in the closing division septum. To avoid this problem, most bacteria use their SpoIIIE/FtsK DNA transporters located in the septum. These proteins are found in many bacteria. They consist of a transmembrane domain at the amino terminus connected by a linker of varying length to a DNA-translocating (ATPase) domain at the carboxy terminus. In *B. subtilis*, the SpoIIIE protein is specially important because, when these cells form spores, they divide asymmetrically–the division septum always closes on a chromosome. Thus, for the smaller cell called the forespore to have a complete chromosome, SpoIIIE must somehow transport some 3 million base pairs of DNA (Wu and Errington, 1994; 2008).

Use of Light by Bacterial Protein to Power Flagella

Proteorhodopsin is a bacterial light-driven proton pump that was identified in 2000 from ocean water samples. It absorbs green light and pumps protons out of the cell, but nothing is known about whether or not it serves as an alternative or complementary energy source. Walter *et al.* (2007) postulated that bacteria which have proteorhodopsin make use of light energy to generate a proton motive force (PMF) which powers their flagella under conditions where normal cellular respiration is impaired (*e.g.*, in a toxic environment). Proteorhodopsin-containing *Escherichia coli* was treated with azide, a respiratory poison that reduces the PMF and slows down the flagellar motor. Under these conditions, green light increased the velocity of proteorhodopsin-containing bacteria. The cells slowed down to their previous speed when the light was removed. These results demonstrate that proteorhodopsin enables bacteria to withstand respiratory poisons and oxygen depletion by augmenting the cellular PMF. Walter *et al.* pointed to the possibility of creating a variety of light-powered bacteria to study PMF-dependent processes, such as ion transport, as well as to optically control the behaviour of synthetically engineered organisms.

Magnetotactic Bacteria (MTB) and Magnetosomes

MTB are motile, Gram-negative bacteria that synthesize nanometer-size (nm = billionth of a metre), membrane-bound magnetic iron oxide (magnetite, Fe_3O_4) or iron sulphide (greigite, Fe_3S_4) (Spring and Bazylinsky, 2002). They are found in the oxic-to-anoxic-transition-zone (OA TZ) of water column, where the oxygen concentration is neither too high nor too low. The bacteria exploit the magnetic properties of magnetite or greigite for navigating in the earth's magnetic field to reach the levels of suitable oxygen concentration in stratified water columns. The magnetites or greigites are arranged in linear chains (needles) inside the cell and serve as tiny magnetic compasses. MTB are veritable-living lodestones that live in the northern and southern hemispheres (north-seeking and south-seeking MTB), besides the equatorial waters (see Sukumaran, 2005).

Membrane-bound crystals of magnetite or greigite within living MTB are termed magnetosomes. They enable the bacteria to orient and migrate along geomagnetic field lines. All MTB possess magnetosomes enclosed within organic membranes. Those from marine, sulphidic environments generally synthesize greigite magnetosomes, while freshwater MTB usually contain magnetite magnetosomes. Genus *Magnetospirillum* represents MTB. It has been isolated and mass cultured. When the MTB die, their magnetite particles settle through the ambient water column and deposit on the bottom sediments; such sediment magnetites of biogenic origin are termed magnetofossils.

Bacterial Communication and Nanovesicles

Gram-negative proteobacteria such as *Pseudomonas aeruginosa* communicate with each other by secreting chemical messengers, and they attack their enemies. The pathogenic *P. aeruginosa* releases a hydrophobic molecule, the "pseudomonas quinolone signal" (PQS) to send messages to other bacteria of the same species. Surprisingly, rather than being secreted as single molecules, PQS is released in bubble-like 'vesicles' that also contain antibacterial agents and probably toxins aimed at host tissue cells as well.

Various groups of bacteria use diffusible chemicals to signal to their own kind. *P. aeruginosa* is a Gram-negative proteobacterium and, like many proteobacteria, it secretes several acylhomoserine lactones (AHLs) whose concentration increases with cell population density, enabling the bacteria to sense how many of their kind surround them–a phenomenon known as quorum sensing. Genes for bioluminescence, DNA transfer, pigment production, and genes required for infecting plants, animals and humans are switched on only at high cell densities. But AHL molecules are readily broken down by other bacteria, and some AHL signals are poorly soluble in water, so they cannot travel far in an aqueous environment–both factors that limit their potential as communication signals. PQS, although not an AHL-type signal, is also poorly water-soluble and readily broken down by other bacteria. So the issue is how these molecules convey the message from one bacterium to another?

Many Gram-negative bacteria release nanovesicles (about 50 nm in diameter) from their outer membranes as a means of delivering toxins to host cells and other bacteria. The vesicles have a lipid bilayer surrounding an aqueous core; they transport lipid-soluble toxins on their surface and protein toxins in their core, releasing their poisonous cargo by fusing with the lipid bilayer of target cells. These vesicles can also release the PQS signal to other bacteria, thereby sidestepping the problems of poor water-solubility and degradation (Winans, 2005).

PQS is one of the many different quinolines made by *P. aeruginosa*, some of which have potent antibacterial activity. The membrane vesicles, in addition to signalling to kindred *P. aeruginosa* cells, show potent antibacterial activity against the Gram-positive bacterium *Staphylococcus aureus* the toxicity is due to quinolines rather than to the protein component of the vesicles.

P. aeruginosa is an opportunistic human pathogen that affects the lungs of cystic fibrosis patients. Greater understanding about how these bacteria communicate with each other and attack host cells could lead to novel treatment strategies for *P. aeruginosa* infections.

Bacterial Caveolae

The uptake of bacterial pathogens as well as their fate inside a host cell is influenced by interactions between bacterial adhesin molecules and host cell receptors. These interactions activate signal transduction cascades in the host cell. In so doing they stimulate preexisting endocytic pathways, leading to internalization of the adherent microbe. Internalization not only provides an invading microbe with a safe haven but also facilitates its dissemination within and across host tissue barriers. At the same time internalization is also an effective component of the host defense system–pathogens engulfed by active phagocytic cells usually enter a phagolysosome pathway where they may be rapidly destroyed. The ultimate fate of engulfed bacteria depends on which endocytic pathway they end up in (Mulvey and Hultgren, 2000).

Mast cells (a type of phagocytic cell) recognize and internalize *Escherichia coli* bacteria expressing the FimH adhesin. Found on the surface of *E. coli* and many other bacteria, FimH is a component of filamentous adhesive organelles called type 1 pili. The FimH of *E. coli* interacts with its receptor, CD48–a glycosylphosphatidylinositol (GPI)-anchored protein–on the surface of mast cells. As CD48 and other GPI-anchored proteins lack a transmembrane or cytoplasmic domain, it was not clear how they transmit intracellular signals that direct internalization of adherent *E. coli*. Shin *et al.* (2000) proposed that the uptake of *E. coli* bound to mast cell CD48 depends on specialized host membrane microdomains called caveolae. Such internalization tends to promote survival of the microbes within mast cells.

The term caveolae means little caves. These structures are distinct from clathrin-coated pits, rank in size from 50 to 100 nm and have been identified in numerous types of cells. Caveolae are enriched in glycosphingolipids, cholesterol, and caveolin, an integral membrane protein. They resemble other detergent-insoluble membrane domains commonly referred to as lipid rafts–microdomains within eukaryotic cell plasma membranes that may be platforms for concentrating and sorting signal proteins and other molecules. Lipid rafts do not always contain caveolin or show the flasklike morphology ascribed to caveolae. The incorporation of caveolin into lipid rafts, along with the presence of cholesterol, possibly facilitates the formation of caveolae.

The functions of caveolae are not clear. Certain proteins involved in signal transduction appear to be localized in caveolae-like domains, and caveolin itself interacts directly with many signaling molecules. This suggests that caveolae are centers from which multiple signaling pathways originate. They have also been implicated in endocytic events, including the transcytosis of macromolecules across cell layers and potocytosis, a potential method for 'pumping' small molecules into cells (Anderson, 1998). Caveolae also may regulate the internalization of particulate agents such as viruses and bacteria.

Morphologically distinct caveolae have not been observed in mast cells, T and B cells, macrophages or other hematopoietic cells but the identification of caveolin in some macrophage cell lines suggests that caveolin-rich, caveolae-like membrane domains may be important in certain types of effector immune cells. Shin *et al.* (2000) report that caveolae-like elements within mast cells are recruited to plasma membrane sites where *E. coli* cells are attached through FimH. These elements then fuse with other caveolae-related domains at these attachment sites, forming massive vesicles that encapsulate and internalize the adherent bacteria. The mast cell CD48 receptor also appears to be intrinsically associated with caveolin- and cholesterol-rich domains during the internalization process (Mulvey and Hultgren, 2000). It appears that caveolae-like regions in the mast cell plasma membrane are highly dynamic elements and that certain anchored receptors can transmit signals and activate host cell endocytic pathways through association with caveolae-like membrane domains.

Caveolae-related domains are also implicated in the pathogenesis of certain bacteria. CD48 and other anchored proteins belong to an expanding class of receptors for various viruses, bacteria, and bacterial toxins. The entry of simian virus 40 into host cells and the uptake of the bacterium *Campylobacter jejuni* by cultured intestinal epithelial cells both depend on caveolae-like membrane domains. As caveolae do not appear to fuse with endocytic vesicles, internalization of bacteria through caveolae-like domains probably facilitates their intracellular survival.

Bacterial Chemotaxis

Chemotaxis means the directed motion of an organism toward a favourable environmental condition and/or away from unsuitable surroundings. Flagellated bacteria such as *Escherichia coli* move through sequential smooth-swimming runs punctuated by intermittent, fleeting tumbles that last only a fraction of a second; even such a short period is sufficient to effectively randomize the direction of the next run. Runs vary in length which can extend from a fraction of a second to several minutes (Webre *et al.*, 2003). Being only a few microns long, *E. coli* cells behave essentially as point sensors. They cannot measure gradients by comparing head-to-tail concentration differences. But they have a good memory and so can compare current and past chemical environments. In fact, the probability that a smooth swimming *E. coli* cell will stop its run and tumble is dictated by the chemistry of its immediate surroundings as compared to the chemistry it encountered a few seconds previously. When the bacterium senses that conditions have become bad, its tendency to tumble increases; this can happen when attractant concentrations decrease or repellent concentrations increase. Conversely, tumbling tendency decreases and cells keep running when they perceive that surroundings are improving–when a bacterium runs up a gradient of attractants or down a gradient of repellents, it continues on course (Webre *et al.*, 2003).

Signal Transduction

The above behaviour is achieved through a signal transduction system that involves two families of proteins found in microorganisms which work in pair-wise fashion to mediate chemotaxis, as well as other regulatory processes ranging from cell differentiation and development to antibiotic resistance and fruiting; *E. coli* haarbours some 30 different 'two-component' regulatory systems.

One of the families of proteins that mediate two-component signalling involves histidine protein kinases, which catalyse the transfer of ¡-phosphoryl groups from ATP to one of their own histidine residues. Another family consists of 'response regulator' proteins, that are activated by the transfer of phosphoryl groups from the kinase phosphohistidines to one of their own aspartic acid residues (see Stock and Surette, 1996; Levit and Stock, 2002).

The response regulators tend to diffuse around the cytoplasm, and aspartate phosphorylation generally increases the ability of a regulator to bind to DNA, or in the case of the chemotaxis response regulator, to bind to motor proteins and regulate the probability of a tumble (Webre *et al.*, 2003).

The histidine protein kinase CheA mediates chemotaxis responses whereas CheY regulates the chemotaxis response. These two are mainly involved in chemotaxis of *E. coli*. In fact, the *E. coli* chemotaxis system includes five different transmembrane proteins. Tar determines responses to aspartate, glutamate and maltose; Tsr mediates responses to serine; Trg to ribose and galactose; Tap to dipeptides; and Aer to oxygen. Aspartate and serine bind directly to Tar and Tsr, respectively. Tar senses maltose through the periplasmic maltose binding protein; and Trg and Tap sense ribose, galactose and dipeptides via periplasmic binding proteins for these nutrients. Aer senses O_2 through the redox state of a bound flavin (Webre *et al.*, 2003).

One interesting finding is the individuality shown by genetically identical bacteria within a population; this aspect of the function of chemotaxis receptor arrays points to general lack of any simple, fixed relationship between stimulus and response. For instance, when Tsr happens to be dimethylated, even a few serine molecules can effectively inhibit the total CheA activity in the cell. But under other conditions where Tsr is highly methylated, thousands of serine molecules are required to inhibit CheA; when cells grow in media containing very low concentration of serine and methylation is low, they are attracted to even weak sources of serine. But since serine increases Tsr methylation, under conditions of high ambient serine concentrations, cells respond only to large changes in serine concentration. In general the change in stimulus concentration that a bacterium can detect tends to be a constant fraction of the background stimulus intensity. This relationship, known as Weber's Law of psychophysics, is a general property of many animal sensory systems, and seems to apply also to bacteria (Webre *et al.*, 2003).

A Nanobrain?

In some ways, bacterial chemotaxis receptor complexes function as rudimentary brains, only a few hundred nanometers in diameter. If we consider a brain as an organ that uses sensory information to control motor activity, then the bacterial nanobrain satisfies the definition.

Neurobiologists feel that bacteria are too small and too primitive to have brains–brains are fairly large, complex, multicellular assemblies containing neurons. But neurobiologists do not argue against the plea of artificial intelligence and machines that function as brains. A consideration of the evolution of machine intelligence suggests that size and complexity cannot be a good measure of processing capacity–the small computers of today are far more powerful than their larger and superficially more complex predecessors (Webre *et al.*, 2003).

The view that bacteria are primitive is also wrong. Bacteria have been evolving for billions of years longer than animals, their short generation times and enormous population sizes make bacterial systems more highly evolved than anything in the animal kingdom.

Compensatory Adaptation in *Escherichia coli*

Deleterious mutations in a small population can be substituted by random drift by hitchhiking with a beneficial mutation in asexual populations or as a consequence of changing environments (such that a formerly beneficial mutation becomes deleterious). A deleterious substitution has several possible fates: it may (i) cause extinction of the population, (ii) persist indefinitely, (iii) revert to its former state, or (iv) be compensated for by another mutation elsewhere in the genome which ameliorates its deleterious effect. This last possibility has been termed compensatory adaptation (CA) (Moore *et al.*, 2000).

In the field of molecular evolution, compensatory mutations are defined as two alleles which are independently deleterious but are neutral when occurring together (Kimura, 1990). The concept of compensation is both older and more general than this usage. Wright (1982) invoked CA in promoting the spread of major mutations by means of modifier alleles which diminish the deleterious side-effects of the major mutations. In this context, a compensatory mutation is any mutation that masks the deleterious effect of another mutation. As a consequence of the potential for compensation, genotypes which carry deleterious mutations tend to be fitter than genotypes which are at or near a fitness peak. This property is used as a means of operationally defining CA. The relative fitness of an unmutated parent strain is 1.0, while deleterious mutations reduce fitness to lower values. Following some genetic adaptation, the final fitness of an evolving lineage will generally be higher than its initial fitness. If the fitness values of all lineages increase to the same relative extent, regardless of their initial fitness, then that is merely proportional adaptation. However, if lineages that were initially less fit experience relatively larger fitness gains, then that indicates CA (provided, of course, that no reversion has occurred).

CA points to epistasis which underlies such diverse evolutionary theories as fitness landscapes with multiple peaks and the mutational deterministic hypothesis for the evolution of sex (Kondrashov, 1988). Yet, despite such importance, very few data exist on the frequency and form of epistatic interactions (Fenster *et al.*, 1997). Several experiments have demonstrated CA by selecting epistatic modifiers which reduce the costs of resistance to insecticides, parasites or antibiotics. Elena and Lenski (1997) showed that many pairs of deleterious mutations are epistatic. Indeed, it seems that epistasis is not confined to deleterious mutations which may help shape the evolution of genetic systems but which are presumably not the main building blocks for most adaptations. Lenski (1988) demonstrated large fitness costs of mutations which confer the resistance of *Escherichia coli* to the virus T4, but after 400 generations in the absence of T4, these costs were greatly reduced even though resistance persisted. In bacteria CA can quickly overcome the pleiotropic costs which arise from resistance to various antibiotics (Cohan *et al.*, 1994)

In a study which did not involve any resistance mutations, Burch and Chao (1999) found that one particular deleterious mutation in an RNA virus could be compensated for to varying degrees by several different mutations.

To investigate CA, Moore *et al* (2000) used genotypes of *E. coli* which were identical except for one or two deleterious mutations. They compared CA for (i) deleterious mutations with large versus small effects, (ii) genotypes carrying one versus two mutations, and (iii) pairs of deleterious mutations which interact in a multiplicative versus synergistic fashion. In all they studied 14 different genotypes, plus a control strain which was not mutated. Most genotypes showed CA during 200 generations of experimental evolution. They observed greater CA for mutations of large effect than for those of small effect, which is explained by the greater benefit to recovery in severely handicapped genotypes given the dynamics of selection. The rates of CA were similar for double and single mutants whose initial fitnesses were approximately equal. CA was faster for synergistic than for multiplicative pairs, presumably because the marginal gain which results from CA for one of the component mutations was greater in that case. The most surprising result was that compensation could be so readily achieved in an organism which is haploid and has little genetic redundancy. This finding points to an unusually high degree of versatility in the *E. coli genome.*

Bacterial Suicide Through Stress

In nature bacterial cells are constantly exposed to various stresses. An ability to resist those stresses is essential to their survival. The degree of stress required to bring about cell death varies with growth phase, exponential-phase cells being more sensitive to stress than stationary-phase ones. A novel hypothesis termed the suicide response, has recently been advanced to explain this difference in sensitivity (see Aldsworth *et al.*, 1999).

Essentially, bacterial cells are predicted to kill themselves through production of free radicals as a result of uncontrolled metabolic activity subsequent to the application of a relatively mild (sublethal) stress. The suicide response proposes that when actively growing and aerobically respiring bacterial cells are subjected to a sublethal stress, their growth is arrested but their metabolism continues to function for some time. The consequence of this uncoupling of energy production from energy utilisation is that a burst of excess free radicals occurs, and it is this burst of free radicals which proves lethal to the cell, rather than the action of the stress *per se* (Dodd *et al.*, 1997).

The suicide response hypothesis unifies a variety of previously unrelated empirical observations, for instance induction of superoxide dismutase by heat shock, alkyl-hydroperoxide reductase by osmotic shock, and catalase by ethanol shock. The suicide response also has important implications for current food processing methods.

The suicide response has grown from observations with thermally stressed cells chiefly of *Salmonella* and *Escherichia* and all of the supporting evidence given so far relates to thermal stress but there are indications that exposure to other forms of stress (*e.g.* ethanol shock) is also experienced by bacterial cells as an oxidative stress. The accumulation of adenylylated nucleotides (termed alarmones) in bacterial cells after exposure to heat shock or ethanol shock prompted the suggestion of a common mode of action for diverse inimical treatments. The accumulation of these possible alarmones in bacterial cells in response to an oxidative stress indicates what that common mode of action might be: that bacterial cells experience diverse stresses as an oxidative stress (Lee *et al.*, 1983; Aldsworth *et al.*, 1999).

These observations all reinforce the idea that bacterial cells experience differing stresses such as thermal, osmotic shock or ethanol shock as an oxidative stress. There is also evidence that the lethality of some stresses depends on metabolic function. For instance, hydrogen peroxide (H_2O_2) can have a lethal action in one of two ways, depending on its concentration: mode I killing occurs at low concentrations of H_2O_2, whilst mode II killing occurs at high concentrations. Whereas mode II killing is due to gross damage of cellular structures by the peroxide and is just as effective against stationary phase cells as exponential phase ones, mode I killing requires the target cells to be metabolically active since starved cells are more resistant than starved cells grown in the presence of glucose (Imlay and Linn, 1986). That is, mode I killing may not operate through the action of the H_2O_2 *per se*, but through the metabolic action of the cell during and after exposure to low concentrations of H_2O_2 (Aldsworth *et al.*, 1999).

The suicide response hypothesis has strong implications for food processing technologies. The use of modified atmosphere packaging is gaining popularity among manufacturers as a means of stabilizing the organoleptic properties of food products (sandwiches and salads, for example). However, the suicide response hypothesis predicts that any food-borne pathogens which have survived processing in an injured state are much more likely to recover in a low oxygen atmosphere than in an aerobic atmosphere (see Rees *et al.*, 1995). Exploitation of the concept of suicide response could allow novel approaches to bacterial eradication to be made. Thus, encouraging bacteria to enter a metabolically active phase by the provision of a short growth period prior to processing is contrary to current ideas but should increase the efficacy of any applied stress. Such ideas may allow more minimal processing of foods to produce safer products in the future (Aldsworth *et al.*, 1999).

Quorum Sensing in Gram-negative Bacteria

Bacteria can perceive and respond to other bacteria. This capability is important in the colonization of animal and plant hosts by symbiotic or pathogenic bacterial species. Different bacterial groups have different mechanisms for monitoring their own abundance in a, local environment. One mechanism that is common in gram-negative bacteria is that found in the luminescent bacterium *Vibrio fischeri.* This phenomenon was originally termed autoinduction and is now known as quorum sensing and response, and has been extensively reviewed recently (see Fuqua *et al.*, 1996; Kaiser and Losick, 1997).

Luminous bacteria like *V. fischeri* contain high levels of luciferase only when cultures reach the late-logarithmic phase of growth. *V. fischeri* and another luminous species, *Vibrio harveyi,* produce diffusible compounds (termed autoinducers) that accumulate in the medium during growth. These autoinducer signals can accumulate to sufficient concentrations only when there is a critical mass of cells in a confined environment. The signals from *V. fischeri* and *V. harveyi* are specific as they do not cross-react. *V. fischeri* occurs at very high densities (10^{10} to 10^{11} cells per ml) as a specific symbiont in light organs of certain fish and is also found free in seawater but at much lower densities (perhaps 5 cells per ml). Autoinduction allows it to sense its elevated density in the light organ and express the luminescence system there, where it is required for the symbiosis, but not in seawater, where luminescence, which is energetically expensive, would be frivolous (Greenberg, 1998).

The concept that bacteria produce pheromones and communicate with one another was met with considerable skepticism by many microbiologists. Although *V. fischeri* does not produce a signal that activates the luminescence genes in *V. harveyi,* many other gram-negative marine bacteria do. This observation of Greenberg and coworkers indicated that cell-to-cell signaling within and perhaps between bacterial species might be a common phenomenon.

In the 1980's it was established that luminescence requires high cell density. The structure of the autoinducer signal, *N*-3-(oxohexanoyl) homoserine lactone, was solved. This molecule was shown to move out of and into cells by passive diffusion. The genes for luminescence were cloned from *V. fischeri* into *E. coli.* Fortunately, the genes for autoinduction are linked to the luminescence structural genes. *E. coli* cells containing this *lux* gene cluster emit light in a cell density-dependent fashion. Thus, quorum sensing could be analyzed with the tools of *E. coli* genetics.

The regulatory region that enables autoinduction of luminescence consists of two genes: *luxR*, which encodes an autoinducer-responsive transcriptional activator, and *lux1*, which encodes a protein required for autoinducer synthesis. The region between *luxR* and *lux1* contains the regulated *lux* promoter elements. Because *lux1* is positively autoregulated, basal levels of luminescence operon transcription lead to low rates of autoinducer production, and quite high densities of cells are essential for activating the luminescence genes. Once activation has occurred, the rate of autoinducer synthesis increases, and cell density must drop considerably before the transcription level of the luminescence operon returns to the basal level (Greenberg, 1998).

luxR requires activation by cyclic AMP and the cAMP receptor protein.

Since the late 1980s, more details of the mechanism of autoinduction have become known. Lux1 is an autoinducer synthase that catalyzes the formation of an amide bond between its two substrates, a six-carbon fatty acyl-acyl carrier protein (acyl-ACP) and S-adenosylmethionine. Studies of Lux1 mutants have revealed that the active site in which amide bond formation is catalyzed is roughly in the region of residues 25-110 of this 193-amino-acid protein. A region in the C terminus may be involved in selection of the appropriate acyl-ACP from those existing in the cellular pools (Greenberg, 1998).

In the early 1990s, key discoveries led to our current view that quorum sensing is common to many Gram-negative bacterial species. First, LuxR homologs were discovered in *Pseudomonas aeruginosa* and *Agrobacterium tumefaciens,* and several bacterial species were found ot produce *N*-3-(oxohexanoyl)homoserine lactone.

Some bacteria that produce acylhomoserine lactones are *Vibrio fischeri, V. harveyi, Pseudomonas aeruginosa, Agrobacterium tumefaciens, Erwinia* spp., *Rhizobium leguminosarum, Psuedomonas solanacearum, P. fluorescens, Rhizobium meliloti* and *Yersinia enterocolitica.*

Quorum sensing in *A. tumefaciens* controls conjugal transfer genes, probably to ensure that the catabolic Ti plasmid is present when cells are at a high density in a crown gall tumor. *P. aeruginosa* is an opportunistic human pathogen. It uses quorum sensing to regulate expression of a battery of extracellular virulence genes, including enzymes and exotoxins. Autoinduction of extracellular enzymes is a common theme.

Non-coding Regulatory MicroRNAs

Small non-coding RNAs with important regulatory roles occur both in eukaryotes and prokaryotes. Recent studies have led to the identification of numerous small regulatory RNAs in *Escherichia coli* and other bacteria. These non-coding (nc) RNAs are known as microRNAs and short interfering RNAs (siRNA). They act as novel regulators of development, cell death and chromosome silencing and are an extremely useful tool–a tool that can be used to inactivate specific messages. Non-coding RNAs also regulate replication and maintenance of prokaryotic extrachromosomal elements (Gerhart *et al.*, 1998). Many small ncRNAs have been identified as crucial regulatory elements in bacterial stress responses and in bacterial virulence (Gottesman, 2004, 2005). A single ncRNA can regulate multiple genes and exert strong effects on cell physiology.

As in eukaryotic cells, a major class of these small RNAs acts by base-pairing with target mRNAs, resulting in changes in the translation and stability of the mRNA. Roles for these non-coding pairing RNAs in several bacteria have been demonstrated. Because these ncRNAs act post-transcriptionally, they impose a regulatory step that is independent of and epistatic to any transcriptional signals for their target mRNAs (Gottesman, 2004, 2005).

Unlike open reading frames, ncRNAs are not easily revealed by sequencing of the genome, and because they are quite short (generally ~ 100 bp), they are missed in searches for mutations. Nonetheless, several ncRNAs could be detected either by physical means or serendipitously in *Escherichia coli*, and these inspired global searches for others having similar properties. Over 60 ncRNAs have been confirmed in *E. coli* (~1-2 per cent of the number of known protein-coding genes); a similar estimate has been made for the abundance of ncRNAs in eukaryotes (Xu *et al.*, 2004).

Some bacterial ncRNAs act by pairing. The largest class of bacterial ncRNAs are defined by their strong interaction with the RNA chaperone Hfg. They regulate gene expression by base-pairing with mRNAs and change the translation or stability of the mRNA. While they resemble eukaryotic microRNAs and siRNAs in their biological effects, they differ significantly in the details of how they are processed and presented to express those effects.

Whereas eukaryotic microRNAs are first synthesized as hairpins embedded in longer transcripts and initial processing down to the hairpin itself precedes exit from the nucleus, the bacterial ncRNA transcripts are usually 80-100 nt in length and are not generally processed. In the (eukaryote) cytoplasm, the hairpin is processed further by the Dicer nuclease to generate the active 22-nt RNAs (Xu *et al.*, 2004).

The structure of bacterial pairing RNAs ensures stability (at least until used). They frequently begin with a sequence that can fold into a stable stem-loop, and transcription terminates with a rho-independent transcription terminator, a stem-loop which also stabilizes the molecule. Most of these nc RNAs are significantly more stable than mRNAs (Vogel *et al.*, 2003).

In both prokaryotes and eukaryotes, the final active RNA needs to be presented to the target mRNAs it will base-pair with, and protein cofactors perform crucial roles in such presentation. In eukaryotes, the microRNAs and siRNAs are bound to a protein complex (the RISC complex) and are then presented to their pairing targets generally at the 3′ untranslated region (UTR) of genes. In contrast, the bacterial regulatory RNAs, although bound to protein, are presented to the RNA chaperone Hfq which interacts profoundly with the small nc RNAs at single-stranded AU-rich regions (Valentin-Hansen *et al.*, 2004). This interaction stabilizes many of the small RNAs.

An important difference between the bacterial regulatory RNAs and the microRNAs and small interfering RNAs of eukaryotes is the placement of targets in the mRNA. Pairing of the bacterial regulatory RNAs is most frequently with the 5′ end of the message, usually occluding the ribosome-binding site and/or starting codons, and is imperfect, sometimes including several short (8-9 nt) regions of pairing. Eukaryotic small RNA pairing to the 3′ UTR favours translational silencing when pairing is imperfect and mRNA degradation when it is perfect, but it is not known whether the bacterial RNAs obey a similar rule (Gottesman, 2005).

One notable result of small regulatory RNAs in *E. coli* which has not yet been reported for eukaryotic microRNAs is positive regulation of translation and mRNA stability. At least two different small RNAs stimulate translation of an important transcription factor, RpoS. Stimulation takes place when the small RNAs pair with a region of RNA ~70 nt upstream of the translation start site–this abrogates the formation of a *cis*-inhibitory RNA structure (Repoila *et al.*, 2003).

Small RNAs perform important roles in the bacterial response to stress and regulation of factors involved in virulence. The recognition of these effects has not come from the traditional route of isolation of mutations in small RNA leading to understanding of its role. Mutations in these RNAs have not been described for bacteria, possibly because the regulatory RNAs are too small targets for mutagenesis, and many point mutations do not inactivate them; also because in some cases redundant small RNAs and other regulatory mechanisms mask the effects of inactivating a single RNA. The shortage or lack of identified mutations in regulatory RNAs means that understanding the roles of a small RNA will require other approaches.

Positive Regulation of a Transcription Factor

Stimulation of translation by small RNAs has so far been unique to prokaryotes, as seen in the stimulation of translation of RpoS, a major transcriptional regulator of *E. coli* and many other bacteria, by the small RNAs DsrA and RprA. RpoS is an alternative sigma factor and makes up the promoter-specificity subunit of the prokaryotic RNA polymerase. The genes dependent on RpoS for transcription allow the cell to enter into the stationary phase. Various nutrient depletion conditions and other stresses (Hengge-Aronis, 2000, 2002) create an increased need for RpoS which is met primarily by increased translation and less degradation of the protein. The default situation for translation of RpoS remains switched off; the message upstream of the translation start folds back to occlude ribosome entry. Small RNAs that pair with the inhibitory stem of the hairpin are conducive to high levels of RpoS translation. The first of these to be found was DsrA, made at highest levels at low temperatures.

Another small RNA, RprA, was identified as a multicopy suppressor of the low RpoS levels in a *dsrA* mutant. RprA functions in a similar way to DsrA, but is produced in response to activation of the RcsC-ResB phosphorelay, which responds to cell surface perturbation or stress (Majdalani *et al.*, 1998, 2002). Both RprA and DsrA pair with the same region of the rpoS message *in vivo*. In both cases, mutations which reduce pairing of the small RNA with the RpoS mRNA block activity; compensating mutations restore it (Majdalani *et al.*, 1998, 2002)

Under some stress conditions producing more RpoS is not always desirable. RpoS does help the cell in dealing with mild oxidative stress but if oxidative stress is strong enough, the specific repair genes of the OxyR regulon are induced, including another small RNA, OxyS. This latter inhibits, rather than stimulates, RpoS translation. The OxyR-specific repair genes thus attract priority over the more general RpoS-dependent stress response genes (Gottesman, 2005).

The regulation of RpoS translation throws light on how the cell exploits multiple small RNAs to fix regulatory priorities. Under some specific stress condition (low temperature or cell surface stress), the appropriate small RNA can be made quickly to activate translation of RpoS. The various different stress conditions that require increased levels of RpoS can be tackled individually, by several specifically regulated small RNAs.

Another notable property of small RNAs is that they can rapidly block the synthesis of many proteins, thereby redirecting essential cell nutrients. A good example is the small RNA RyhB, formed under iron-limitation conditions. It pairs with the mRNAs of at least five operons encoding iron-binding proteins, including *bfr*, encoding bacterioferritin, *sdh* (succinate dehydrogenase) and *sodB* (superoxide dismutase) and leads to rapid degradation of their message (Masse and Gottesman, 2002). Hfq binding enables remodeling of the *sodB* message to promote pairing between the mRNA and RyhB. Degradation of both the target mRNA and RyhB itself depend on RNase E cleavage sites which are quite similar to Hfq-binding sites (Moll *et al.*, 2003). The outcome of destroying the mRNAs and therefore stopping synthesis of these iron-binding proteins is less cellular need for Fe, and hence increased availability for essential proteins.

The Functome

The availability of complete genome sequences has spurred analyses of entire complements of proteins, RNAs, metabolites, and other cellular constituents, generating a need for a terminology to encompass the global scale of the data (Fields and Johnston, 2002). A sensible suffix ('omes) for this purpose has proliferated : genome, proteome, transcriptome, interactome, functome, even phenome; and many more 'omes are being coined. It is not entirely coincidental that 'ome, is also the anglicized form of 'oma, used for such unwelcome intrusions as sarcoma, lipoma, and fibroma. This unchecked growth of the 'ome is generating some imprecision and confusion.

In contrast to the long history of related suffix 'some, which has been used to name such intracellular particles as "chromosome", "ribosome", "lysosome", "nucleosome", "replisome", "spliceosome" and "proteasome", the suffix 'ome is quite recent in origin–the term genome was first used in English in 1930 and none of the other 'omes has a history of more than a few years.

There are two unresolved problems with the 'ome terminology. Whereas the extent of the genome is quite clear, what constitutes a transcriptome is not so clear. The precise constituents of an 'ome are often not well specified. A second, more serious problem is the conditional nature of some 'omes. The genome–notwithstanding the occasional jump of a transposon or rearrangement of an immunoglobulin gene–is a relatively fixed entity, and most workers agree on its definition But the proteome present in a cell at one moment will differ drastically from that in the same cell even a few seconds after it is heated to 65°C. Or, suppose a cell's glycosylome is defined at time zero, and seconds later the cell dies, its carbohydrate moieties are likely to give up the ghost nonuniformly, with some persisting to the last. At what point in this process should the glycosylome be defined?

To circumvent such difficulties, Fields and Johnston proposed some simple rules. First, much precision is gained by a more circumscribed representation of the 'ome's constituents, for example, phospholipidome rather than lipidome. The minimum number of similar cellular constitutents that constitute an 'ome should be clearly defined. Thus, there can be no "nucleicacidome" but there can be a "nucleotideome" (A, T, G, C, U, I, plus myriad modified purines and pyrimidines); no "actinome," but definitely a "tubulome" (multiple a and β tubulin isotypes).

It is also better if the state of the cells for which an 'ome is defined were reflected in the nomenclature. If initially we use basic parameters like temperature, pH, and subcellular localization, we could define, "the 37°-7.4-G1-Golgi-N-but-not-O-linked glycosylome". As a less cumbersome alternative, an Enzyme Commission (E.C.)–style nomenclature may be designed that allows the incorporation of as many specifications as needed.

According to Fields and Johnston, as biologists attempt to a define the various machines that carry out life's basic processes, it should be possible to define the ultimate 'ome, the collection of all of these machines : the "someome", or the "omesome", since it defines the machine comprising all the 'omes. Either of these two terms represents a great improvement over their imprecise and prosaic synonym currently being widely used: the cell (Fields and Johnston, 2002).

Bacterial Proteomes

Protein function has traditionally been associated with a well-defined three dimensional fold. This view has suffered a setback by recent discoveries that intrinsically disordered proteins do exist and that disorder probably has some important role in protein interactions. Catalysis would seem more structurally demanding. Although it is increasingly recognized that dynamics contributes to enzyme activity, most biochemists assume that this occurs in the context of a folded protein; nevertheless, catalytic activity has been observed in an engineered enzyme with molten globule properties, where function seems to be coupled to substrate-induced folding. Bemporad *et al.* (2008) showed that the partially folded *Sulfolobus solfataricus* acylphosphatase is functional and that this does not derive from a global substrate-induced folding. Molecular dynamics simulations showed that an ensemble of partly folded molecules is characterized by substantial structural flexibility of the catalytic region; the remainder of the protein forms a scaffold that restricts the conformational space of the flexible region so that in a large fraction of the ensemble, residues important for catalysis are usually found close together. It was suggested that scaffold regions in proteins might allow functional regions to mutate without disturbing overall stability, so facilitating the emergence of new activities. Conceivably, catalysis in the absence of well-defined folded structure may be a more common but hitherto not well studied enzyme property.

Kearns *et al.* (2008) have identified a protein that allows the soil bacterium *Bacillus subtilis* to quickly stop its propeller-like propulsion and thus stick to a surface. This protein called, EpsE, appears to function like a clutch rather than a brake. It leaves the rotors that drive the bacterium's flagella unpowered but spinning freely rather than slowing them down.

The researchers labelled EpsE with a fluorescent protein and noted that EpsE is associated with flagellar motors. They then attached bacteria to a surface by a flagellum. The cells rotated passively even when they produced EpsE, which would have prevented them from swimming.

According to Stouthamer (1973), biosynthesis of a typical bacterial cell (*e.g. Escherichia coli*), with organic compounds as sources of energy and carbon, requires about 20 to 60 billion high-energy phosphate bonds. A substantial fraction of this energy budget is used for biosynthesis of amino acids, the building blocks of proteins. The fueling reactions of central metabolism provide precursor metabolites for synthesis of the 20 amino acids incorporated into proteins (N.B. The number 20 has recently increased to 22). Synthesis of an amino acid involves a dual cost (a) energy is lost by diverting chemical intermediates from fueling reactions, and (b) additional energy is needed to convert precursor metabolites to amino acids. The energetic requirement for amino acid biosynthesis for a given codon in the genome is the product of the cost per encoded amino acid and the number of times the codon is translated. Among amino acids, costs of synthesis vary from 12 to 74 high-energy phosphate bonds per molecule. If a significant fraction of amino acids are synthesized in bacterial cells and energy limits survival or reproduction, then the amino acid composition of proteins coded for in the genome should be biased toward less energetically costly amino acids. The extent to which amino acid composition is biased to reduce metabolic costs would be expected to be a positive function of the numbers of proteins synthesized, per generation, from each gene. The energetic advantage of coding for a less costly amino acid in a highly expressed gene can exceed 0.025 per cent of the total energy budget. Akashi and Gojobori (2002) showed that amino acid composition in the proteomes of *E. coli* and *Bacillus subtilis* reflects the action of natural selection to enhance metabolic efficiency. They employed synonymous codon usage bias as a measure of translation rates and observed increases in the abundance of less energetically costly amino acids in highly expressed proteins.

Mutational processes and relationships between primary structure and function are the major determinants not only of amino acid composition but also rates of protein evolution (Li, 1997). Natural selection either preserves or enhances protein specificity, activity, or stability by favouring codons that code for particular amino acids in gene regions corresponding to critical locations in the primary structure of protein (Akashi and Gojobori, 2002). At locations that are less constrained, both mutation pressure and genetic drift account for encoded amino acids. Differences in mutational biases explain some of the variation in amino acid composition among bacterial species (Singer and Hickey, 2000).

Critical analyses and calculations made by Akashi and Gojobori (2002) give strong statistical support to increasing usage of less energetically costly amino acids in abundant proteins in both *B. subtilis* and *E. coli.* These patterns probably do not result from relationships between functional and expression classes of proteins, mutational biases, or biases in estimates of translation rates. In highly expressed genes, the average cost of amino acid biosynthesis falls by about 2.5 and 2.0 ~P per amino acid per protein synthesized in *B. subtilis* and *E. coli*, respectively. Selection to enhance metabolic efficiency is likely to be weak at a given codon but the energetic savings for biosynthesis of the proteome could be in the hundreds of millions of ~P per generation.

Functional Versatility of *Agrobacterium* VirE2

Agrobacterium has evolved a mechanism to genetically engineer host cells to create a favourable niche for itself. This natural ability has been coopted by *Agrobacterium* researchers to engineer plants for agricultural purposes.

Of particular interest is the mechanism that transfers DNA across the inner and outer bacterial membranes and the plasma membrane of the host. It appears that an exporter transfers the T-strand across the bacterial membranes in a process analogous to conjugation. *Agrobacterium* is the prototypical example of a pathogen using this exporter. Exporter proteins are involved in bacterial conjugation, and are virulence proteins in several important human pathogens including *Bordetella pertussis* (whooping cough), *Helicobacter pylori* (gastric ulcers), *Legionella pneumophila* (Legionnaire's disease), and *Rickettsia prowazekii* (epidemic typhus) (Covacci *et al.*, 1999; Ward and Zambryski, 2001).

Dumas *et al.* (2001) suggested that a single protein, VirE2 forms a membrane channel that transfers the T-strand through the plant plasma membrane. This result has implications for our understanding of *Agrobacterium* mediated transformation and, potentially, for delivery of DNA in gene therapy.

VirE2 is one of the most abundant Vir proteins. It binds the T-strand cooperatively, without sequence specificity, protecting it from degradation. VirE2 and VirD2 contain nuclear localization sequences (NLS) which promote nuclear uptake of the T-complex. VirE2:ssDNA complexes microinjected into plant cells give rise to nuclear accumulation of the ssDNA that can be blocked by nuclear import inhibitors. VirE2 may also assist nuclear uptake of the T-complex by keeping the T-strand in an unfolded state (Ward and Zambryski, 2001). It is established that VirE2 associates with the T-strand in the plant cell; but there is controversy over whether this also occurs in the bacterium. VirE2's strong cooperative T-strand binding and abundance suggest association in *Agrobacterium*. It also appears, however, that VirE2 can enter the plant cell independent of the T-strand (Ward and Zambryski, 2001).

VirE2 interacts with lipids and, interestingly, forms large, anion-selective, voltage-gated channels selective for transport of ssDNA. Formation of pores large enough to allow passage of ssDNA may have deleterious effects in plant cells containing VirE2. It seems that the opening/closing of the channels is regulated *in vivo*. The specific chaperone, VirE1, conceivably prevents the formation of pores in *Agrobacterium*. VirE1 is required for export of VirE2 from *Agrobacterium* and may inhibit T-strand binding.

The results of Dumas *et al.* (2001) may assist development of gene therapy technologies. VirE2 mediated transformation may avoid problems inherent in the use of viral delivery systems. Once transported across the plasma membrane of the recipient cell, however, additional factors are probably needed for nuclear uptake and integration of the DNA–factors that are lacking in non-plant hosts.

VirE2 seems to perform an unusually large number of functions. Dumas *et al.* also reported that Vir E2 forms channels in lipid bilayers. Thus, VirE2 may insert in the plant plasma membrane and facilitate passage of the T-strand. VirE2 protein performs multiple critical functions at several points in T-strand transfer (Ward and Zambryski, 2001).

Noncultured Microbes

Some 99 per cent of microorganisms from the environment cannot be/have not been cultivated in the laboratory. It is estimated that existing microbial species number around 10^5 to 10^6 but only several thousand have been isolated in pure culture because very few microorganisms from environmental samples grow on nutrient media in petri dishes (see Colwell and Grimes, 2000). Ribosomal RNA analysis suggests that these uncultivable organisms are found in nearly every prokaryotic group. Several divisions have no known cultivable representatives. It has recently been possible to grow in the lab several strains of previously unculturable beach-growing bacteria–an achievement that may be a new way of exploring the vast diversity of microbial species. The feat was accomplished by transplanting not just the organisms but their whole sandy neighborhood along with them, thereby mimicking the natural conditions.

In view of the difficulty of culturing most microbes, researchers have usually explored microbial diversity by looking for RNA signatures in the environment that point to the presence of novel active genes. But this kind of information does not help in identifying and characterizing the organisms. It is only by studying the organisms in culture that we can learn about microbial evolution and ecology; this may also yield novel useful compounds, such as antibiotics or enzymes (see Greene, 2002).

Kaeberlein *et al.* (2002) designed a diffusion chamber that allows the growth of previously uncultivated microorganisms in a simulated natural environment. Colonies of representative marine organisms could be isolated in pure culture. These isolates did not grow on artificial media alone but formed colonies in the presence of other microorganisms. This observation may help explain the nature of microbial uncultivability (Kaeberlein *et al.*, 2002).

Gene-expression Sorting of Metagenome Libraries

Readily cultured microorganisms represent less than about 1 per cent of the organisms in most habitats. This has aroused interest in nonculture-based methods for studying microbial life (Pace, 1997). Uchiyama *et al.* (2005) described an elegant screen that rapidly sorts out thousands of environmental DNA clones for the purpose of identifying those having a desired metabolic activity. The screen employs a reporter system to trap genes encoding biodegradative pathways; it is based on the prediction that operons encoding these pathways may be induced by the substrate of the pathway (Handelsman, 2005).

Metagenomics refers to the genomic analysis of an assemblage of microorganisms. It opens a window on the uncultured world. It requires cloning of DNA extracted directly from an environmental sample (Handelsman *et al.*, 1998). Screening of metagenomic libraries has traditionally followed two routes: sequence-based, and function-based. Some libraries have been screened by hybridization or PCR to detect genes showing homology to known genes. This approach is limited to discovering genes in previously described families.

Functional analysis allows one to discover entirely new classes of genes for useful functions but requires expression of genes from exotic organisms in such tame bacteria as *Escherichia coli*. It depends on low-throughput screens, *e.g.*, visual detection of growth inhibition of indicator bacteria, and selection, such as resistance to antibiotics or complementation, of necessary biochemical pathways (Diaz-Torres *et al.*, 2003; Henne *et al.*, 1999). Low-throughput screens often detect a wide range of activities, but their utility is usually limited by the fairly low frequency of active clones.

The high-throughput screen designed by Uchiyama *et al.* for catabolic pathways is termed substrate-induced gene expression 'screening' or SIGEX. What Uchiyama *et al.* did was to design an 'operon trap' in which the metagenomic DNA is cloned upstream of the *gfp* gene, in such a way as to place green fluorescent protein (GFP) expression under the control of promoters in the metagenomic DNA. They then looked for clones in which GFP expression is regulated by the substrate of interest.

SIGEX relies on promoter activity, rather than sequence similarity or phenotypic expression. This removes several limitations of previous metagenomic screens. SIGEX provides access to genes that are silent or too rare to be revealed by other methods (Handelsman, 2005). Unfortunately, SIGEX also suffers from a constraint–the regulatory machinery which can recognize the presence of promoter and substrate, and also that they are functional in the host cell. The host cell must also take up the substrate According to Handelsman, the SIGEX template facilitates the challenging task of discovering infinite variation in the functions, the regulatory molecules to be employed, and the reporter proteins to be detected. The high-throughput nature of the screen should make SIGEX a landmark in metagenomic analysis.

As pointed out by Daniel (2004), most microorganisms in the environment resist cultivation, prompting microbiologists to directly clone useful genes from environmental metagenomes. Two screening methods being used for the metagenome approach are nucleotide sequence-based screening (Okuta *et al.*, 1998) and enzyme activity-based screening (Henne *et al.*, 1999). A third option has been optimized for isolating novel catabolic operons–substrate-induced gene expression screening (SIGEX). SIGEX is based on the assumption that catabolic-gene expression is usually induced by relevant substrates and commonly controlled by regulatory elements situated proximate to catabolic genes. For making SIGEX high throughput, they constructed an operon-trap *gfp*-expression vector available for shotgun cloning; this vector makes possible the selection of positive clones in liquid cultures by fluorescence-activated cell sorting. The utility of SIGEX has been demonstrated by the cloning of aromatic hydrocarbon-induced genes from a groundwater metagenome library and subsequent genome-informatics analysis (Uchiyama *et al.*, 2005).

Designing a Bacterial Signalling Network

Cellular biochemical networks function in a noisy environment by using imperfect components. Any errors in signal transduction can upset development and behavioural decisions, and result in growth impairment or, in multicellular organisms, cancer. Networks involved in gene regulation or signal transduction have only small output tolerances, and the underlying network structures have probably evolved for inherent robustness against perturbations (Kitano, 2004). Kollmann *et al.* (2005) investigated an optimal design for the signaling network of bacterial chemotaxis, a most thoroughly studied biological signaling network. They experimentally determined the extent of intercellular variations in the expression levels of chemotaxis proteins and used computer simulations to quantify the robustness of several hypothetical chemotaxis pathway topologies to such gene expression noise. Among these topologies the experimentally established chemotaxis network of *Escherichia coli* was shown to have the smallest sufficiently robust network structure, allowing accurate chemotactic response for almost all individuals within a population. It appears that this pathway has evolved to show an optimal chemotactic performance while minimizing the cost of resources associated with high levels of protein expression. Moreover, the underlying topological design principles compensating for intercellular variations seem to be highly conserved among bacterial chemosensory systems (Szurmant and Ordal, 2004).

Survival of Bacterial Spores

Species of *Bacillus* and *Clostridium* form dormant spores which can survive for centuries or longer (Kennedy *et al.*, 1994; Nicholson *et al.*, 2000). The physiological properties of these spores markedly differ from those of other known cells. *Bacillus* and *Clostridium* spores in water apparently do not metabolize endogenous or exogenous compounds. They survive treatments with wet and dry heat, UV and gamma radiation, desiccation and toxic chemicals that rapidly kill growing bacteria (Setlow and Setlow, 1996; Tennen *et al.*, 2000; Setlow, 2006, 2007). Several factors are involved in spore resistance and long-term survival, but minimization of damage to spore DNA is crucial. Two mechanisms that minimize spore DNA damage are protection and repair. Protection is preferable for spores because DNA repair can be mutagenic and repair of dormant spore DNA damage has to wait until spore germination is completed and spore outgrowth begins. Spores seem to contain only a single chromosome–this rules out homologous recombination early in outgrowth as a mechanism to repair damage in dormant spore DNA, although there are several spore-specific DNA repair proteins (Wang *et al.*, 2006). While DNA repair contributes to spore DNA resistance to some treatments, protection of DNA is the chief factor that makes spore DNA resistant (Nicholson *et al.*, 2005).

Spores of *Bacillus* and *Clostridium* species have attracted serious concern in view of their novel properties and involvement as agents of food spoilage and food-borne disease. This concern in spores has enhanced greatly in recent years in view of the use of *Bacillus anthracis* spores as a bioterrorism weapon. The resistance of spore DNA is a primary factor determining the ability of spores to cause disease, warranting critical research aimed at elucidating mechanisms of DNA protection in spores. Dormant spores of *Bacillus, Clostridium* and related species can survive for many years because spore DNA is well protected against damage by many different agents. Such protection results partly from the high level of Ca^{2+}-dipicolinic acid in spores and DNA repair during spore outgrowth, but is primarily caused by the saturation of spore DNA with a group of small, acid-soluble spore proteins (SASP). These proteins are synthesized in the developing spore and degraded after completion of spore germination. The structure of both DNA and SASP changes when they associate. This causes marked changes in the chemical and photochemical reactivity of DNA (Setlow, 2007).

The DNA is located in the core of the spore in a region of lower water content (25-50 per cent wet weight) than the protoplast of a growing cell or the outer spore layers. The pH of the spore core is ~6.5, *i.e.*, ~1 pH unit lower than in growing cells, and the core has a huge amount of pyridine-2.6-dicarboxylic acid [dipicolinic acid (DPA); ~20 per cent of core dry weight] (Setlow, 1994). DPA in spores is present as a 1:1 chelate with divalent cations, mostly Ca^{2+}, giving Ca-DPA. In the core, spore DNA is saturated with α/β-type SASP, named after the major proteins of this type in *Bacillus subtilis* spores (Setlow and Setlow, 1996; Tennen *et al.*, 2000; Setlow, 2006, 2007). The α/β-type SASP are synthesized in developing spores late in sporulation, and suffer degradation early in spore outgrowth (Setlow, 2007).

Among all the core components the one that confers the greatest protection on spore DNA is the α/β-type SASP. There are quite high levels of these proteins in spores of *Bacillus* and *Clostridium* species. In *Bacillus megaterium* and *B. subtilis* spores there is enough α/β-type SASP to saturate spore DNA. These proteins have been experimentally shown to bind to DNA in spores and are crucial for spore DNA protection; this is shown by the sensitivity of DNA to many DNA-damaging agents in *B. subtilis* spores that lack ~85 per cent of their α/β-type SASP (termed $\alpha^-\beta^-$ spores), and the protection of DNA *in vitro* against damage upon binding of α/β-type SASP. Lower levels of α/β-type SASP also reduce the resistance of *Clostridium perfringens* spores to wet heat and UV radiation (Raju *et al.*, 2006). α/β-type SASP and Ca-DPA together are also indispensable for survival of spore DNA–spores that lack both of these components rapidly lose viability during sporulation because of DNA damage (Setlow *et al.* 2006).

It now seems established that the α/β-type SASP are the chief factor protecting spore DNA from various damaging agents. These proteins are unique and, considering their role in protecting DNA in a dormant organism where an entire chromosome is transcriptionally silent, such uniqueness is not surprising. On the other hand, presence of these proteins in growing cells is extremely deleterious.

Humanizing Yeast for Producing Sialylated Glycoproteins

Except for antibodies, the half-life and therapeutic efficacy of most glycoproteins depends on the presence of terminal sialic acid. As pointed out by Stockert (1995), the exposure of other terminal sugars *e.g.* mannose, *N*-acetylglucosamine, and galactose on a glycoprotein leads to clearance by sugar-specific receptors or lectins. Most therapeutic glycoproteins require sialylation and have been produced on mammalian hosts, which can perform humanlike N-glycosylation and also add terminal sialic acid. Yeast and some filamentous fungi are in some ways more suitable as recombinant protein expression systems than mammalian cell culture–yeast gives higher recombinant protein titers, requires shorter fermentation time, and can grow in chemically defined media. But wild-type yeast glycosylate proteins with high-mannose type N-glycans, shorten half-life and compromise therapeutic function (Gemmill and Trimble, 1999). Choi *et al.* (2003) and Hamilton *et al.* (2003, 2006) have engineered human glycosylation pathways into fungal hosts, including the yeast *Pichia pastoris*. The human glycosylation reactions require the sequential removal of mannose by two distinct mannosidases (*i.e.*, α-1,2-mannosidase and mannosidase II), the addition of *N*-acetylglucosamine (by N-acetylglucosaminyltransferase I and II), the addition of galactose (by β-1,4-galactosyltransferase), and the addition of sialic acid by sialyltransferase. Sialylation is the final step of human glycosylation. It is very difficult to accomplish in yeast, because wild-type yeast lacks all the following prerequisites: (i) ability to produce the N-glycosylated precursors terminating in β-1,4-galactose, (ii) biosynthetic ability to make the sugar nucleotide precursor cytidine monophosphate (CMP)-sialic acid, (iii) transporter to shuttle CMP-sialic acid into the Golgi, and (iv) a sialyltransferase which can carry sialic acid to terminal galactose on the nascent glycoprotein. All these requirements need to be met for achieving high enough efficiency to produce sialylated glycoproteins. Also organelle-specific targeting of several elements is a must if these functions are to occur in concert (Hamilton *et al.*, 2006).

Hamilton *et al.* enhanced the utility of yeast as a recombinant protein expression system by engineering *P. pastoris* to secrete human glycoproteins with fully complex terminally sialylated N-glycans. Four genes were knocked out to eliminate yeast-specific glycosylation, and 14 heterologous genes were introduced. This made it possible to replicate the sequential steps of human glycosylation. The reported cell lines produced complex glycoproteins with over 90 per cent terminal sialylation. Also, the utility of these yeast strains was demonstrated by the production of recombinant erythropoietin.

The yeast cell lines of *P. pastoris* developed by Hamilton *et al.* have a significantly reengineered secretory pathway. These lines secrete terminally sialylated, complex, glycoproteins as well as some other recombinant proteins that were tested. The availability of such yeast cell lines could possibly abolish the need for mammalian cell culture and facilitate efficient production of therapeutic glycoproteins in a nonmammalian host. This can significantly reduce production time and viral containment issues and improve product uniformity and overall production economics.

Dimorphism in Yeast

The baker's yeast (*Saccharomyces cerevisiae*) is a good model to examine the eukaryotic way of life at the molecular level. It has shed valuable light on fundamental mechanisms in fungal pathogenesis because it adopts two distinct morphologies: a spherical or avoid (yeastlike) form, which proliferates by budding, and a filamentous form. For some fungal pathogens, *e.g.*, *Candida albicans*, the ability to undergo this dimorphic change is intimately correlated with invasion of host tissue and virulence (Sudbery *et al.*, 2004).

When growing on a rich medium, *S. cerevisiae* proliferates as separate spherical cells. But when exposed to nitrogen limitation, the cells become thin and oblong, and when they produce a daughter bud, it cannot detach from the mother cell but continues to adhere and remain connected end-to-end with the mother. The switch from budding to pseudohyphal growth is regulated by at least three signalling modalities (Gagiano *et al.*, 2002). Each input occurs through the action of different classes of protein kinases. The centerpiece of one pathway is a three-tiered mitogen-activated protein kinase (MAPK) cascade (Mösch, 2000). Each kinase mediates the transfer of phosphate from ATP to either serine, theonine, or tyrosine residues in the next kinase which acts like a switch in the amplification of signal propagation. The MAPK cascade cooperates with a second input involving cyclic adenosine monophosphate (cAMP)-dependent protein kinase (PKA); this aspect has been widely studied in yeast with respect to its role in glucose metabolism in vegetatively growing cells. When the glucose level is low or exhausted, a third input, acting through 5'-AMP-activated protein kinase (AMPK) is pressed into service so that filamentous growth on fermentation end products and other nonglucose carbon sources continues (Palecek *et al.*, 2002). During the above series of reactions, there is a signal transduction from the plasma membrane to the nucleus.

In recent years it has emerged that in some signalling pathways, many components are connected in complicated networks. A small set of similar or even identical components is assembled in different ways in distinct cell types to control completely different biological responses (Ray *et al.*, 2004). The MAPK cascade (certain protein kinases that are sequentially activated by phosphorylation) also acts in three distinct signalling pathways in yeast: a pheromone sensing pathway (Wang and Dohlman, 2004), a pathway that monitors extracellular osmotic conditions (Westfall *et al.*, 2004), and a nutrient-sensitive pathway that converts yeast into a connected filamentous form (Truckses *et al.*, 2004). These pathways even share one identical member of the MAPK cascade, yet specificity is usually faithfully maintained (Ray *et al.*, 2004).

Diploid yeast grows as filaments when it is deprived of nitrogen. Haploid yeast undergo a similar dimorphic switch ("invasive growth") but only when glucose is limiting. In either case, the MAPK cascade and PKA are involved. Yeast can sense a decrease in the nitrogen or glucose supply.

Work is underway to understand the genetic mechanism underlying this sensing and also to learn integration of signalling pathways, and signalling specificity in complex networks by continued study of the filamentous growth process in yeast (Truckses *et al.*, 2004).

Lifespan in Yeast

Much research aimed at discovering the genetic bases of longevity has focused on *Saccharomyces cerevisiae* (Jazwinski, 2002; Anderson *et al.*, 2003). It is hoped that after identifying genes that promote longer life in yeast cells, it may be possible to spot analogous genes for more complex species, including humans. Unfortunately, yeast researchers use a definition of longevity not applied to other species. Minois *et al.* (2005) proposed a method that makes it possible to estimate for yeast the same measures of longevity calculated for other species. They showed that the conventional method (equating longevity with the number of offspring) is only an approximate measure of true chronological lifespan. This method will allow results for yeast to be compared more correctly with those for other species.

The commonest measure of longevity reported in yeast studies is budding lifespan (Lin *et al.*, 2002), *i.e.* the number of buds or daughter cells produced by a mother cell. To avoid colony formation and to ensure that the same cell is followed throughout the experiment, buds or daughter cells are removed from the mother cell with a micromanipulator. But this is a most unusual definition of longevity–it is actually a measure of fertility rather than longevity (Minois *et al.*, 2005). Budding lifespan is a measure of the quantum of fertility. It will only be the same as true longevity (defined in time elapsed from birth) if certain assumptions are true: this definition assumes that there is a directly proportional relationship between the number of buds and the time taken to produce them. Moreover, it also assumes that there is no postreproductive life. Clubbing fertility and longevity into one measure is flawed. For example, it is impossible to tell whether a mutation that appears to increase longevity is in fact doing so. It may be increasing fertility, without prolonging life.

An alternative approach for studying the lifespan of yeast is to use the definitions preferred for other species: reproductive lifespan and postreproductive lifespan. The former is the length of time taken for a cell to produce the number of offspring reported in the budding lifespan, *i.e.*, the time elapsed from birth to the appearance of the last daughter cell. Postreproductive lifespan is the time elapsed from the appearance of the last bud to the death of the cell. Reproductive lifespan is calculated from the same experimental observations made for budding lifespan, reporting the time elapsed rather than the number of offspring. But estimation of postreproductive lifespan requires a methodological innovation, as normal experimental methods do not directly observe death in all circumstances. To make this possible, Minois *et al.* incorporated the vital dye Phloxine B, into the food on which the cells were kept. All cells absorb the dye, but metabolically active cells pump it out and remain colourless. Dead cells are dyed: they passively incorporate the dye and stain red. Phloxine B is used in studies of both budding and fission yeast, both on cells mounted for fluorescence microscopy and added in the food to detect, for instance, temperature-sensitive colonies. Observations of single cells have been made with the dye incorporated in the food. By measuring elapsed time and using the dye, one can estimate both reproductive and postreproductive lifespan. Adding them together gives the overall lifespan.

Minois *et al.* found that budding lifespan is not an accurate predictor of reproductive lifespan, and that unqualified use of budding lifespan as a measure of longevity is not justified. They also showed that postreproductive lifespan differs between strains.

Fungi for Weed Control in Crop Fields

In row-crop fields, weeds are the chief pest constraint. They impose heavy variability costs of crop production whether mechanically by cultivation (at high costs in erosion and energy) or chemically (with erosion prevention and high-energy efficiency but with considerable potential for environmental problems). Over 60 per cent of pesticides used in the developed world are herbicides for weed control. Certain pathogens and insects infect/infest weeds, and are used to replace mechanical and chemical control, although they have not proven very successful on row crops. They have been more successful in pastures and managed forests, as well as in natural ecosystems (Amsellem *et al.*, 2002). Successful strategies have been based on the biotechnological control of alien weeds by introducing the pathogens or insects that can be the wild species in balance in its center of origin ("classical biocontrol"). But many important weeds of major row crops are globally distributed, considerably evolved from their wild cohorts at the centers of origin, and so cannot be controlled by classical methods. However, some pathogens have been isolated that attack these major global weeds. These pathogens can help to control only when high levels of inoculum are applied (inundative biocontrol), but they raise the costs greatly.

Pathogens are effective at a 100 per cent humidity level (near the dew point) for 6-18 h after application; this requirement is quite difficult to meet. The major organism being marketed is a strain of *Colletotrichum gleosporioides* (COLLEGO, Encore Technologies, Plymouth, MN), for use in rice paddies (where the humidity for establishment of the mycoherbicide is adequate to control *Aeschynomene virginica* (northern jointvetch), a weed that is difficult to control chemically (Amsellem *et al.*, 2002).

The lack of success with potential inundative mycoherbicides has prompted recourse to genetically engineering hypervirulence into weed-specific pathogens (Kistler, 1991; Gressel, 2002), while adopting suitable biosafety mechanisms to prevent the spread of the transformed hypervirulent organisms as well as to mitigate introgression of transgenes into related organisms (Gressel, 2001).

Sex Life of *Candida albicans*

The opportunistic yeast, *Candida albicans* causes thrush and serious systemic disease and arouses serious concern among physicians, but it has attracted great interest of microbial geneticists by revealing the amazingly varied sex lives of fungi.

Sex drives evolution by recombining parental genomes and eliminating deleterious mutations. Sex has various different forms but all share three common steps: (a) meiosis, which halves the number of chromosomes per cell; (b) gamete production; and (c) cell-cell fusion. In fungi, sex usually involves morphologically similar cells of opposite mating type (**a** and α) which reciprocally produce sex-stimulating pheromones. Some fungi are self-fertile, and lone individuals can reproduce sexually, as exemplified by the yeast *Saccharomyces cerevisiae*. In this yeast, a single cell divides, one of the resulting cells switches mating type, and mating yields **a**/α diploid cells, with a double complement of chromosomes, that no longer mate but do undergo meiosis. Alby *et al.* (2009) reported a conceptually similar but mechanistically different self-mating phenomenon in the pathogenic yeast, *Candida albicans*.

C. albicans lives in the human microbiota, *e.g.*, in the gastrointestinal tract, and commonly causes infections. It has been supposed to be strictly asexual, but its mating-type gene locus has been discovered, and most isolates turned out to be **a**/α diploids that, like *S. cerevisiae*, are sterile (Bennett and Johnson, 2005). Mutants with only one mating type (**a**/a or α/α) can mate in the laboratory or during infection but only at a low frequency; this raised the issue of whether sex occurs in nature. This issue has been resolved by various discoveries (Huang *et al.*, 2009) and by the observation that mating is enhanced enormously when cells switch from 'white' to an 'opaque' type specialized for mating (see Figure 5.1).

In opposite-sex mating in *C. albicans*, cell-cell fusion generates tetraploid cells with four copies of each chromosome (a/**a**/ α/α). These tetraploids undergo a 'parasexual' phase (Bennett and Johnson, 2003), which restores the diploid state so that some genetic recombination occurs but neither involves overt meiosis nor produces spores. (The parasexual recombination that does occur requires Spo11 (Forche *et al.*, 2008); this is an enzyme which functions as a recombinase during meiosis in other organisms. The implication: sex in *C. albicans* possibly involves cryptic meiosis or novel paraxesual roles for Spo11).

Alby *et al.* reported an interesting aspect of *C. albicans* reproduction in which cells mate with each other via same-sex mating, not with the opposite mating type. They discovered this by analyzing *C. albicans* mutants lacking barrier protease. Barrier is encoded by the *BAR1* gene in a mating-type-specific manner, and acts on α-pheromone, enabling **a** cells to choose mates producing the most pheromone (Heitman, 2009). Interestingly, *C. albicans* a cells express genes coding for both **a**- and α-pheromone; **a** cell *bar1* mutants lacking barrier give rise to abnormal wrinkled colonies which respond to pheromone. Surprisingly, **a-a** cell fusion can occur both between *bar1* mutants and between wild-type and *bar1*-mutant cells. Crosses with mutant strains revealed that same-sex mating occurs when α-pheromone is produced

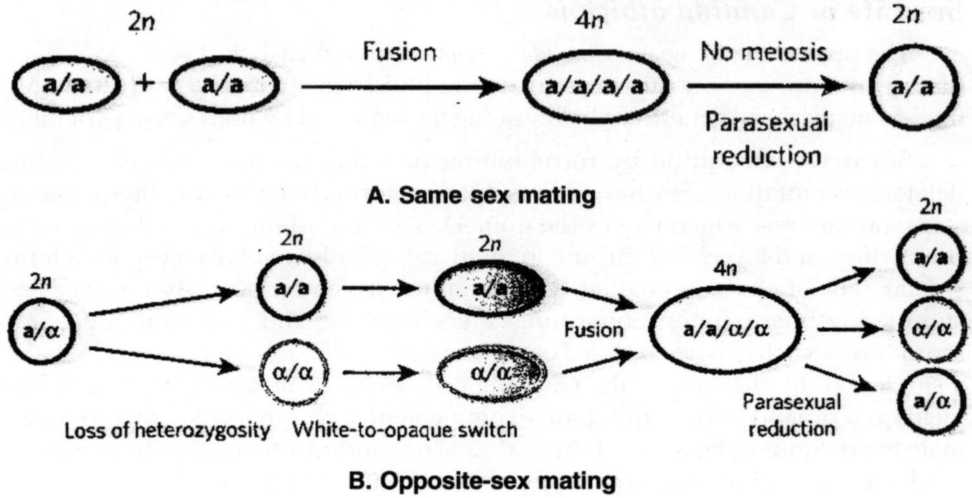

Figure 5.1: *C. albicans*. Same-sex mating (A), and opposite-sex mating (B). Self-fertility may result in population features (such as inbreeding and clonality) that are associated with pathogenesis. B. *C. albicans* is diploid (2*n*): most strains are heterozygous, having both mating-type gene forms (a/α), and are sterile. Opposite-sex mating involves loss of heterozygosity, yielding a and α diploid cells that then switch from the white type to the opaque type specialized for mating. Cell-cell fusion yields tetraploid (4*n*) cells in which paraxesual chromosome reduction without recognized meiosis yields diploid progeny (after Heitman, 2009 and Alby *et al.*, 2009). (Reprinted by permission from Macmillan Publishers Ltd. (*Nature* 460: 807-808, 2009).

by one **a** cell and is sensed by α-pheromone receptors on both the producing and the partner a cell (Heitman, 2009).

Same-sex mating is less efficient than opposite-sex mating, but can be easily detected. *C. albicans* does not require both mating-type locus versions for the paraxesual cycle. This contrasts with their requirement for meiosis in *S. cerevisiae*.

The notable point is that the yeast can mate by both opposite-sex/outcrossing and same-sex/inbreeding. In *S. cerevisiae*, the natural population consists of isolates that switch mating type and self-mate, and others that cannot switch but are fertile with the opposite mating type. Thus, a well-defined species can have individuals with different modes of sexual reproduction. Strikingly, however, two closely related yeasts (*C. albicans* and *S. cerevisiae*) have evolved such distinct modes of self-fertility. The more evolutionarily divergent fission yeast, *Schizosaccharomyces pombe*, is also known to switch mating type and can self-mate, like *S. cerevisiae*, but it independently evolved a machinery to do so. Such a diversity in forms of sexual reproduction attests to the remarkable plasticity of fungi, which have evolved to exploit different environments (Heitman, 2009).

Biological Nitrogen Fixation–Ecological Dimensions

Although biological nitrogen fixation (BNF) has traditionally been considered as a process restricted to the rhizosphere and carried out by only a few free-living organisms, recent work has revealed several new representatives of diazotrophs as well as the presence of known nitrogen-fixing organisms in new habitats. These data have enhanced knowledge of diazotrophs' ecology and capabilities and thrown new insights into the biogeochemical cycle of nitrogen and its implications.

Much is known about BNF in the Leguminosae-*Rhizobium* symbiotic interaction. This scenario has been extended to the Graminaceous (Poaceae) plants, where a symbiotic interaction also occurs, but through a different process from that of the legume-*Rhizobium* (Dobereiner, 1961). Several free-living bacteria and cyanobacteria are also recognized as nitrogen-fixing organisms (diazotrophs).

The enzyme (nitrogenase) involved in BNF has been studied extensively. The nitrogenase complex is sensitive to O_2 and hence can be irreversibly inactivated by O_2. This means that diazotrophs must employ mechanisms which concomitantly permit the supply of O_2 required for obtaining chemical energy and also protect nitrogenase from this negative effect. Some microbial strategies that limit the access of O_2 to nitrogenase are the following:

1. Avoiding O_2 by living in strict anaerobic environments.
2. Balancing nitrogenase inactivation with the expression of the enzyme.
3. Creating a physical barrier that prevents O_2 from diffusing to the enzyme. In obligate aerobic organisms, this barrier should not completely exclude O_2 and should not affect the diffusion of nitrogen (N_2). This selective sequestration is accomplished by heterocystous cyanobacteria, which can fix N_2 and simultaneously carry out oxygenic photosynthesis. Generally, bacteria from the genus *Rhizobium* are protected from contact with oxygen by the structural nature of the legume plant's symbiotic nodule.
4. Metabolic elimination of O_2 to reduce its concentration to acceptable levels around the enzymatic complex.

   Robson and Postgate (1980) postulated that uncoupled respiratory activity protects nitrogenase against O_2. *Azotobacter* can adjust its respiratory rate throughout a wide range of O_2 concentrations. *A. vinelandii* has a ramified respiratory chain and the role of its terminal cytochrome oxidase–responsible for the O_2 reduction to H_2O–plays a crucial role in avoiding damage to nitrogenase. A mutant strain altered in its cytochrome can fix nitrogen only under anaerobic conditions. This is called respiratory protection to the nitrogenase (Montagna and Torres, 2008).

   When the respiratory protection is not functioning well in *Azotobacter chroococcum*, the O_2 reduction by nitrogenase can cooperate to minimize its inactivation through self-protection. Nitrogenase can reduce O_2 when the cells have high energy and reduction potential levels. The partial reduction of O_2 generates free-radicals whose concentration depends on the levels of O_2 and Fe-protein; O_2^- and H_2O_2 can be eliminated by superoxide dismutase and catalase. The activity of these enzymes rises sharply when cells fix nitrogen (Dingler and Oelze, 1987).

5. Nitrogenase may be so modified as to make it resistant to inactivation by O_2. Under O_2 stress or carbon limitation, nitrogenase activity stops but when the O_2 concentration falls, the nitrogenase activity recovers without any fresh synthesis of enzyme (Scherings *et al.*, 1983).

Soil

Table 5.1 shows the occurrence and distribution of diazotrophs in a wide range of temperatures, from the very cold to the very wet and hot soil of rain forests. Diazotrophs have also been reported in coastline soil with higher salinity and a higher level of oxygen as compared to forest soil due to the compaction of the latter. Organisms previously described from the rhizosphere of sugar cane, in free-living conditions (*Beijerinckia* sp.) have also been found growing in coastline soil.

Table 5.1: BNF Occurrence in Soil (after Montagna and Torres, 2008)

Soil Type	*Organisms*	*Detection**	*Detected Gene*
Acid soil	*Frankia, Azospirillum, Clostridium*	NGD, AR	*nifH*
Coastline soil	*Beijerinckia derxii*	NGD, AR	*NifH*
Siberian conifer forest	*Methylocella, Methanococcus*	NGD, IG	*nifH, nifD*
European continental temperate river basin	*Methylosynus*	NGD, IG	*nifH, nifD*
Soil and floodplain of Amazon	*Nostoc, Calothrix, Cylindrospermum, Fischerella, Anabaena*	AR	–

* NGD: *Nif* gene detection; AR: Nitrogenase activity measured by acetylene reduction; IG: isolation and growing of diazotrophs.

Water

Table 5.2 lists some diazotrophs reported from aquatic habitats. In fresh water, there are 13 unidentified species of diazotrophs or at least, 13 species of organisms able to carry out BNF, plus another 14 identified species, with only 5 previously described (Montagna and Torres, 2008).

The non-heterocystous cyanobacterium *Trichodesmium* dominates N_2-fixing organisms in the tropical oceans whereas heterocystous species dominate N_2 fixation in freshwater (see Capone *et al.*, 1997; Staal *et al.*, 2003).

Deep sea hydrothermal vents also contain methanogenic archaea which can reduce N_2 to NH_3 at up to 92°C and still express *nifH* messenger RNA (Mehta and Baross, 2006).

Table 5.2: BNF Occurrence in Water (after Montagna and Torres, 2008)

Environment	*Organisms*	*Detection**	*Detected gene*
Freshwater			
Alkaline lake	*Phormidium, Pseudomonas, Desulfovibrio, 13 unidentified species*	IG, AR	*nifH*
Antarctic freeze lake	*Pseudomonas stutzeri, Azospirillum brasilense, Aquaspirillum*	NGD, NC	*nifH*
Marine			
North Pacific-subtropical (25 m depth)	*Trichodesmium*	NGD, NC	*nifH, nifD*
	Trichodesmium, Cyanothece, Richelia, Synechocystis	NGD, NC	*nifH, nifD, nifK*

* NGD: *Nif* gene detection; AR: Nitrogenase activity measured by acetylene reduction; NC: Nitrogenase activity measured by N^{15} consumption; IG: Isolation and growing of diazotrophs.

The major input of nitrogen comes from cyanobacteria, either free-living or symbiotic with protozoans or protistans. The wide dispersal of cyanobacteria throughout tropical and subtropical seas points to a substantial input of nitrogenous compounds to the base of the ecologic chain in this environment.

Intracellular

Several diazotrophs grow as intracellular symbionts and play important role in the nitrogen metabolism of the host which is enabled to grow without a nitrogen supply other than N_2, showing that these intracellular diazotrophs fix nitrogen. Because of paucity of experimental data, the importance of marine protozoans' biomass for the marine nitrogen cycle is not clear but it seems that cyanobacteria and Archaea are the major diazotrophs in a marine environment (Montagna and Torres, 2008).

Endophytic

Besides legumes and grasses, many other plant groups in some environments harbour diazotrophs. Almost all plant tissues and organs host diazotrophs. De Hoff and Hirsch (2003) cited several works that relate diverse strategies of interaction and dependence between the plant and the bacteria, including the possibilities of hormonal signaling between them (see Table 5.3)

Insects

Certain enterobacteria that fix nitrogen colonize termite and dipteran gut tract (Table 5.4). Not only common enterobacteria such as *Klebsiella* spp. fix nitrogen under specific conditions, but some other enterobacteria also share this ability. These bacteria are widely disseminated as free-living organisms in nature and as gut tract residents in many species of terrestrial vertebrates. These bacteria have been demonstrated to fix nitrogen *in vitro* but we do not know whether they also fix N_2 *in vivo* in the gut tract. Conceivably, their ability to fix nitrogen in vivo might improve their autonomy when outside the gut tract, contributing to their disseminating capability. Like marine protozoans with intracellular symbiotic diazotrophs, which are able to grow without a nitrogen source other than N_2, more data are needed to test this possibility. Another analogous possibility is to extrapolate the termite and dipteran gut tract colonization to other herbivorous insects. Since diazotrophs (especially enterobacteria) are found in grasses and cereals, it is possible that they reach the insect gut tract. The issue is whether or not these diazotrophs would preserve their BNF ability once they reach the gut tract? If so, they might contribute significantly to nitrogen uptake by the host. Insects being ubiquitous, it seems reasonable that if some representatives of these phyla do show nitrogen-fixing enterobacteria, others also might be doing so. Since insects make up a significant fraction of terrestrial biomass, their BNF assumes relevance to the global terrestrial nitrogen cycle and quantitative importance to the process as a whole.

Table 5.3: Selected Examples of Plants and Animals in which *Nif* Genes or Nitrogenase Activity have been Detected (After Montagna and Torres, 2008)

Plant	*Organisms*
Graminaceae roots (9 spp.)	*Enterobacter, Klebsiella, Pseudomonas*
Graminaceae–aerial parts	*Azospirillum, Pseudomonas, Klebsiella, Enterobacter, Serratia marscens, Azoarcus, Herbaspirillum, Acetobacter*
Graminaceae (*Oryza*)–roots	*Azoarcus*
Gossypium hirsutum, Ipomoea batatas, Coffea arabica	*Acetobacter diazotrophicus*
Ananas comosus	*Acetobacter diazotrophicus*
Oryza	*Azoarcus, Herbaspirillum, Ideonella, Azospirillum*

Table 5.4: Bacteria Colonizing Insect Gut in which *nifH* Gene has been Detected

Bacteria	*Insect*
Citrobacter, Enterobacter, Desulfovibrio	Termites
Enterobacter, Klebsiella	Diptera *Ceratitis capitata, Anastrepha ludens, Bactocera tryoni, Rhagoletis pomonera*

Lichens

The BNF ability recognized in lichens was previously attributed to cyanobacteria and not much quantitative importance was assigned to this association due to its low occurrence. BNF was later demonstrated also in fungi and green algae associations. Nitrogen-fixing bacteria were demonstrated to participate in this association.

Lichenized fungi are widespread in nature. They represent about 20 per cent of all fungal species. About 85 per cent of the lichenized fungi comprise associations with green microalgae, 10 per cent with cyanobacteria and 4 per cent with both cyanobacteria and green algae. The implication is that potentially almost all lichens contribute to BNF relying on that association to face severe conditions of restricted nutrient availability (Liba *et al.*, 2006).

Novel Aspects

"Classical" diazotrophs have now been recorded from novel habitats and likewise non-classical nitrogen fixing diazotrophs have been found in unsuspected, new habitats. An attractive hypothesis that may explain the wide distribution of BNF has emerged from the strong similarity reported among nitrogenases as a consequence of lateral gene transfer between unrelated organisms (Hanson *et al.*, 2004).

Rhizobial Systematics

Biological nitrogen fixation (BNF) is an important feature of symbiotic associations of legume hosts with rhizobia. To achieve maximum BNF out of any legume-rhizobium association, proper characterization and identification of rhizobia is essential before they can be made commercially available for field applications. The advent of molecular tools is proving helpful for this purpose. The polyphasic approach is proving reliable in classifying microorganisms and has led to the reorganization of the already described species of rhhizobia as also identification of new species from the nodules of already explored as well as from a large number of hitherto unexplored legume plants (Sahgal and Johri, 2003).

According to Postgate (1998), symbioses between leguminous plants and soil bacteria (rhizobia) are responsible for an estimated 180 x 10^6 tonnes per year of biological nitrogen fixation worldwide. Symbiotic BNF contributes about 120 x 10^6 metric tones year^{-1} to global nitrogen economy and represents about 65 per cent of the nitrogen used in agriculture (Graham, 1988); it is much greater than the input of nitrogen from N fertilizers, estimated at 65 x 10^6 tonnes per annum.

Recent times have seen an alarming, ongoing land degradation in India, where annual rate of abandonment of dry lands due to land degradation is between 9 to 11 m ha. Legume-*Rhizobium* associations prove beneficial in restoring such degraded lands. Considering the potential of legume-*Rhizobium* associations, rhizobial inoculants can be used to improve plant and soil health, provided that the problem of variability in field performance and successful establishment of introduced strain(s) from of competition with the indigenous rhizobacterial population is addressed. Correctly identified and characterized bioinoculant isolates showing high growth promotion and survival in soil can be exploited for technology development and commercialization.

Effective plant growth-promoting rhizobacteria (PGPR), including rhizobia, have recently been characterized by molecular fingerprinting and other modern tools. Tripathi *et al.* (2002) grouped salt-tolerant (3 per cent NaCl) bacteria of rice rhizosphere in four clusters assigned to *Alcaligenes xylosoxidans*, *Ochrobactrum anthropi*, *Pseudomonas aeruginosa* and *Serratia marcescens*. All these four are known potential human pathogens. Likewise, *Ralstonia taiwanensis* and *R. eutropha* have been isolated from root nodules of *Mimosa*. Several species of *Ralstonia* are opportunistic pathogens of humans and plants. The isolation of pathogenic bacteria from root nodules of legumes has generated considerable concern and calls for due precautions in the contest of field application of rhizobial inoculants; it has underscored the necessity of using only well-identified and well characterized non-pathogenic strains.

Fred *et al.* (1932) recognized six species in the genus *Rhizobium*, *viz*. *R. japonicum* (hosts: *Lathyrus*, *Lens*, *Pisum* and *Vicia*), *R. lupine* (*Lupinus*), *R. meliloti* (*Melilotus*, *Medicago*, *Trigonella*), *R. phaseoli* (*Phaseolus*) and *R. trifolii* (*Trifolium*), based on their host range. Using growth rate as the criterion, bacteria were grouped as fast growers and slow growers, but both categories being placed in the genus *Rhizobium* till Jordan (1982) created the new genus *Bradyrhizobium*. A single species, *Bradyrhizobium japonicum*, was described for isolates of *Glycine max*. *In* general, alkali-producing slow growers were associated with tropical legumes whereas acid-producing fast growers were associated with temperate legumes. Some exceptions to this are known, *e.g.*, the temperate legumes *Corollina* and *Lupinus* are infected by slow-growing rhizobia whereas tropical *Acacia*, *Leucaena* and *Sesbania* can be infected by fast-growing rhizobia, also both fast and slow-growing rhizobia have been isolated from the same legume *Glycine max*, or even from the same plant *Acacia*, *Lupin* and *Prosopis*. Thus, classifying rhizobia on the basis of host range and physiological properties does not reflect the true phylogeny of the group (Sahgal and Johri, 2003).

Sequence comparisons of 16SrRNA genes and genetic fingerprinting methods based on the use of PCR have changed the above picture. Sequencing of full 16SrRNA genes and use of different PCR tools yields reliable information about the phylogeny and relationships of rhizobial isolates.

Yanagi and Yamasoto (1993) divided rhizobia into three genera: *Azorhizobium*, *Bradyrhizobium* and *Rhizobium* based on 16SrRNA sequence alignment. Later, *Rhizobium* was further divided into two new genera, *Mesorhizobium* and *Sinorhizobium*. However, the distinctions between *Rhizobium* and *Sinorhizobium* are not very clearcut.

In some cases characterization of rhizobial populations nodulating *Leucaena leucocephala*, *Mimosa affinis* and *Sesbania herbacea* through PCR-RFLP of 16SrRNA genes revealed that isolates from a single legume species can belong to different species of the same genus or even different genera. A population of 150 isolates of *L. leucocephala* was clustered into 18rDNA types corresponding to *Mesorhizobium*, *Rhizobium* and *Sinorhizobium* (see Sahgal and Johri, 2003).

The current approach to taxonomy of root nodule bacteria is that DNA-DNA homology exceeding about 70 per cent, coupled to distinctive phenotypic characters, can suffice to define various species. Table 5.5 lists some rhizobial species and their generic assignment as recognized on the basis of the polyphasic approach.

On the basis of various modern molecular biology and biochemical criteria, Sahgal and Johri (2003) classified rhizobia to fall in four different branches, as shown in Figure 5.2.

Carbon Fixation Pathways in Bacteria and Archaea

Autotrophs are those organisms that can grow by using CO_2 as their only source of carbon. Autotrophs include plants, algae, cyanobacteria, purple and green bacteria, and some other bacteria and archaea that do not derive energy from light. Autotrophs not only produce the biomass on which all other organisms thrive but also have a crucial role in Earth's nitrogen and sulfur cycles. Autotrophic organisms fix carbon by five mechanisms (see Berg *et al.*, 2007).

Table 5.5: Some Genera and spp. of *Rhizobia* (After Sahgal and Johri, 2003)

Genus	*Species*	*Host*
Azorhizobium	*Az. caulinodans*	*Sesbania rostrata*
Bradyrhizobium	*B. elkanii*	*Glycine max*
	B. japonicum	*G. max*
Mesorhizobium	*M. chacoense*	*Prosopis alba*
	M. ciceri, M. mediterraneum	*Cicer arietinum*
	M. loti	*Loti*
	M. plurifarium	*Acacia, Leucaena*
Rhizobium	*R. etli*	*Phaseolus vulgaris*
	R. galegae	*Galega*
	R. gallicum, R. giardinii	*P. vulgaris*
	R. huautlense	*Sesbania herbacea*
	R. leguminosarum	*Trifolium, Vicia*
	R. mongolense	*Medicago ruthenica*
	R. phaseoli	*P. vulgaris*
	R. tropici	*Leucaena, P. vulgaris*
	R. trifolii	*Trifolium*
Sinorhizobium	*S. arboris*	*Acacia senegal, Prosopis chilensis*
	S. fredii	*G. max*
	S. kostiense	*A. senegal, P. chilensis*
	S. medicae	*Medicago* spp.
	S. meliloti	*Medicago sativa*
	S. saheli	*Sesbania*
	S. terangae	*Acacia, Sesbania*

The first autotrophic CO_2 fixing pathway was elucidated by Calvin half a century ago. In this, CO_2 reacts with a five-carbon sugar, yielding two carboxylic acids, from which the sugar is regenerated in a cyclic process. The Calvin cycle occurs in plants, algae, and cyanobacteria (all these groups perform oxygenic photosynthesis), also in autotrophic proteobacteria, some of which do not tolerate oxygen (anaerobes). The key enzyme of the cycle–RuBisCO is also present in several other bacteria and some archaea, but these either lack another enzyme crucial for the cycle and/or do not show autotrophic growth (Thauer, 2007).

The green sulfur bacterium *Chlorobium* uses a second cycle (reductive citric acid cycle) for autotrophic CO_2 fixation (Evans *et al.*, 1966). This cycle also takes places in

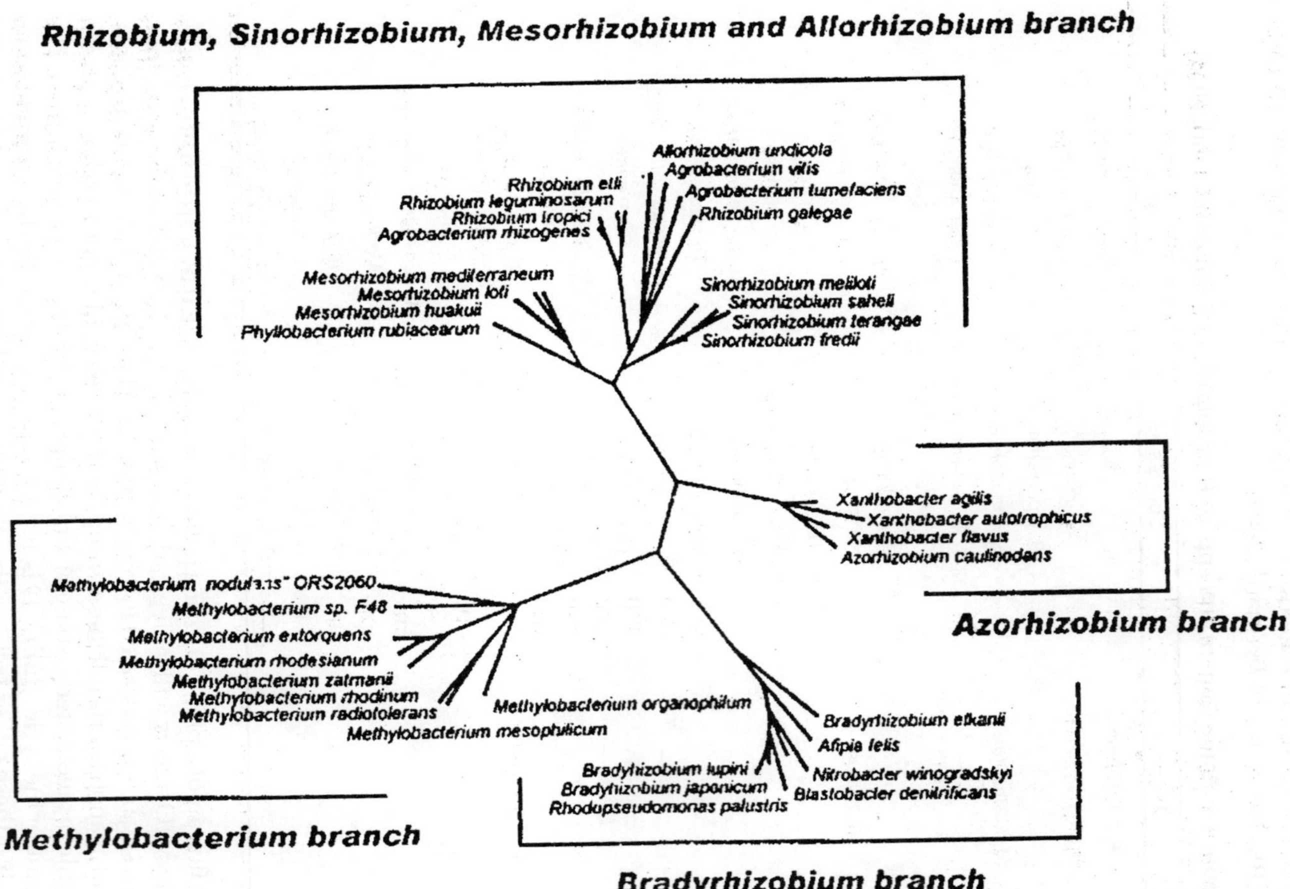

Figure 5.2: Unrooted Phylogenetic Tree (Neighbour-joining) Showing Different Rhizobial Branches in the α-subdivision of Proteobactera (After Sahgal and Johri, 2003).

several other classes of bacteria and archaea. It involves oxygen-sensitive enzymes, so it is only active in anaerobes or in microaerophiles. A third pathway was discovered in certain Gram-positive bacteria and methane-forming archaea in which one CO_2 molecule is reduced to CO and one to methanol (bound to a carrier); thereafter, acetyl-coenzyme A (CoA) is synthesized from CO and methanol (Ragsdale, 1991). This reductive acetyl-CoA route also operates in some other bacteria and archaea and involves one of the most oxygen-sensitive enzymes known–it is only found in strict anaerobes.

The fourth pathway was discovered in the green nonsulfur bacterium *Chloroflexus*. CO_2 fixation starts with the carboxylation of acetyl-CoA; the CO_2 acceptor is then regenerated in a cyclic process, with 3-hydroxypropionate and malyl-CoA as typical intermediates (Herter *et al.*, 2002). The cycle appears to be restricted to *Chloroflexus* species. None of the enzymes involved in this cycle is sensitive toward oxygen.

The assimilation of CO_2 into organic material is quantitatively a most important biosynthetic process. What Berg *et al.* found is that autotrophic *Metallosphaera sedula* fixes CO_2 with acetyl-coenzyme A (acetyl-CoA/propionyl-CoA carboxylase as the key carboxylating enzyme. In this system, one acetyl-CoA and two bicarbonate molecules are reductively converted via 3-hydroxypropionate to succinyl-CoA. This intermediate is then reduced to 4-hydroxybutyrate and converted into two acetyl-CoA molecules via 4-hydroxybutyryl-CoA dehydratase. In fact, the key genes of this pathway are present not only in *Metallosphaera* but also in species of *Sulfolobus*, *Archaeoglobus*, and *Cenarchaeum*. The fact that the Global Ocean Sampling database contains half as many 4-hydroxybutyryl-CoA dehydratase sequences as compared with those found for another key photosynthetic CO_2-fixing enzyme, ribulose-1,5-bisphosphate carboxylase-oxygenase, points to the importance of this enzyme in global carbon cycling (Berg *et al.*, 2007).

The new autotrophic CO_2 fixation pathway has some of the same intermediates as the 3-hydroxypropionate/malyl-CoA cycle (*i.e.*, the 4th pathway). Succinyl-CoA is also formed from acetate and 2 CO_2 molcules via 3-hydroxypropionate. The difference is that the enzymes involved are not phylogenetically related, indicating convergent evolution. From succinyl-CoA on, the two pathways are different.

The novel cycle operates in *Metallosphaera* growing on H_2 and O_2 as the energy source. The genes for this cycle are also found in other archaea. All these organisms are either microaerophiles or strict anaeobes. The cycle involves 4-hydroxybutyryl-CoA dehydratase, a radical enzyme sensitive to oxygen (Martin *et al.*, 2004).

Of the different pathways, only the Calvin cycle made it into the aerobic world of plants–possibly because it does not use enzymes that are inactivated by O_2 or by light (Thauer, 2007).

Bacterial Source Tracking

This is a new field of microbiology that can prove useful for cleaning up polluted waterways. In this, waterborne bacteria and pathogens are tracked back to their origins.

Since long, public health officials have measured water quality by monitoring levels of faecal coliforms; these bacteria live in animal guts and survive in the environment when expelled along with faeces. Although the coliforms may not cause disease, they can be accompanied by certain pathogens, including microbes that cause hepatitis, cholera, and gastrointestinal illnesses. They are therefore used as a marker to routinely close certain waterways, wells, and shellfish beds when coliform counts rise above certain levels (Malakoff, 2002).

Unfortunately, the approach has not kept waterways clean. Modern sewage treatment has no doubt eradicated the worst problems, but still there are millions of km of streams and coastal waters having bacterial loads that exceed health standards. Microbiologists have been called upon to suggest suitable methods to pinpoint the sources of bacterial contamination. Researchers have proposed ideas ranging from DNA fingerprinting of different species of bacteria to techniques that identify viruses unique to specific animals.

The most prominent techniques are "library-dependent" methods, which require researchers to match a bacterium found in a waterway to one included in a previously created library of bacteria from known sources. This can be done *e.g.*, by means of a technique called antibiotic resistance analysis (ARA) which assumes that the strains of bacteria in people, farm animals, and wildlife respond differently to antibiotics. To catalog these signatures, the researcher first combs a small watershed for excrement from every potential source such as people, cows, or other animals. This is followed by culturing *Enterococcus* bacteria from the samples, exposing the microbes to various antibiotics, and recording the results. This can reveal signatures in *Enterococcus* taken from the stream signatures that may match catalogued signatures produced by cattle-borne bacteria.

Several genetic approaches also depend on such libraries, which are used by researchers who wish to match the unique DNA or RNA profiles of bacteria from waterways with those from bacteria associated with different sources.

Some scientists feel, however, that the ARA method is too crude to reliably segregate a group of *Escherichia coli* bacteria–the most common faecal bacteria–into useful groups. Certainly the present version of ARA may need some improvements.

One pressing question facing the researchers is whether a library constructed for one study can be used for others. Waterborne bacteria tend to vary widely by place, season, and time of day, and even diet can shift the dominant strains found within an individual. In the face of this variability, some BST researchers fear that they will have to construct new libraries for every new watershed, which will push up expenses. One cheaper option may be to use, *e.g.*, certain *Bacteroides* bacteria, which have host-specific genetic markers that vary little from place to place. Other techniques seek to detect widespread, species-specific antibodies that adhere to shed bacteria (Malakoff, 2002).

For fully developing and testing such methods, however, researchers will need to adopt standard methods and share data. It is likely that more than one method will be needed to track down solutions.

Chapter 6
Microbes, Infectious Disease and Antibiotics

Introduction

For every cell in our body, we are supporting about 10 bacterial cells that produce vitamins, trigger hormones, and may even influence how obese we are (Salyers, 2005). By taking penicillin or other antibiotics we affect these microbes. Microbes that we once thought were eliminated by antibiotics can still thrive in the body. Serious issues arise about the behaviour of microbes in the human body and how they should be treated.

One such bacterium (*Chlamydia*) is pathogenic and can persist in human cells even after treatment with penicillin. This microbe cannot be grown outside the cells. Some quiescent varieties of *Chlamydia* seem to have a role in chronic ailments not traditionally thought to be related to this infectious agent. In the 1990s, two types of *Chlamydia–C. trachomatis* and *C. pneumonia* were found in the joint tissue of patients with inflammatory arthritis, and in 1996, researchers fished *C. pneumonia* out of the brain cells of Alzheimer's victims. Since then, the genetic fingerprints of *C. pneumonia*, and of several kinds of common mouth bacteria, in the arterial plaque of heart attack patients have been worked out. Hidden infections are now known to be responsible for several stubbornly elusive ills like chronic fatigue syndrome, Gulf War syndrome, multiple sclerosis, lupus, Parkinson's disease, and some kinds of cancer (Sachs, 2005).

To deal with these killers, some physicians turn to lengthy or lifelong courses of antibiotics. But others find that bacteria we think are bad for us may ward off other diseases and keep us healthy. Using antibiotics to disturb this complicated and little-understood population can potentially and irrevocably change the microbial ecology in an individual and speed up the spread of drug-resistant genes to the general public.

The enigma regarding the role of bacteria is two-faced. Louis Pasteur was the first to show that bacteria can cause disease but he also felt that bacteria normally present in the body are essential to life. It is now clear that trillions of bacteria line up not only our intestines but also our skin and much of our respiratory and urinary tracts. The vast majority of them are harmless, if not beneficial. Bacteria are everywhere, in abundance–they outnumber other cells in our body by 10 to one. The genetic fingerprints of hundreds of new bacterial species have recently been reported in the mouths, stomachs, and intestines of healthy volunteers. The puzzling fact is that whereas researchers are turning up stealth infections everywhere, these infections cause problems only in *some* people *sometimes*. This goes against Koch's postulates–Robert Koch had postulated long ago that any microbe that causes disease should be found in every case of the disease and always cause the disease when it is introduced into a new host. This view prevailed until the middle of the 20th century, but is not universally accepted today and it is emerging that Koch's postulates simply do not apply. With new technologies such as PCR, researchers are reporting stealth infections everywhere, yet they cause problems only in *some* people *sometimes*, often many years after the infection.

Notwithstanding these mysteries, however, antibiotics continue to be prescribed freely and frequently. Many rheumatologists even prescribe long-term or lifelong course of antibiotics for inflammatory arthritis, even though it is not known whether the antibiotics actually clear away bacteria or reduce inflammatory arthritis through some other unknown mechanism.

Antibiotics are also frequently used to treat heart disease. Two large-scale studies in April 2005 reported that use of antibiotics does not reduce the incidence of heart attacks or eliminate *C. pneumonia*, that is sometime associated with the formation and accumulation of plaque in arteries.

Meanwhile, alarms are being voiced. Unfortunately, many physicians still consider antibiotics as benign. We are just beginning to learn how our normal microflora does such an excellent job of preventing our colonization by disease-causing microbes. And we are just beginning to understand the medical consequences of disturbing that with antibiotics.

Hospital patients get the broadest spectrum, most powerful antibiotics but they also happen to be in an environment where they are exposed to the nastiest, most drug-resistant pathogens. Powerful antibiotics can be very dangerous in this setting–they kill off benign bacteria that compete with drug-resistant colonizers which can then proliferate. The result is that hospital-acquired infections are a major cause of death in intensive care units.

Bacterial Flora of the Human Body

More than 100 trillion bacteria inhabit our body. They digest our food, make vitamins, and protect us from pathogens (Liver, gall bladder, brain thymus, blood and lower lungs are all normally sterile). A brief outline is given below.

1. *Eyes*: Natural antibiotics in tears kill most organisms, but the eyes are home to a few resistant forms–mostly harmless strains of *Staphylococcus*, such as *S. epidermis*, and *Streptococcus*–that keep more virulent strains, such as *Chlamydia trachomatis*, at bay.
2. *Ears*: Waxy secretions contain antibacterial components, yet over 200 bacterial species normally live in the outer ear.
3. *Nose*: Many people carry a virulent strain of *Staphylococcus aureus*. When a cut lets it into the blood stream, it can become serious and even fatal. All people contain less harmful strains of *Staphylococcus*, *Neisseria*, and *Corynebacterium* that provide a buffer against colonization by such pathogens as *Streptococcus pneumonia*.
4. *Mouth*: About 500 microbial species are found in the mouth out of which only about 150 have been cultured in laboratories. On our teeth, *Actinomyces viscosus* secrete plaque, which traps volatile sulfur producers and acid-leaking *Streptococcus nutans*, the cause of bad breath and cavities.
5. *Skin*: Relatively low moisture, a low pH, and high salinity make most areas of the skin inhospitable to all except a few species.
6. *Armpits*: The majority of skin bacteria thrive in the moist armpits and groin, where urea protein, salts, and lactic acid leak out of sweat ducts and collect around hair follicles.
7. *Stomach*: Once supposed to be too acidic to contain any life, it actually harbors the bacterium *Helicobacter pylori*, which can cause ulcers.
8. *Small Intestine*: Bile and antimicrobial mucus keep the small intestine sparsely populated, but *Bacteroides*, streptococci, bifidobacteria, and clostridia can live. One of these, *Bacteroides thetaiotaomicron*, sends signals needed for the blood vessels of the bowels to develop properly after birth.
9. *Colon*: Our colon contains about two pounds of bacteria which make up a third of human faeces by weight. Mainly composed of the anaerobic members of the phyla Bacteroides and Firmicutes, these bacteria metabolize bile acids, break down indigestible parts of our food, and produce vitamins K and B_{12}. A strain has been identified that contributes to obesity by disrupting the appetite-regulating hormone ghrelin.
10. *Urinary Tract*: The urethra is normally sterile except for the half inch near the exit. Urinary tract infections occur when certain strains of colon-dwelling *Escherichia coli* succeed in colonizing the opening and move upward.
11. *Reproductive Tract*: Various species of *Lactobacillus* help to maintain the vagina at a slightly acidic pH ranging from 4 to 5. If these bacteria are killed, the pH rises, encouraging the overgrowth of the fungus *Candida*.
12. *Feet*: Several species of moisture-loving bacteria thrive between the toes. Some ferment acids which emit the smell of sweaty feet.

A thorough study of our bacteria and their role in antibiotic resistance has been undertaken by microbiologist Abigail Salyers at the Univ. of Illinois, Urbana-

Champaign. Her research showed that decades of antibiotic use have bred a high degree of drug resistance into our intestinal flora. The resistance remains harmless so long as the bacteria are confined to their normal habitat. But it can be deadly when those bacteria happen to infect an open wound or cause an infection after surgery (Salyers, 2005). She focused on the genus *Bacteroides* that includes the notorious *B. theta* and about a quarter of the bacteria in the human gut. Salyers observed strong increases in the prevalence of several genes and suites of genes coding for drug resistance. She particularly explored tetQ, a DNA sequence that confers resistance to tetracycline.

When faecal samples were taken in the 1970s, less than 25 percent of human-based *Bacteroides* carried tetQ. Two decades later, that rate had crossed the 85 per cent mark, even among strains isolated from healthy people who had not used antibiotics for long. Salyers also documented the spread of several *Bacteroides* genes conferring resistance to macrolides, which are widely used to treat skin, respiratory, genital, and blood infections.

When drug-resistant genes become common in the gut bacteria they are more likely to pass on their information to more dangerous organisms that only move periodically through our bodies. Even distantly related bacteria exchange genes with one another by various techniques, from direct cell-to-cell transfer, called conjugation, to transformation, in which a bacterium releases snippets of DNA that other bacteria pick up and use.

Salyers showed that *Bacteroides* may have picked up erythromycin-resistant genes from distantly related species of staphylococcus and streptococcus. Although neither microbe colonizes the intestine, they are routinely inhaled and swallowed, providing a period of up to two days in which they can commingle with intestinal flora before existing.

Salyers also discovered that antibiotics like tetracycline actually stimulate *Bacteroides* to begin swapping its resistance genes. When *Bacteroides* is exposed to other bacteria such as *E. coli*, the disinhibiting influence of antibiotics can be tracked through the step-by-step process by which the bacteria excise and transfer the tetQ gene from one species to another.

But *Bacteroides* is by no means the only intestinal resident showing such abilities. In June 2002, epidemiologists discovered that hospital-bred strains of the gut bacterium *Enterococcus* harbor a gene that makes them resistant to vancomycin. Since then, the bacterium may have even passed this gene to the much more dangerous *Staphylococcus aureus*, the commonest cause of fatal surgical and wound infections.

Microbial Virulence

Bacterial Lifestyles and Regulatory Systems

There are many bacteria which have the ability to adopt different lifestyles. When they live free, they are virulent and cause disease; when living in a surface-attached community, their virulence decreases. Goodman *et al.* (2004) identified a regulatory system in *Pseudomonas aeruginosa* that determines whether it causes disease or remains unnoticed. This bacterium can infect people whose immune system is damaged, who have sustained serious burns or an eye injury, or who suffer from cystic fibrosis. Inactivating a two-component regulatory system in *P. aeruginosa* confers in a strain a greatly decreased ability to cause disease, while increasing its ability to form surface-attached, persistent communities or biofilms (see O'Toole, 2004).

The bacterial two-component regulatory systems constantly sample the external environment and transmit this information to the bacterial interior, allowing the organism to adapt to an everchanging environment. A novel protein component of such a regulatory system has been reported by Goodman *et al.* who named this protein RetS.

A RetS-deficient strain was better than a wild-type strain at forming a biofilm on both an abiotic surface (glass) and a biotic surface (cultured hamster cells). But the RetS-deficient bacteria caused less damage to the hamster cells they colonized. *P. aeruginosa* readily forms biofilms on such abiotic surfaces as catheters, and it might also possibly produce biofilms on tissues within a host, in diseases such as otitis media (earache) and cystic fibrosis, as reported by Post *et al.* (2004). Conceivably, Goodman *et al.* probably have identified a control element which allows this bacterium to switch between a virulent, pathogenic state and a biofilm state in a mammalian host. The biofilm state, though less virulent, probably allows the bacterium to persist for longer periods (O'Toole, 2004).

Analysis and identification of various genes of P. *aeruginosa* by Goodman *et al.* indicated that in the Ret-S deficient strain, the expression of certain genes needed by *P. aeruginosa* to cause a short-term infection was greatly reduced. The expression of other genes associated with virulence and with pili formation was also reduced. It appears that RetS is normally required for the full expression of the factors required to produce an acute infection. In contrast, certain genes involved in the formation of the sugar-rich matrix that encloses a biofilm were markedly upregulated in the mutant bacteria, suggesting that RetS usually turns off the genes needed to make a biofilm (O'Toole, 2004).

Bacteria make choices all the time on the basis of the information available. These choices–*e.g.*, whether or not to form a persistent biofilm are determined by local environmental cues, and are effected through altered gene expression. *P. aeruginosa* tends to form biofilms on catheters or contact lenses when an energy source on some nutrients are readily available (Singh *et al.*, 2002). In other conditions, it chooses to remain free-living. It appears that *P. aeruginosa* also is called upon to decide when in the context of a mammalian host: does it cause a short-term infection or does it persist in a biofilm state? An acute infection is conducive to bacterial propagation, whereas in a biofilm the organism is lying low and so is less likely to be recognized and attacked by the immune system (O'Toole, 2004).

Disease and Defence

Pathogenic impacts of microorganisms on animals and plants have evolved through an arms race of attack and defence. This race reflects successive developments in virulence and disease resistance. The leap from a saprophytic mode of life on dead animal or plant tissues to the colonization of a living host has been followed by the adaptation of pathogenicity to cross evolving mechanisms of defence so as to colonize new host species. Effector proteins delivered through the type III secretion system (T3SS) are the key players in bacterial colonization. They probably played a leading role in the establishment of a parasitic lifestyle (Mota and Cornelius, 2006; Grant *et al.*, 2006). Stavrinides *et al.* (2006) discussed genetic divergence associated with the emergence of pathogenicity that led to identification of differences between the genomes of pathogens and related non-pathogens that could possibly have changed lifestyle or host range. The genomic diversity approach is retrospective and predicts past events from bioinformatics data. To develop effective methods for disease control there is need to know not only about the weapons available to pathogen and host, but also the factors involved in the generation of new genotypes and their selection within microbial populations (Arnold *et al.*, 2007).

Filamentous protein complexes called adhesive pili are found on bacterial surfaces and mediate the adhesion of pathogenic bacteria to host tissues. Pili are good models for studying ordered macromolecular assembly reactions at the bacterial cell membrane. Nishiyama *et al.* (2008) have described the complete *in vitro* reconstitution of an assembly and secretion system for adhesive pili from purified pilus proteins, using type 1 pili from uropathogenic *Escherichia coli*. The reconstitution reveals how a protein catalyst can accelerate the ordered assembly of a supramolecular protein complex.

The bacterium that causes anthrax, *Bacillus anthracis*, exists as a small spore until it gets into a mammal, but once it does enter a mammal, it produces a host-killing toxin and self-protecting capsules in response to higher bicarbonate levels than normally found in the environment. M. Perego *et al.* (2008) discussed how it recognizes this compound. Perego *et al.* searched the literature for other organisms that use bicarbonate, and came across some work on a photosynthetic cyanobacterium, *Synechococcus elongatus* which shuttles bicarbonate through its cell membrane by means of a multiprotein structure known as an ABC transporter. When *B. anthracis'* genome was scanned for genes that encode similar proteins, a section named BAS2714-12 was revealed. Deleting this region rendered the pathogen harmless to mice.

Genome sequencing of microbial pathogens has thrown light on the evolution of virulence but factors determining the gain and loss of genes and of pathogenicity islands (gene clusters), which contribute to the emergence of new disease outbreaks, are not known. Recent experiments with the bean pathogen *Pseudomonas syringae* pv. *phaseolicola* illustrate how exposure to resistance mechanisms acts as the driving force for genome reorganization. Arnold *et al.* argued that the antimicrobial conditions generated by host defences can speed up the generation of genome rearrangements which confer selective advantages on the invading microbe. Similar exposure to environmental stress outside the host may also drive the horizontal gene transfer that has led to the evolution of pathogenicity towards both animals and plants.

Primary defences in plant leaves against nonpathogenic bacteria are reflected by localized alterations to the plant cell wall. Pathogens can suppress such wall-based defences but cannot do so in resistant varieties of their host plant in view of the hypersensitive resistance reaction, HR. The HR involves programmed cell death (PCD) which, through some tissue sacrifice, aggravates antimicrobial conditions and restricts colonization of the host at the initial site of infection (Pitman *et al.*, 2005).

One major focus of recent research has been to identify enzymes that bacteria need for their survival. It is hoped that molecules that inhibit these essential enzymes would be found in chemical libraries and turned into drugs. An important pre-requisite for such search, however, is to determine what effect losing the enzyme might have on the bacterium. After the bacterium's genome has been sequenced, the genes encoding certain enzymes can then be disabled to see if the bacterium can survive without them.

One strong challenge being faced in this effort is the formidable barrier to entry presented by the bacterial cell wall--even if a small molecule that inhibits a crucial bacterial enzyme is discovered, it proves useful only if it can successfully enter the cell. Rather than seeking weak spots in pathogens, a wiser alternative would be to identify novel microbes that produce new antibiotics. Organisms growing in unusual habitats are more likely to produce antibiotics that have not been discovered before.

Researchers are exploring new methods of killing or disabling pathogenic microbes. Three such potentially promising strategies are being considered. One is based on *defensins*–pore-forming peptide tubes exert lethal effect by puncturing a bacterial cell membrane. Defensins are small, naturally occurring proteins and perform a similar function in vertebrates to defend against microbes. Attempts are underway to develop synthetic protein fragments (peptides) that would self-assemble into tubes within bacterial membranes. A second strategy exploits *narrow targeting*. As is well known, bacteriophages infect a bacterium, typically preferring only one host. Phages have long been studied for possible use against pathogenic bacteria, but they also exemplify the principle underlying new "narrow spectrum" drugs, which target only a single pathogen, and do not harm human cells or friendly bacteria. The third method aims to *subdue without killing*, *i.e.*, sparing the pathogen but affecting its ability to cause illness as a method to treat disease without promoting antibiotic resistance. *E. coli* has been genetically engineered to imitate cells in the human gut. When the harmless *E. coli* cells are consumed, they soak up deadly Shiga toxin produced by another microbe.

Yet another approach to mining new ecological conditions for useful drugs is to examine microbes that participate in mutualisms–interspecies interactions in which both parties benefit. A good example is that of the southern pine beetle which harbours a mutualistic fungus that digests the inside of pine trees the beetle invades. The beetle protects its fungal mutualist from a second, antagonistic strain of fungus that competes with the first strain for food. It contains a second mutualistic microbe–an actinomycete–that produces a powerful and hitherto unknown antifungal agent. This molecule, called mycangimycin, kills only the antagonistic fungus but spares the mutualistic one (Walsh and Fischbach, 2009).

Injection of Effector Proteins into Host Cells by the T3SS

The T3SS is involved in the transfer of bacterial proteins into eukaryotic cells. Core proteins of the T3SS architecture that make a channel across the bacterial membranes are highly conserved between pathogens of animals and plants but notable differences occur in external T3SS appendages. Pathogens of plants use an extending pilus as an effective passage to cross the plant cell wall. Several of the proteins delivered through the T3SS suppress defence responses in animals and plants, but their enzymatic activities are not very clear (Arnold *et al.*, 2007).

Two types of disease-causing bacteria utilize a special injection system to deliver proteins into host cells (see Roy *et al.*, 2008). These proteins have regions known as 'Anks' (ankyrin repeat homology domains), which often form scaffolds that enable other proteins to interact.

Roy *et al.* (2008) reported that *Legionella pneumophila*, the bacterium that causes Legionnaires' disease, injects four Ank proteins into mammalian cells via a complex called a 'type IV secretion system'. Another microbe, *Coxiella burnetti*, which causes Q fever, injects eight such proteins.

One of the *L. pneumophila* proteins, AnkX, prevents host vesicles (bags of membrane-bound fluid that contain the bacteria) from moving towards the lysosome, where the bacteria would be destroyed. This may contribute to the bacterium's virulence.

Antimicrobial Defence Systems

Antimicrobial secondary metabolites of low molecular weight such as phytoalexins and phytoanticipins play an important role in challenged plant tissues (Lewis and Ausubel, 2006). One fairly common response of plants attacked by pathogenic microbes involves the secretion of agglutinating glycoproteins (Huffaker *et al.*, 2006). Plants and animals all use free radical oxygen species, antimicrobial peptides (defensins) and microbial cell wall-degrading enzymes (lysozyme, chitinase) to defend themselves. Plants lack antibody-based immunity (Radtke *et al.*, 2006). Some common antimicrobial denfences in plants and animals are:

For Plants

Phytoalexin accumulation (involves synthesis of different types of antimicrobial secondary metabolites such as the isoflavonoid phaseollin from bean); localized formation of active oxygen species such as H_2O_2; produced by NADPH oxidase or peroxidase activity at infection sites; agglutination by glycoproteins (binds bacteria to cell walls in the intercellular spaces of plant tissues); secretion of chitinase and glucanase which damage microbial cell walls; production of defensin peptides *e.g.*, thionines; and release of phytoanticipins from damaged plant cells (a wide range of antimicrobial compounds such as saponins and alkaloids present in plant tissues are released upon infection (Arnold *et al.*, 2007).

For Animals

Antibacterial peptides: *e.g.*, defensins, hepcidin, cecropins, and diptericins; lysozyme: degrades cell walls containing peptidoglycan; humoral 'cascade' reactions (complement pathways involving assembly of membrane attack complex that kill bacteria in vertebrates, and the phenol oxidase-mediated melanization cascade in invertebrates; adaptive immunity: recognizing invading bacteria by immunoglobulin family antibody molecules mediates many immune responses such as opsonization for phagocytosis, the complement pathway and antibacterial peptide release; phagocytosis: several immune cell types take up bacteria which have been recognized by pattern recognition receptors such as lg-family or innate immunity lectins. Upon being phagocytosed, the bacteria are killed by active oxygen species and killer proteases; iron limitation: soluble Fe^{2+}, which is essential for the formation of many key metabolic enzymes, is sequestered by host molecules such as transferrins; and fever: the pyrogenic response in homeothermic animals and behavioural fever in insects increases the host body temperature, exerting thermal stress on invading pathogens (Adamo, 2006; Arnold *et al.*, 2007).

To succeed as a pathogen, a microbe needs to avoid or suppress basal defences which are activated in response to microbial components called elicitors, or pathogen- (or microbe-) associated molecular patterns (PAMPs or MAMPs) (Asubel, 2005). Many effector proteins suppress PAMP-mediated defences in both animals and plants. However, plant pathogens produce several effectors that can suppress the same aspect of basal defences (Grant *et al.*, 2006).

Some other PAMPs that activate defence responses are outlined below, with examples of molecules found in many bacteria that are known to activate defences in animals and plants.

- ☆ *Lipopolysaccharide (LPS)*: the commonest PAMP recognized by animals through Toll-like receptor 4 (TLR4);
- ☆ *Flagellin* is a structural component of the bacterial flagellum. It is recognized by TLR5 and the flagellin-sensing 2 receptor kinase, respectively. *Elongation factor-Tu (EF-Tu)* is a most conserved and abundant protein found in the bacterial cytosol and the component Elf peptide is recognized in plants through the elongation factor receptor (EFR);
- ☆ *Cold-shock protein* has been identified from preparations of *Micrococcus lysodeikticus*. It caused rapid alkalinization of the extracellular medium of cultured tobacco cells. Elicitor activity was found only in plants of the Solanaceae family.
- ☆ *Peptidoglycan (PGN)* is produced as a thick layer in the cell wall of Gram-positive bacteria and as a thin layer in Gram-negative bacteria. It is recognized by animals and insects and stimulates immunity. Muramyl components of peptidoglycan can be recognized intracellularly by certain nucleotide oligomerization domain (NOD) proteins.
- ☆ *N-acyl homoserine lactones (AHLs)* are formed by various different bacteria as signaling molecules believed to control gene expression in a cell density-dependent manner. They seem to cause TLR- and NOD-independent physiological effects in a wide range of animal cell types (Kravchenko *et al.*, 2006).

Valid comparisons between animal and plant pathogenesis are difficult in view of differences between defence responses in the two types of host. Apparent increases in virulence in animals might not be related to increased bacterial colonization during the early phase of infection as observed in plant-pathogen interactions, but to increased response of challenged tissues to bacterial toxins (see Maurelli *et al.*, 1998). During disease progression from acute to chronic gastritis, *Helicobacter pylori* suffers various genomic rearrangements, but there is no clear pattern of defence-driven change (Oh *et al.*, 2006).

Whole genome sequences of several closely related animal pathogens with different host ranges has revealed large-scale gene losses that accompanied changes in virulence and expansion of host range. Extensive genomic degradation took place in the evolution of the specific human pathogen *Bordetella pertussis* from *B. bronchiseptica*; the latter has a much larger genome and broader host range (Bentley and Parkhill, 2004). Conversely, it appears that the more general mammalian pathogen, *Mycobacterium bovis*, has evolved from a specifically human pathogen, *M. tuberculosis*, again by repeated gene deletions (Hewinson *et al.*, 2006).

The emergence of novel plant pathogens from nonpathogenic relatives is not known although the periodic emergence of more severe infections, such as in Dutch elm disease or horse chestnut canker, suggests that it is a continuing but ill-defined happening. According to Toth *et al.* (2006), the genomic history of plant pathogens *e.g.* enteric *Erwinia carotovora* points to the importance of horizontal transfer in the emergence of a current pathogenic genome. Soil borne plant pathogenic bacteria such as *E. carotovora* have considerable opportunity for interactions with other microbes.

Advantages of a Stressful Life Existence

Adverse conditions in the external environment seem to promote gene exchange within diverse microbial communities (Sorensen *et al.*, 2005). Once pathogenicity to one host has been established, exposure to eukaryotic defences can trigger the deletions and rearrangements that maintain virulence and broaden host range. There is a strong need to define the precise nature of the stress to which microbes are exposed–it can vary from exposure to a wide range of physiochemical factors, such as low temperature and nutrient deficiency outside the host, to attack by antimicrobial compounds within the host (Arnold *et al.*, 2007). Stress seems to be beneficial to microbes if they are to acquire the ability to parasitize living animal and plant tissue, and if they are to reap the advantages of parasitism, feeding without competition on their specific host.

A Virulent Bacterial Pathogen: Response to Light

The ubiquitous environmental signal of light regulates many developmental and behavioral responses in plant, fungal, bacterial, and animal cells. Photosensitive proteins are widespread and common in the bacterial kingdom, even though their cellular functions are often obscure. Swartz *et al.* (2007) identified a functional role for a new type of light sensor in bacteria–light, oxygen, or voltage (LOV) histidine kinase. In the serious pathogen *Brucella abortus*, light enhances the activity of this kinase, and that in turn significantly increases virulence of the bacterium. Related LOV histidine kinases are conserved across various bacterial taxa, suggesting that this virulence pathway may well be a new photosensory pathway that regulates bacterial physiology (Kennis and Crosson, 2007).

LOV protein domains were initially described as the light-absorbing region in plant serine-threonine kinases called phototropins (Taylor and Zhulin, 1999; Christie and Briggs, 2005). The LOV domain is a compact protein module of ~110 amino acids. It is rendered sensitive to blue light through a noncovalently bound flavin cofactor. When it absorbs a blue photon, the flavin cofactor forms a covalent bond with a conserved cysteine residue in the LOV domain. This triggers a conformational change within the domain. The change communicates photon detection to regions outside the domain's core. Such a unique signaling mechanism is involved in several well-known photomorphogenic and circadian responses in plants and fungi (Dunlap and Loros, 2006).

Flavin-binding LOV domains are widely distributed in the plant, fungal, and bacterial kingdoms. Being fused to a range of "output" domains including bacterial diguanylate cyclases and phosphodiesterases, zinc-finger transcription factors, and histidine kinases, this flavin-binding protein fold allows the light environment to control a wide range of cellular phenomena.

Surprisingly, several bacterial taxa that encode LOV domains in their genomes happen to be nonpigmented heterotrophs (Crosson *et al.*, 2003) with no known or expected response to visible light. So, what is the function of this widely distributed light-sensing domain in these species (LOV domains are encoded in 16 per cent of sequenced bacterial genomes)? One very enigmatic family of bacterial LOV proteins is the LOV histidine kinases. These kinases, along with a response regulator family of proteins, form the two-component signaling systems that constitute the predominant mode of signal transduction in bacteria (Stock *et al.*, 2000).

Swartz *et al.* (2007) showed that light increases the enzymatic (autophosphorylation) activity of a *Brucella* LOV histidine kinase. Visible light increases the virulence of *B. abortus* in a macrophage infection model: Wild-type cells of *Brucella* become up to 10 times more virulent after exposure to visible light than cells never exposed to light. Deletion of the gene encoding LOV kinase from *Brucella* abolished this light-mediated virulence response and decreased bacterial multiplication in light-exposed macrophage cells to the level observed in the dark. These results suggest a clear connection between *B. abortus* LOV histidine kinase and the light-induced increase in virulence. Furthermore, mutagenizing the conserved cysteine in the LOV domain abolished the light response, showing that cysteinyl-flavin bond formation in the LOV domain is needed for light-mediated virulence. It appears that upon release from an infected animal host, sunlight acts as an environmental signal that enhances the virulence of *B. abortus*, and so exposure to visible radiation in natural reservoirs outside the host could well increase the chance of infecting another host (Kennis and Crosson, 2007). Swartz *et al.* also showed that light regulates histidine kinase activity across many other bacterial taxa.

Listeria Infections

Pathogens use many ways of subverting molecular machinery of their hosts. A striking example of this has emerged from studies of *Listeria*, the bacterium that causes listeriosis.

Listeria monocytogenes infections pose a serious threat to public health. Ingestion of contaminated food can have very serious consequences for pregnant women, newborns, the elderly, and individuals with compromised immune systems (Lam and Brumell, 2008).

Typically, these bacteria are engulfed by macrophages, that are versatile agents of the immune system. They are then confined in a cellular compartment called a phagosome which, upon becoming acidified, fuses with a compartment (the lysosome) containing digestive enzymes and the invader is destroyed in the resulting phagolysosome. However, *L. monocytogenes* can avoid destruction by escaping through the phagosomal membrane into the cell cytosol, where it undergoes replication and infects neighbouring cells. The bacterial virulence factor, listeriolysin O (LLO) performs the central role in the escape process. Singh *et al.* (2008) showed that *L. monocytogenes* exploits a host factor to promote LLO activity during infection.

Listeriolysin O belongs to a family of cholesterol-dependent cytolysins which are able to oligomerize and form pores in host membranes. It seems to mediate *L. monocytogenes* escape from phagosomes in two steps. First, in response to acidification of the phagosome, LLO undergoes structural reconfigurations that make the protein stable and active (Schnupf and Portnoy, 2007). It then makes tiny pores in the phagosome and blocks acidification of this compartment. This prevents its fusion with lysosomes (Schaughnessy *et al.*, 2006, 2007) and a 'window of opportunity' is opened for the second step of bacterial escape, involving the combined actions of LLO and two phospholipase enzymes. But LLO also carries some potential to kill the host cell through unchecked pore formation in the cell's membrane. This is undesirable for the bacterium as the latter depends on the cell for its own existence. Schnupf and Portnoy (2007) reviewed the mechanism employed by *L. monocytogenes* to limit LLO activity in the cytosol.

The phagosomal environment also modulates the activity of LLO, NADPH oxidase and nitrogen oxide synthase and two phagosomal enzyme complexes that produce antimicrobial agents (reactive oxygen and nitrogen species) which inhibit escape of *L. monocytogenes* (Myers *et al.*, 2003). This points to the possibility that the redox (reduction-oxidation) environment in the phagosome might control LLO activity. A single, sulphur-containing cysteine amino-acid residue in the protein could be the target of oxidation. This possibility receives support from the fact that reducing agents are needed for optimal LLO activity *in vitro* (Portnoy *et al.*, 1992).

According to Singh *et al.*, a host protein enables the bacteria to overcome LLO oxidation and escape into the cytosol. This host protein enzyme, γ-interferon-inducible lysosomal thiol reductase (GILT), normally ensures the proper degradation of cysteine-containing peptides as part of another macrophage function–presentation of foreign antigen at the macrophage surface to stimulate antibody production. However, it seems that *L. monocytogenes* also exploits GILT to promote LLO activity in the phagosome (Lam and Brumell, 2008).

What Singh *et al.* (2008) have shown is that GILT regulates LLO activity, and they also highlighted the importance of the redox environment in bacterial escape. These bacteria require multiple levels of regulation on LLO. It is conceivable that precise control over the time and place that LLO is active is critical in allowing bacterial escape from a specific stage in the phagolysosomal pathway. There is also some potential of GILT as an attractive target for therapeutic intervention.

Biological Control of Plant Pathogens

Bacterial resistance due to widespread and indiscriminate misuse of antibiotics is a serious global issue and alternative methods are being developed that might decrease the use of antimicrobials in farms and crop fields. Bacteriophage therapy is a possible new way of controlling the growth of plant-based bacterial pathogens.

Not only has the use of antibiotics resulted in antimicrobial resistance in pathogenic bacteria that infect humans, animals and plants, but antibiotic resistance in animal and plant bacterial pathogens can also be transferred to humans via the food chain. Animals and plants can be a potential reservoir of antibiotic-resistant bacteria for humans (and *vice versa*). This warrants very careful prescription or use of antibiotics in medical, veterinary and agricultural practices. The emergence of multiple antibiotic-resistant bacteria has aroused considerable interest in the use of bacteriophages (phages) to kill pathogenic bacteria (Sulakvelidze *et al.*, 2001; Goodridge 2004).

Besides the interest in using phage-based therapy as a potential treatment of human infections, the application of phages as a biocontrol method for bacterial species of agricultural importance has received much recent attention. Biocontrol is the intentional introduction of specific organisms–usually predators or parasites–into such agricultural environments as croplands, to suppress pest populations.

Phages pass most of their life while engaged in extracellular search for host bacteria. Once a susceptible target organism is located, the phage adsorbs to the surface of the bacteria via proteins that show high affinity to bacterial cell surface receptors. Following bacterial-specific adsorption, the phage nucleic acid is transferred to the bacterial cytoplasm where it either transforms the cell into a bacteriophage 'factory' (lytic phage) leading to the production of thousands of phages and subsequent lysis of the bacterial cell, or it integrates into the bacterial chromosome as a prophage and is then replicated in keeping with the bacterial chromosome division (lysogenic phage). Eventually, the prophage excises itself from the bacterial chromosome (usually when the bacterial cell becomes stressed), leading to lytic growth of the phage. It is the life cycle of a given phage that usually defines its usefulness in a biocontrol method : lysogenic phages deliver potentially harmful genes (*e.g.* those encoding the Shiga toxins of Shigatoxinogenic *Escherichia coli*) to their target bacteria. Lysogenic phages can transform non-pathogenic bacteria into pathogens, which renders these phages unsuitable from a biocontrol point of view. So, phages that can undergo lytic growth may be the vehicles of choice for any phage-based biocontrol method (Goodridge, 2004).

Burkholderia pseudomallei–A Bacterial Time Bomb

The notorious bacterium, *Burkholderia pseudomallei*, which causes the deadly disease melioidosis, can kill most of its victims. Melioidosis used to be an obscure malady two decades ago, it now often proves fatal. Experts are intrigued by the ability of *B. pseudomallei* to alter its form in such a way that it can survive in environments as disparate as soil, distilled water, and the human body (Stone, 2007).

Melioidosis is largely confined to Southeast Asia and northern Australia. But its properties make it a deadly threat as a bioweapon. These bacteria remain hidden in the body for decades but once the time bomb explodes, its various symptoms allow melioidosis to masquerade as other ailments. Many patients are hospitalized with acute disease. Others have symptoms resembling tuberculosis or cancer. Misdiagnosis can be fatal because *B. pseudomallei* is impervious to all except a few antibiotics.

Though not as frightening as anthrax or small pox, meliodosis still is a significant terror threat. If soil becomes contaminated, *B. pseudomallei* is very difficult to get rid of.

Whereas viruses are known to be very smart, bacteria are normally not so smart. The important point about *B. pseudomallei* is that it acts like a virus in its deviousness.

The first case of melioidosis came from opium addicts in Burma in 1911. In 1947, two cases involving prisoners of war held in Siam were reported. But it was not until the Vietnam War, when American soldiers returned home with the disease, that melioidosis attracted attention in the West.

In the 1960s, *B. pseudomallei* was isolated from soils throughout Thailand and it emerged that about a third of Thai soldiers sampled had antibodies to it. For several years, many cases of melioidosis were being overlooked or misdiagnosed as TB or other ailments. Why the bacterium eluded detection was because it grows quite slowly in culture.

Since then, a highly perplexing picture of the malady has emerged. We do not know whether *B. pseudomallei* lives freely in soil, in association with rhizomes that stud roots, or in some host like an amoeba. It infects humans by accident. Conceivably, *B. pseudomallei* enters the body through cuts, abrasions, or the lungs.

Once the bacterium enters its human host, illness often sets in within a few hours. Many patients usually deteriorate very fast. But in other people, the bacterium lies low for years. The bacterium seems to lurk in asymptomatic people in such hiding places as the lymph nodes, reticular epithelial cells, the spleen, and bone marrow. Antibodies do not seem to confer protection.

In many patients, some chronic illness such as diabetes or kidney disease is the spark which ignites a latent infection or allows infection from a fresh exposure. Person-to-person transmission is rare, and relapses are common.

Researchers are working hard to unravel the genetic machinery that underlies *B. pseudomallei*'s exceptional hardiness and virulence. Its ability for morphing (*i.e.*, phenotype switching) seems to account for its versatility. Researchers have identified seven morphotypes, each with a distinct gene-expression pattern. Type 1 is most often cultivated from people and soil. If we make it switch over to another morphotype by starving it, for example, many other things can switch on or off.

Its genome was sequenced in 2004 by Matthew Holden and published in Proceedings of the National Academy of Sciences, 28 September 2004, p. 14240. The genome is a hefty 7.25 million base pairs on two chromosomes (most bacteria have one), and its chromosomes are dotted with "genomic islands" that contain genes acquired from other organisms. Much of the genome codes for functions that equip the bacterium to thrive in the environment, and that includes a well-stocked store of virulence factors.

Like *Salmonella* and *Shigella*, *B. pseudomallei* also uses a molecular syringe to invade host cells, but its virulence mechanisms are still unknown. Also perplexing are the mutational hot spots (short sequence repeats) that are present in two-thirds of its genes. Indeed, *B. pseudomallei* is a pastmaster at adapting.

A major concern today is that terrorists may exploit this resourceful bug. In fact one of the few biowarfare agents ever deployed is a cousin of *B. pseudomallei'*, *viz.*, *B. mallei*, which cause glanders, a disease that primarily affects horses but can also cause an illness in people that significantly resembles melioidosis.

Melioidosis is much more sinister than glanders, differing from most other potential bioweapons in that melioidosis is actually an important human disease and there is no vaccine. Because it takes several months on antibiotics to eradicate the bacterium, treating many patients in an emergency can not only be extremely expensive and difficult to manage but also an enormous burden.

A vaccine against this pathogen is a distant dream. Because the presence of antibodies does not appear to be protective, a vaccine must also stimulate the cellular immune response.

Another possible option is to let loose competing microbes (such as the beneficial bacteria found in some commercial biofertilizers) which can knock down *B. pseudomallei* soil concentrations to levels that cause few or no infections. (A high density is correlated with a high infection rate).

Antibiotic Resistance in Bacteria

The history of antibiotics and antibiotic resistance is quite short: antibiotics have been used for the treatment of infectious diseases and for applications in agriculture, animal husbandry and aquaculture during the last half century only. This was a highly successful period in medical history during which many of the important diseases could be mostly controlled in the industrial nations. Unfortunately, this situation is now threatened by the increasing incidence of antibiotic resistance. The microbes that were once sensitive to antibiotics are becoming more and more difficult to treat.

The treatment of infectious disease can be greatly compromised by the development of antibiotic-resistant strains of microbial pathogens. A variety of biochemical processes are involved that may keep antibiotics out of the cell, alter the target of the drug, or inactivate the antibiotic. Studies have shown that resistance determinants arise by either of two genetic mechanisms–mutation and acquisition. Antibiotic resistance genes can be spread among bacterial populations by several processes, but principally by conjugation. Thus the overall problem of antibiotic resistance is one of genetic ecology. A better understanding of the contributing parameters is necessary to devise rational approaches to reduce the development and spread of antibiotic resistance and thereby avoid a critical situation in therapy–a return to the pre-antibiotic era (Mazel and Davies, 1999).

There is considerable evidence to suggest that prior to the introduction and use of antibiotics, antibiotic-resistant microbes were absent from human or animal flora; this comes from studies of pre-antibiotic bacterial populations and also from analysis of the development of bacterial resistance to antibiotics such as tetracycline following their introduction into the environment. In all cases studied, a direct correlation over time between antibiotic use and the increase in proportion of resistant to non-resistant strains can be recorded. The dramatic appearance of antibiotic resistance in pathogenic *Shigella* spp. during epidemics of intestinal infections in Japan in the 1950s is yet another example of the rapid response of bacteria to the threat of antibiotic use (Mitsuhashi, 1993).

Resistance Gene Capture

One important aspect of the development of antibiotic resistance in bacteria concerns the tandem assembling of resistance genes within single, mobile genetic elements to generate multi-drug-resistance clusters. Multi-drug resistance can include the following scenarios (see Mazel and Davies, 1999). A single plasmid or transposon encodes resistance to several different antibiotics; *e.g.*, clusters. A single gene encodes a biochemical mechanism that engenders resistance to a class of related antibiotics. A single resistance gene encodes resistance to a group of structurally unrelated antibiotics. A non-antibiotic, such as a disinfectant or organic salt (*e.g.* mercury), selects for the establishment of genetically linked antibiotic resistance genes. Mutations occur in multiple, independent target genes (as in *M. tuberculosis*).

The commonest mechanism by which tandem arrays of antibiotic resistance genes are assembled in the Enterobacteriaceae is gene capture by integrons: many resistance determinants are carried on such elements. All known integrons are composed of three essential elements: a gene encoding an integrase, a primary recombination site, and a strong promoter. The integron-associated integrases belong to the site-specific recombinase family and can recombine discrete units of DNA, now known as gene cassettes, providing them with a promoter for their expression. Most of the resistance cassettes contain only a single resistance gene associated with a specific recombination sequence known as the 59-base-pair (bp) elements or *attC*.

Evolution of Pathogenesis in Bacteria

Bacterial pathogenesis and antibiotic resistance appear to be unrelated, yet there is a common thread which links them–their evolution usually responds to a selective pressure. In the former case, it is the immune system of the host, comprising the specific humoral or cellular responses and all the non-specific mechanisms contributing to removing the bacteria from host tissues or cells. In the case of antibiotic resistance, the selection pressure is exerted by the antibacterial drugs either administered by physicians or found naturally in foodstuffs and the environment. Because they can be essential to bacterial survival, genes that encode pathogenic factors or antibiotic resistance are easily spread among bacteria by such processes as conjugation, transformation, transduction or transposition. Selection pressure coupled with gene transfer leads to the emergence of micro-organisms with increased pathogenicity and/or antibiotic resistance.

The evolution of bacteria is associated with continuous generation of novel genetic variants. The major driving forces in this process are point mutations, genetic rearrangements, and horizontal gene transfer. A large number of human and animal bacterial pathogens have evolved the capacity to produce virulence factors that are directly involved in infection and disease. Further, many bacteria express resistance traits against antibiotics. Both virulence factors and resistance determinants are subject to intrastrain genetic and phenotypic variation. They are often encoded on unstable DNA regions. Thus, they can be readily transferred to bacteria of the same species or even to non-related prokaryotes.

Whereas microevolution leads to the emergence, in the short term, of new pathogenic or resistant variants, macroevolution over long periods of time, leads to the stable fixation in the pathogen genome of sequences encoding improved or new adaptive properties (Piffaretti and Frey, 1999).

Different classes of repetitive DNA sequences are found in bacterial genomes and have a role in microbial pathogenesis and evolution. Besides providing a modulatry mechanism for the expression of genes, short-sequence repeat (SSR) variability can be used to study the evolution of genetic diversity among bacteria.

Strains of *Escherichia coli* that produce shiga toxin exemplify the evolutionary mechanisms driven by genetic transfers–the DNA sequences encoding the major virulence factors of these micro-organisms may probably be the product of horizontal genetic transfers and recombination events involving alien DNA.

The problem of antibiotic resistance is of growing concern not only to health workers but also to the general public. Indeed, we may be approaching a critical scenario where patients are infected with multiresistant pathogens that do not respond to any of the antibiotics on the market.

There is a strong ecological and global dimension of drug resistance exemplified by the role of farm animals and food as reservoirs for resistance genes. The growing application of antibiotics in agriculture is a major factor responsible for the spread of drug resistance among food-borne pathogens. Methicillin-resistant strains of *Staphylococcus aureus* are causing concern in many hospital environments: these strains are resistant to all b-lactam antibiotics and, in addition, they often encode many other resistance determinants.

Multidrug efflux pumps represent a powerful and adaptive bacterial response to the presence of antibiotics in the environment. These systems were probably originally intended as secretion machineries for cellular products and as defence mechanisms against harmful substances–they evolved to develop the capacity for pumping antibiotics out of the bacterial cytoplasm (Piffaretti and Frey, 1999). These systems often show broad substrate specificity.

Even though no new family of antibiotics has been introduced onto the market in the last three decades, recent technologies associated with genomic sequencing of pathogens have the potential to facilitate identification of additional bacterial targets which might lead to the development of new antibacterial drugs.

Our current vision of the bacterial genome is that of a mosaic genetic structure composed of different DNA sequences originating not only from closely related micro-organisms, but also from distant taxa. Genes are moving constantly in the global environment and the importance and incidence of particular useful encoding sequences may be greatly enhanced by strong selection pressures, contributing to the generation of highly pathogenic and resistant strains. These facts must be considered when deciding on measures for the control of the emergence of such bugs (Piffaretti and Frey, 1999).

Pathogenicity Islands

Vibrio cholerae has two chromosomes whose base sequences are now known. There is a total of 3885 open reading frames (ORFs, lengths of DNA that encode proteins): 2770 on the larger chromosome 1 (2.9 million base pairs) and 1115 on the smaller chromosome 2 (1.1 million bp). Slightly more than 50 per cent of these ORFs encode proteins homologous to proteins of known function; the remainder encode proteins without ascribed functions or that are not homologous to any known protein (DiRita, 2000).

Cholera is a scourge throughout much of the world with seven global epidemics (pandemics) recorded since 1817. Between epidemics, *V. cholerae* lives in aquatic environment, often in association with marine invertebrates. Although most isolates of *V. cholerae* do not cause disease, some carry genes that have enabled the microbe to adapt to humans and to produce virulence factors. Upon infection of humans, these strains colonize the gut mucosa and produce cholera toxin (CT), which stimulates secretion of water and electrolytes by gut epithelial cells leading to severe diarrhoea that is usually fatal.

In many bacteria there are found segments of DNA called pathogenicity islands. These are enriched for genes encoding the proteins that contribute to virulence. These proteins include secreted toxins, factors and signalling devices that help the bacterium to colonize its host, and even some special types of molecular apparatus that inject bacterial products into host cells. But how the bacteria acquired these pathogenicity islands during evolution? Karaolis *et al.*, (1999) describe a fascinating interplay between two pathogenicity islands–the first allowed the bacterium *Vibrio cholerae* to be infected with the second, which is a bacteriophage that carries the cholera-toxin genes.

Over a century ago, *V. cholerae* was identified as the bacterial species that causes the deadly, infectious disease cholera. More recently, the potent cholera toxin was found to be encoded by a filamentous bacteriophage called cholera-toxin phage (CTXΦ), which exists within *V. cholerae*. The CTXΦ contains single-stranded DNA, and it uses the bacterium's main colonization pilus–an appendage used to infect the host's intestine–as its receptor. This pilus, termed toxin co-regulated pilus (TCP), is a fibre of polymerized pilin protein termed TcpA. It promotes colonization by enhancing interactions between bacterial cells and, perhaps, directly with the host (Taylor, 1999).

The *V. cholerae* pathogenicity island phage (VPIΦ) interacts with a recipient *V. cholerae* cell, resulting in infection and insertion of VPIΦ DNA into the bacterial chromosome. This allows production of the TCP, which, in turn, allows infection by the CTXΦ. The bacterium now has a full complement of virulence genes. Their expression is controlled by complex interactions between regulators encoded within the ancestral chromosome and those encoded on the newly acquired genetic elements. For example, the *toxR* gene product activates the expression of other genes encoding proteins in the bacterium's outer membrane.

Expression of the genes that encode the TCP parallels expression of the toxin genes: these genes are expressed only when the bacterium finds itself *e.g.*, in the epithelial surface of the intestine. The genes encoding the TCP are located on a pathogenicity island known as the *V. cholerae* pathogenicity island known as the *V. cholerae* pathogenicity island (VPI). All strains of *V. cholerae* that can cause epidemic cholera contain a VPI, which is thought to be transferred between strains in the environment. It appears that this genetic element might be encoded by a bacteriophage (Taylor, 1999).

According to Karaolis *et al.*, the VPI is, indeed, a filamentous bacteriophage. Termed the VPI phage, it can transmit itself between certain strains of *V. cholerae*, and it carries ssDNA in a similar way to CTXΦ. The protein that surrounds the DNA (the coat protein) is TcpA–the implication is that the same structure may be both a colonization pilus and a bacteriophage particle. The bacteriophage may be formed and bud off from the bacterium while acting as a colonization pilus. If true, it indicates that the VPIΦ is the receptor for yet another phage–the CTXΦ. But it is strange how an infectious phage particle, which must be released from the bacterial cell to infect another cell, can also act as a colonization factor and a phage receptor (Taylor, 1999).

Previously it was believed that virulence genes directing the interaction between this pathogen and its human host have been acquired by pathogenic *V. cholerae* from bacteriophage and from large gene clusters called pathogenicity islands by horizontal gene transfer. Two major virulence factors, CT and a colonization factor called toxin-coregulated pilus (TCP), are each encoded on genetic elements in chromosome 1 that are not universally found in *V. cholerae* strains; yet they are regulated by genes that seem to predate entry of these elements into the genome. The genes encoding CT are acquired from the genome of bacteriophage CTXf (see DiRita, 2000); those encoding TCP are carried on an element called the *Vibrio* pathogenicity island (VPI), also thought to be of bacteriophage origin. The genome sequence data is at variance with the idea that the VPI is of bacteriophage origin because none of the VPI genes encode phage structural or morphogenetic proteins. An integron island (a system for gene capture and dissemination) on chromosome 2 may also be important for survival of *V. cholerae* in human populations since it codes for several genes for antibiotic resistance as well as potential pathogenicity determinants.

Besides CT, the genome sequence shows some unknown toxins that probably contribute to the side effects of current vaccine strains.

There exists considerable potential for combining genome information with functional genomic approaches to identify essential genes and characterize patterns of gene expression during infection. Conversely there is paucity of information about how *V. cholerae* behaves in natural environments outside of its human host. Although interactions of the free-living microbe within biofilms and with marine invertebrates and zooplankton have been described, the factors involved in these associations are not known. Possibly enzymes that digest chitin (a component of the exoskeleton of zooplankton), genes for which have now been identified in the *V. cholerae* genome, are involved.

It appears that bacteriophages will continue to surprise us with their multifaceted roles in the evolution of pathogenic and other bacterial species (Taylor, 1999).

A Unique Linear Plasmid?

Until over a decade ago, it was believed that γ-proteobacteria had a single circular chromosome. Although extrachromosomal elements were well understood, except for some linear bacteriophage, all DNA molecules were thought to exist as closed rings to replicate within *Escherichia coli* and its cousins. This dogma suffered a major setback when Trucksis *et al.* (1998) discovered that *Vibrio* spp. had two, rather than one, circular chromosomes. Another notable exception to the genome structure rules emerged when the presence of a linear, extrachromosomal and autonomously replicating element was reported in a *Salmonella enterica* sub sp. Whereas previously described linear bacteriophge genomes from *E. coli*, *Klebsiella oxytoca* and *Yersinia enterocolitica* all had covalently closed hairpin ends (similar to the ends of linear *Borrelia* DNA, Casjens *et al.*, 2004), this new element is more like a plasmid than a bacteriophage; its covalently capped terminal inverted repeats are similar to those of *Streptomyces* linear replicons and lack any open reading frames (ORFs) with even remote similarity to bacteriophage DNA. A central low GC region–from which predicted ORFs appear to diverge–could possibly serve as the origin of replication, and only eight of thirty other predicted ORFs are similar to any sequences in the GenBank database (see Baker *et al.*, 2007; Okeke, 2008).

Bacterial linear elements that are not phage are thought to have eukaryotic or eukaryotic virus ancestry (Stewart *et al.*, 2004; Okeke *et al.*, 2008).

Bacterial Antibiotic Resistome

Antibiotics play an important role in the battle against bacterial infections, but several antibiotics which are natural products of microorganisms and the ecological roles of these antibiotics in the wider environment are not known (Dantas *et al.*, 2008). A rampant spread of multiple antibiotic resistance in clinically relevant pathogenic microbes has taken place and indeed a substantial environmental reservoir of antibiotic-resistance determinants, termed the *antibiotic resistome*, has been discovered (Riesenfeld *et al.*, 2004; D'Costa *et al.*, 2006). Efflux pumps, target gene-product modifications, and enzymatic inactivation of the antibiotic compound are some notable microbial antibiotic-resistance mechanisms (Alekshun and Levy, 2007). Many common mechanisms are shared by several species of pathogens and spread by horizontal gene transfer (Davies, 1994). According to Dantas *et al.* (2008), the soil microbiome contains a large reservoir of bacteria that can subsist on antibiotics. Many bacteria growing in extreme environments and which can degrade toxic substrates have been reported; only a few organisms seem to subsist on a limited number of antibiotic substrates (Johnsen, 1977).

Many soil bacteria able to grow on antibiotics as a sole carbon source were isolated by Dantas *et al.* (2008). Of 18 antibiotics tested, representing eight major classes of natural and synthetic origin, 13 to 17 supported the growth of clonal bacteria from each of 11 diverse soils. Bacteria subsisting on antibiotics are phylogenetically diverse, and many are closely related to human pathogens. Furthermore, each antibiotic-consuming isolate was resistant to multiple antibiotics at clinically relevant concentrations. This suggests that this hitherto little understood source of antibiotic-resistance determinants can contribute to the increasing levels of multiple antibiotic resistance in pathogenic bacteria.

It appears very likely that pathogens obtain antibiotic-resistance genes from environmentally distributed super-resistant microbes subsisting on antibiotics (Dantas *et al.*, 2008).

Microcosm, Life and *E. coli*

The bacterium *Escherichia coli* has recently become notorious from the various outbreaks of harmful strains. But for biologists specialized in many different aspects of life, *E. coli* forms a basis from which one can gain a better understanding of almost every facet of biotic life ranging from the inner workings of the cell, through the social life of *E. coli*, to outer space. For long, scientists did not believe microbes could have sex. Joshua Lederberg *et al.* discovered bacterial conjugation and showed that *E. coli* could exchange genes through direct contact with other cells. Not only can *E. coli* have sex, but, like higher organisms, it has a rather complex social life. For example, when food is limited, members of a population signal to one another to enter a stationary phase which improves the chances that at least some members will survive. However, life is not always so friendly for *E. coli*–in addition to cooperation, there is also strife in the bacterial world. To compete with other species, or even its own kind, *E. coli* can produce antibiotics (Zimmer, 2008; Rankin, 2008).

Zimmer even presented *E. coli* in defense of evolution itself. The bacterial flagellum, which is rotated to propel the organism, has figured in the recent controversy surrounding intelligent design whose advocates cite the flagellum as an example of a structure too complex to have evolved through natural selection. Michael Behe testified to that effect for the defense when parents in Dover, Pennsylvania, sued the local school district over the teaching of intelligent design in science classes (see Humes, 2007). The plaintiff's lawyers countered by bringing scientists who demonstrated how it was possible for the flagellum to have evolved from useful, intermediate structures, even if those structures did not function in bacterial locomotion. They argued that complicated structures such as flagella could have evolved through small evolutionary steps, thereby discrediting a central argument used against evolutionary theory.

Biotechnology owes much to *E. coli*. By introducing animal genes into *E. coli*, strains were developed to fill tanks that now pump out gallons of insulin and other drugs every day. *E. coli* genes are now being inserted into animals: researchers have transferred to pigs genes that allow *E. coli* to break down phosphate-bearing compounds. This raises hope that such pigs may some day be used to reduce phosphate pollution in farms.

This minute creature has in fact advanced humanity much further than any other organism. Not only has it inhabited human guts for as long as history, it has benefited almost all areas of the biosciences, from genetic engineering to evolutionary theory (Rankin, 2008).

Combating Antibiotic Resistant Bacteria

Invisible dangers lurk everywhere. Antibiotic-resistant bacteria infest hospital catheters, the hands of health care workers and the dirt on supermarket potatoes. Information about how to genetically manipulate microbes could allow terrorists to cause mayhem with bioweapons.

Successes in the eternal battle against microscopic bugs usually have far-reaching *Domino effects*. For instance, wider use of the influenza vaccine advocated by many with the argument that immunizing all infants against the nasty virus could have profound public health benefits, can have a negative consequence of widespread vaccinations–protecting children against chickenpox might make their grandparents' generation more susceptible to shingles, a painful malady caused by the same virus (Cohen, 2002).

Certain proteins of host and bacteria which are essential for bacterial survival provide new way for developing nonantibiotic-based treatments for infections. The notorious human pathogen *Staphylococcus aureus* has already become even more deadly following its acquisition of antibiotic resistance. A new strain of methicillin-resistant *S. aureus* (MRSA), known as USA300, and its close relatives are not only resistant to antibiotics but are highly virulent and contagious. MRSA has been spreading throughout the world, in hospitals and also in community settings where people come in close contact (Kennedy *et al.*, 2008). In the United States, MRSA infections now cause more deaths each year than AIDS. But some new therapeutic possibilities are being explored to treat such infections. One of these targets the ability of *S. aureus* to thwart attacks that are mounted by the immune system during infection. Bacteria defend against lethal reactive oxygen species (ROS) produced by neutrophils, immune cells that are mobilized to sites of infection (Novick, 2008). Calprotectin, a well-known mammalian calcium-binding protein, chelates manganese (Mn^{2+}), an essential requirement of the bacterium for its growth and for detoxifying ROS.

Staphylococcal infection particularly threatens people with weakened immune systems, impaired circulation (as with diabetics), and surgical wounds. Staphylococcal abscesses appear when the host's immune system senses certain bacterial products such as cell wall components and chemotactic (neutrophil-attracting) enzymes. The host then produces a strong inflammatory response. It walls off the infected area by forming a mesh-work of fibrin protein packed with neutrophils and other immune cells. Neutrophils phagocytose bacteria and kill them with ROS and other lethal agents. Staphylococci in turn defend themselves by deploying cytotoxins and enzymes to degrade host tissue and kill neutrophils and other cells.

Fortunately, however, neutrophils continue their antibacterial activities even after their lysis by releasing bactericidal ROS as well as DNA that forms a network to entrap bacteria (Brinkmann *et al.*, 2004). The bacteria not only secrete potent nucleases to degrade the DNA, but also produce substances that detoxify ROS, including superoxide dismutase and calprotectin (Novick, 2008). Calprotectin, a calcium-binding heterodimer, makes up about 40 per cent of the soluble cytoplasmic protein of a neutrophil. One of its major biological roles is signaling to other immune cells after tissue damage and/or inflammation. Corbin *et al.* showed that it defends against bacterial infection by chelating Mn^{2+} as did those in wild-type mice. It appears that abscess attenuation is due to the effects of Mn^{2+} depletion on bacterial defenses against ROS. Mn^{2+} is also an essential component of bacterial superoxide dismutase A, which inactivates ROS enzymatically (Karavolos *et al.*, 2003).

The situation is somewhat complicated because staphylococci express catalase and other antioxidant enzymes that are repressed by Mn^{2+} through a regulatory protein, PerR (Horsburgh *et al.*, 2002). Calprotectin could possibly enhance the expression of these enzymes, so enabling bacteria to counter the effects of ROS, even when the Mn^{2+} concentration is low.

Infection of Eukaryotic Cells by *Escherichia coli*

Escherichia coli is not only the commonest cause of infections by Gram-negative bacilli but also a commensal of the normal gut microflora (Kaper *et al.*, 2004). It is a versatile pathogen as it can produce diverse types of virulence factors that affect basic host cell functions (Hacker *et al.*, 2003), including the cyclomodulins, which target the host cell cycle and influence whether an infected cell will grow and divide, or die. Nougayrede *et al.* (2005, 2006) reported that certain *E. coli* strains can induce megalocytosis in cultured eukaryotic cells; such megalocytosis involves a progressive enlargement of the cell body and nucleus and the absence of mitosis. This cytopathic effect is reminiscent of the effect of cyclomodulins and manifests upon transient infection of different mammalian cells (HeLa, CHO, A375, and IEC-6). It can be induced by pathogenic E. coli strains isolated from meningitis and urinary tract infections. It is also induced by certain commensal strains but not by laboratory K-12 strains nor by enteropathogenic or enterohemorrhagic *E. coli.* The cytopathic activity is contact-dependent and is not seen in bacteria separated from mammalian cells by a 0.2-µm permeable membrane. Inhibition of bacterial internalization with cytochalasin-D did not eliminate the cytopathic effect. Heat-killed bacteria, gentamicin-killed bacteria, bacterial culture supernatants, and bacterial lysates do not produce cytopathic effect, nor can the effect be explained by the production of any toxins that change the host cell cycle (Oswald *et al.*, 1994; Marches *et al.*, 2003).

Transient infection of eukaryotic cells with certain commensal and extraintestinal pathogenic *E. coli* strains blocks mitosis and induces megalocytosis. This trait is connected to a very common genomic island which codes for giant modular nonribosomal peptide and polyketide synthases. Contact with *E. coli* expressing this gene cluster causes DNA double-strand breaks and activation of the DNA damage checkpoint pathway. This arrests cell cycle and eventually causes cell death. Discovery of such hybrid peptide-polyketide genotoxins in *E. coli* changes our concept of pathogenesis and commensalism and can lead to novel biotechnological applications (Nougayrede *et al.*, 2005, 2006).

E. coli strains containing a genomic island, widely found in both pathogenic and commensal isolates, induce double strand breaks (DSBs) upon transient contact with epithelial cells. This genomic island is present in Nissle 1917, a commensal strain of *E. coli* that readily colonizes in mice and humans and is also used as a probiotic treatment for ulcerative colitis and other intestinal disorders (Kruis *et al.*, 2004). Blocking the cell cycle slows down the renewal of the intestinal epithelium and so could be a bacterial strategy to prolong colonization of the intestinal epithelium, which in turn is expected to have some impact on pathogenicity and commensalism (Nowrouzian *et al.*, 2005). The quantity of the genotoxin made by various strains together with the location and duration of the contact with the target host cells probably determine whether commensalism or pathogenicity is promoted. Because DSBs in DNA can cause genomic instability, the occurrence of bacteria with the pathogenicity island may well be a predisposing factor in the development of intestinal cancer (Lax, 2005). Indeed, synthesis of a bioactive polyketide-peptide in *E. coli* should catalyze progress in the engineering of hybrid peptide-polyketide biosynthetic pathways for creating natural products such as anticancer agents, antibiotics, and immunosuppressants (Weissman *et al.*, 2005).

Bacterial Interactions with Eukaryotic Hosts

Bacteria show wide variety of specialized interactions with eukaryotic hosts. One way to classify these associations is along an axis from pathogenic to mutualistic: that is, how much does the infection help or hurt the host relative to noninfected individuals? While from the host perspective, this distinction is central, for the bacteria themselves, pathogenesis and symbiosis present similar obstacles. Primary among these is the invasion of host cells or tissues in the face of generalized cellular and chemical defenses. Host invasion often depends on certain evolutionarily homologous genes that have been transferred among distinct bacterial lineages (see Moran, 2001). These loci code for a secretion system or complex of about 20 proteins that collectively make up a membrane-spanning structure that transports proteins outside of the bacterial cell. Transport is sometimes induced by contact with host cells, and the effectors are delivered to the host cytoplasm and linked to suppression of host defenses and / or invasion of host cells (see Moran, 2001). Dale *et al.* (2001) showed that *Sodalis glossinidius*, an intra-cellular symbiotic bacterium restricted to tsetse flies (*Glossinia* spp.), possesses the above secretion system which is essential for entry of the bacteria into cultured insect cells and for infection of intact hosts. *Sodalis* is maternally transmitted in invertebrate lineages and appears to be mutualistic or not obviously pathogenic. This is the first report in which an animal symbiont has been shown to use a secretory system associated with pathogenesis for host infection.

The discovery by Dale *et al.* extends the role of this particular transport apparatus to a broader range of interactions between bacteria and animal hosts (see Moran, 2001).

The bacteria harbouring the secretion system are phylogenetically diverse, belong to several major bacterial divisions, with a concentration of known cases in the

Enterobacteriaceae. Among animal pathogens, these secretion systems have been found in *Yersinia pestis* (plague), *Salmonella enterica* (typhus), *Shigella* spp. (dysentery), and the more distantly related genus *Chlamydia* (sexually transmitted and other diseases. In plants, gene clusters encoding the secretion systems not only occur in various *Erwinia* species, *Pseudomonas syringae,* and other pathogenic bacteria, but also in symbiotic *Rhizobium* species that function in nitrogen fixation. These gene complexes are typically encoded within pathogenicity islands, *i.e.* segments of chromosomal DNA that are present in pathogens but absent from reated nonpathogens and that are required for virulence. Most pathogenicity islands seem to have been acquired through horizontal transfer of DNA.

Although systems such as the above seem to have been acquired by bacterial pathogens, *Sodalis* is transmitted mainly through vertical passage from mother to developing progeny and is not obviously harmful to hosts.

The diversity of bacterial residents within insect bodies covers various interaction types and evolutionary patterns. The organisms classified as primary symbionts often descend from ancient infections based on recent molecular phylogenetic studies. For example, *Buchnera aphidicola*, the primary symbiont of aphids, has been vertically transmitted for over 100 million years and shows various genomic changes, some reflecting mutualistic adaptations for provisioning hosts with amino acids lacking from their phoem sap diet. In contrast, phylogenetic studies on secondary symbionts such as *Sodalis* point to a history of horizontal infection in addition to the usual maternal route. The effect of such organisms on host fitness, that is whether they are mutualists or pathogens, is usually not as clear as in the case of primary symbionts. Another widespread group of symbionts present in many insects is *Wolbachia pipientis,* a maternally transmitted symbiont linked to reproductive abnormalities. Horizontal transfer of *Wolbachia* is well established.

The diversity of insect symbionts is particularly high within the gamma Proteobacteria and especially within the Enterobacteriaceae. Whereas *Buchnera* and *Wigglesworthia* fall just outside of the Enterobacteriaceae, a whole assemblage of insect symbionts, including *Sodalis*, are grouped into several independently derived clades within it (see Moran, 2001). The heterogeneity of the insect hosts infected by each of these clades suggests that symbiont lineages move among insect host species, even occasionally, to members of different orders. For example, the closest relative of *Sodalis* is a symbiont of *Sitophilus* weevils feeding on stored grain. Another clade contains the symbiont of the bloodsucking bug *Triatoma infestans,* a symbiont causing reproductive abnormalities in parasitoid wasps, and several secondary symbionts of psyllids. These findings indicate transfer of bacterial symbionts between these ecologically divergent insect groups.

Mutualistic symbionts and pathogens come across similar problems besides that of establishing infections and overcoming defenses. In both pathogens and mutualistic bacteria, limitation of damage to host cells is usually essential for persistence and subsequent spread to new hosts. Selection against virulence is higher in life cycles that require persistence within individual hosts. This pressure for decreasing damage and increasing benefits to hosts will be especially high if the infection is transmitted between successive host generations. Thus, pathogens sometimes evolve into mutualists following the establishment of vertical transmission. However, the transition into effective mutualists requires that the bacterial genome possess the capacity for processes beneficial to hosts, such as provision of nutrients (see Moran, 2001).

Ecofriendly Microbes for Insect Pest Control

Pesticides have long been used as a weapon against insect pests, but the growing concern about their ill effects has led to a change in strategies to manage insect pests in an ecofriendly manner by using biopesticides which leave no negative effects on the environment. Some selected, desirable attributes of microbial control agents and nematodes are: Rapid spread, power of search for host, persistence, safety and acceptability, control to sub-economic level, predictable control, virulence, easy production, low cost, good storage and convenience of application.

Besides this use of *Bacillus thuringiensis* (*Bt*) in insect pest management, the part played by various other microbes such as viruses, bacteria, fungi and protozoans should be explored.

Most of the organisms tested for field insect pest control have caused more than 50 per cent mortality of the pest.

Results on the combinational studies revealed possible synergism of *Metarhizium anisopliae* and optimum level of synthetic insecticides against *Spodoptera litura*. *Bt* and Neem Azal were found to be more effective on *S. litura* whereas *Bt* alone was effective on *Plutella xylostella*. Additive and synergistic interactions of nematodes (*Steinernema carpocapsae*) and *Bt* showed considerable effect on root grub (*Holotrica serrata*) (Ignacimuthu, 2000).

Liquid formulations of *M. anisopliae* proved to be stable and effective against diamond black moth (*Plutella xylostella*).

NPV has been successfully used against teak leaf defoliator (*Hyblae puera*), groundnut red hairy caterpillar (*Amsacta albistriga*), southern armyworm (*Spodoptera litura*), cotton boll worm (*Helicoverpa armigera*) and caterpillars of *Spilosoma obliqua*.

The fungus *Beauveria bassiana* is effective against *H. armigera, S. litura* and stored product pest *Sitophilus oryzae;* so also the fungus *Nomuraea rileyi* is effective against *H. armigera* and *S. litura*.

While using biopesticides, careful consideration should be given to improvement of inoculum load in the formulations, safety to non-target organisms and beneficial insects, genetic changes in the hosts, standard bioassay procedures, combating spurious cultures, and adverse effects to humans during production and use. There is also a strong need for the following:

(1) Technology for large-scale production of microbials; (2) Identification of nodal centres of excellence for technology transfer; (3) Developing a protocol or an effective implementation of programmes in vast and diverse areas; (4) Detachment of scientists from commercialization and involvement of commercial organizations for bioefficacy testing; (5) Anticipatory research in the management of insect resistance; (6) Further

work on insect and fungal pathogens and establishment of identification services; and (7) Identification of important crops for use of microbials and economic threshold level.

Mutasynthesis for Drug Discovery

Mutasynthesis joins the ability of chemical synthesis with molecular biology to derive medicinally valuable natural products. Biotica Technology Ltd., Cambridge has exploited mutasynthesis to generate promising new variants of the polyketide anti-cancer compounds rapamycin and borrelidin (Weissman, 2007).

Many of our most valuable drugs are either natural products or were inspired by their structures (Clardy and Walsh, 2004). Unfortunately, the high complexity underlying this biological activity makes it very difficult to modify these metabolites by semisynthesis, or to assemble analogues from scratch. A better alternative to augmenting natural product diversity is PDB (precursor-directed biosynthesis) in which synthetic chemistry is used to produce analogues of the early intermediates in the metabolic route, leaving the producing organism to accomplish the remaining, more difficult, transformations (Weist and Süssmuth, 2005). The synthetic compounds can be added into the culture medium of the microbe. If these so-called 'mutasynthons' are absorbed by the cells and happen to out-compete the natural building blocks for the biosynthetic machinery, a product with a modified structure results. As competition with endogenous precursors can lower the yield of new molecules: PDB is often coupled with mutagenesis of the producing strains to block the normal pathway at a crucial point. In such mutasynthesis (MBS) experiments, the microbes have no alternative except to utilize the surrogate substrate to initiate the assembly process. MBS has been widely applied to many classes of compounds, such as siderophores, nucleosides, aminocoumarins, non-ribosomal polypeptides and polyketides (Galm *et al.*, 2004; Weist *et al.*, 2004; Weist and Süssmuth, 2005).

The earlier PDB experiments depended on laborious chemical synthesis to create suitable precursor compounds, but today several thousands of complex small molecules are available from commercial suppliers not only as racemic mixtures but also in stereochemically pure forms. If the desired analogues cannot be purchased, it is still possible to elaborate commercial compounds into the target molecules by means of advanced chemical synthesis. Further explosive growth has occurred in sequence information for many natural product pathways. *In silico* analysis of the proteins coded for by the gene clusters, together with gene inactivation *in vivo*, frequently enables each gene product to be unambiguously assigned to a specific role in the biosynthesis. Suitable experiments can identify appropriate candidate genes for making metabolic non-producers; this is a much better alternative to screening large numbers of randomly generated mutants for the desired phenotypes. Inactivating the genes in the native hosts has become straightforward, following the advent of recent molecular biology techniques, such as Red/ET-mediated recombination (also known as λ-mediated recombination) (Zhang *et al.*, 1998). The ability of modern mutasynthesis to produce natural product analogues is highlighted by several recent papers from the company Biotica (Little Chesterford, UK).

"Rapalogues"

Streptomyces hygroscopicus produces rapamycin–a secondary metabolite of mixed polypeptide-polyketide origin. Its binding to the protein mTOR (the mammalian target of rapamycin), a downstream protein kinase within the phosphatidylinositol 3-kinase (PI3K)/Akt (protein kinase B) signaling pathway, inhibits the signals needed for cell cycle progression, growth and proliferation (Brown *et al.*, 1994). Rapamycin has an exciting lead structure for development of anti-cancer, anti-inflammatory and immunosuppressive agents. All rapamycin analogues currently in clinical trials have been modified by semi-synthesis in the portion of the molecule derived from the starter unit, 4,5-dihydroxycyclohex-1-enecarboxylic acid (DHCHC). Since the starter unit of polyketides is specially vulnerable to mutasynthesis, Gregory *et al.* (2005) examined the utility of this approach for generating a series of 'rapalogues'.

In the first of three stages of rapamycin biosynthesis, DHCHC is produced from shikimic acid. It then acts as a starter unit for polyketide chain extension on a modular polyketide synthase (PKS) and results in the release of 'pre-rapamycin' which is then modified by certain post-PKS enzymes through regio- and stereospecific oxidation and methylation. This analysis points to the possibility that the enzymes involved in DHCHC synthesis, or its incorporation, might gainfully be targeted to allow starter unit mutasynthesis; in fact, a suitable mutant was created accidentally in an attempt to generate a strain of *S. hygroscopicus* that produced pre-rapamycin by deletion of an entire set of genes (*rapK1JMNOQL*) that were implicated, by sequence analysis, in processing the core to the fully elaborated structure. Unexpectedly, the mutant strain *MG2-10* did not produce the expected polyketide; however, efficient biosynthesis of pre-rapamycin was restored by supplementing the culture medium of *MG2-10* with the surrogate starter unit, 4,5-dihydroxycyclohexanecarboxylic acid. Selective restoration of the deleted genes revealed that the gene involved in the regulation of starter unit production was *rapK*, till then believed to code for an enzyme responsible for a late-stage oxidation (Weissman, 2007). Further researches and collaboration with Wyeth, enabled Biotica to develop rapalogues as potential drugs (Weissman, 2007).

Borrelidin

Borrelidin, a natural polyketide, is a strong inhibitor of angiogenesis. As angiogenesis is essential for the survival of tumour cells, various inhibitors of the complex metabolic routes involved in this process are being explored as potential agents for cancer chemotherapy. Analogous to rapamycin, borrelidin biosynthesis also occurs in three stages: assembly of the starter unit, *trans*-cyclopentane-(1*R*, 2*R*)-dicarboxylic acid (*trans*-1,2-CPDA); PKS-catalyzed chain extension; and post-PKS processing. Olano *et al.* (2004) cloned and sequenced the borrelidin gene cluster from *Streptomyces parvulus* Tü4055. This enabled them to mutate a crucial gene involved in starter unit production by employing insertional gene inactivation. Importantly, adding *trans*-1,2-CPDA to the mutant strain BIOT-1302 elicited 15-fold higher production of borrelidin than from the wild-type strain. Moss *et al.* (2006) fed over 40 commercially available mono- and dicarboxylic acids to BIOT-1302 cultures for evaluating their ability to initiate the biosynthesis of modified borellidins. Critical and advanced chemical reactions and analysis led them to develop six novel borrelidin analogues.

Structure-activity data obtained on these six analogues suggest that the precise steric arrangement of the starter unit is critical for differentiating the cytotoxic and anti-angiogenic activities of borrelidin (Wilkinson *et al.*, 2006). A cyclobutyl analogue particularly showed a significant increased selectivity *in vitro* for inhibition of angiogenesis as compared with cytotoxicity; this might make it a better therapeutic agent than borrelidin.

The above successes notwithstanding, however, mutasynthesis does have certain limitations as a means to expand natural product diversity. Incorporation of the modified precursors requires a broad substrate tolerance of all catalytic activities downstream from the blocked step and, in practice, only a small subset of analogues will be accepted. Furthermore, the approach is only okay for modifying those parts of the structures which are generated in the earliest stages of the biosynthesis. The hope is that even such minor modifications often result in a new drug–the commercially valuable antiparasitic agent doramectin differs from its parent macrolide avermectin only in the starter unit (see McArthur, 1998). It appears that mutasynthesis is destined to have a bright future in lead optimization and drug discovery, especially when coordinated with attempts to change additional portions of the natural product structures (Weissman, 2007).

Chapter 7
Marine Microorganisms

Microbial Census in the Sea

Most archaea thrive in extreme environments. Some live in saline lakes that are much more saltier than the ocean; some live in anaerobic habitats, where even trace amounts of oxygen are lethal; and some live in so hot geothermal environments that most organisms would be cooked. *Pyrolobus fumarii* can grow in anaerobic deep-sea hydrothermal vents at temperatures around 235 degrees F.

Surveys of the frigid, aerobic Antarctic waters have revealed archaea in unexpectedly high numbers. According to DeLong (2003), cold-adapted cousins of thermophilic archaea flourish in shallow and deep marine waters at polar, temperate, and tropical latitudes. They turn up in the guts of abyssal sea cucumbers and in sediments at the bottom of the sea. In fact, archaea comprise 20 to 30 percent of all the microbial cells in the ocean.

The discovery and improved understanding of novel microbial groups has become possible from examining novel habitats. Besides, microbiologists now have new, high-tech tools for identifying and counting microbial life. In the past the common approach was simply to culture a sample of, say, seawater and see what, if any, species grew. Microbial life has turned out to be much more diverse than indicated by cultured samples. A lot of the newly recognized life in the oceans is so small in size as to be termed picoplankton.

The plankton includes the floating microbes, single-celled and multicellular plants and animals (including immature larval forms) that move primarily by drifting with the currents. Anything smaller than 0.05 millimeter but larger than 2.0 microns, that can pass through fine-mesh nets, is designated nanoplankton. The picoplankton comprises the smallest cells that range between 0.2 and 2.0 microns across. (Note that prefixes "nano" and "pico" do not exactly correspond to the units nanometer or picometer but have been named as per traditions in marine biology).

Until over two decades ago, picoplankton was thought to be insignificant in the marine microbial food web because its biomass seemed much too low to play a primary role. But the advent of epifluorescence microscopy increased the estimated numbers greatly. We now know that the density of microorganisms ranges from tens of thousands per milliliter in the deep ocean to millions per ml in surface waters. Measuring the rate of synthesis of DNA and protein showed that marine picoplankton can double in biomass every day or so and that the cells observed through fluorescence microscopy are alive and metabolically active. The explanation of why the marine populations of picoplankton tend to remain roughly constant is that protist predators actively graze on them at about the same rate as the picoplankton reproduces.

The metabolism in the enormous biomass of picoplankton involves a huge flow of carbon and energy. While some of this carbon is released as carbon dioxide, much of it is locked up in organic molecules that help sustain the rest of the food web. Especially important in the carbon cycle and to the oceanic food web are those microorganisms which grow at or near the ocean's surface (DeLong, 2003). The upper 600 feet of the water column in the oceans constitute a region of intense photosynthetic activity with the resulting photosynthate supporting not only all the other inhabitants of the ocean's surface but also most denizens of the deep sea. Until recently all that productivity has been erroneously credited to eukaryotic algae such as diatoms and dinoflagellates. It was in the 1970s that *Synechococcus*–the first of a new kind of photosynthetic microorganism, the marine picoplanktonic cyanobacterium, was discovered. It was found to be extremely abundant in coastal and open-ocean environments, often reaching densities higher than 100,000 cells per ml. At certain times and places these cells are responsible for as much as half of the primary production of food in the ocean (DeLong, 2003). Then, in the late 1980s, came the discovery of another small (less than a micron in diameter), red-fluorescing organism that was even more abundant than *Synechococcus*. The new cells–in size, also a kind of picoplankton–were eventually cultured and isolated in the laboratory and named *Prochlorococcus*. They are closely related to *Synechococcus*, but the two genera differ in their pigment composition.

Field experiments have revealed that in the open ocean *Prochlorococcus* cells reach concentrations of hundreds of thousands per ml of sea water. Indeed, *Prochlorococcus* constitutes half of the total chlorophyll-based biomass in the ocean. So picoplankton has turned out to be much more central to the carbon cycle. Picoplanktonic cells circulate in so large numbers and are heavily grazed upon by protists, suggesting that they supply nutrients directly to larger organisms. This microbial portion of the food web has been dubbed the "microbial loop" (DeLong, 2003).

Cloning of large DNA fragments from mixed natural microbial communities has been used to create libraries of microbial genetic diversity. These libraries can be scanned for the presence of particular genes–and by extension, the presence of the proteins and metabolic functions that the genes encode. The biomass can also be screened for markers that identify just which species of microorganism the genes belong to. Also proteins encoded by individual genes can be readily produced, enabling study of their structure, function, and role in the natural world. From one such library, unexpectedly, a previously unknown kind of photoprotein was discovered (see DeLong, 2003). The protein molecule came from the genome of a widespread planktonic bacterium; it absorbs light of a characteristic wavelength–much as does rhodopsin in the human eye. Like rhodopsin, it converts light into energy that can be used by the microbial cell. Indeed, the novel form of rhodopsin occurs in natural communities of marine picoplankton. Several marine surveys from Antarctica to Hawaii have revealed that colour variants of the photoprotein exist virtually everywhere. In deep waters the photoprotein is adapted to absorb the blue wavelengths of light most abundant there. In shallower waters, it absorbs the green light available at the surface (DeLong, 2003).

Besides the importance of cyanobacteria in the global carbon cycle, other marine microorganisms also support the biosphere. Bacteria, archaea, and other microorganisms are also essential for the nitrogen cycle as they break down organic nitrogen to produce ammonia; can convert ammonia to nitrate, and recycle nitrate into other nitrogenous compounds in oxygen-poor marine sediments. Some cyanobacteria fix nitrogen to synthesize organic nitrogen compounds. Without microorganisms, nitrogen wouldn't cycle at all–neither would most other elements (DeLong, 2003).

Such transformations depend not on single species but on entire communities of microorganisms. Microorganisms can be viewed as our planetary engineers, they have been at work for at least 3.5 billion years ever since life originated on Earth.

Biogeography of Microbial Communities in a Model Ocean

Quantification and modeling the role of ocean ecosystems in the global carbon cycle can enhance our understanding of the changing earth system. Microbial communities in the surface ocean regulate critical biogeochemical pathways such as the efficiency of export of organic carbon to the deep ocean. Despite the extraordinary diversity in the oceans, the biomass of local microbial communities at any location is usually dominated by only a small number of species whose relative fitness and ecosystem community structure are regulated by various factors, *e.g.*, physical conditions, dispersal, predation, competition for resources, and the variability of the environment (Pedros-Alio, 2006). Pimm and Lawton (1977) examined some models based on this conceptual view in idealized ecological settings and Kleidon and Mooney (2000) applied these to studies of terrestrial ecosystems. Follows *et al.* (2007) adopted this approach in a marine ecosystem model which covers the diversity and genomics of microbes. In this model microbial community structure is generated from a broader group of possibilities and, hence, mimics some aspects of the process of natural selection. The high flexibility of the system allows it to respond to changing ocean environments, helps to interpret the structure and development of marine microbial communities, and to reveal connections between marine ecosystem structure, global biogeochemical cycles, and climate change.

Gregg *et al.* (2003) and Litchman *et al.* (2006) attempted to resolve community structure in ocean models by the clear representation of three or four classes (functional groups) of phytoplankton, but four notable challenges in the effort are (a) subjective and arbitrary specification of functional groups and diversity of the model ecosystem; (b) difficulty in evaluating the parameters controlling such models–quantitative, physiological information from laboratory cultures is extremely limited; (c) paucity of observations of microbial community structure with which to evaluate global-scale models; and (d) model ecosystem structures that are optimized to reflect today's ocean are not dynamic enough to adapt appropriately to a changing climate where strong shifts in community structure might be possible (Follows *et al.*, 2007).

To deal with some of these challenges, Follows *et al.* created a marine ecosystem model which represents several potentially viable phytoplankton types whose physiological characteristics were determined stochastically. These initial taxa interacted with one another and with their environment and evolved into a sustainable ecosystem where community structure and diversity were not imposed, but were emergent properties.

This ecosystem model, seeded with many phytoplankton types, whose physiological traits were randomly assigned from ranges defined by field and laboratory data, generated an emergent community structure and biogeography consistent with observed global phytoplankton distributions. The modeled organisms include types analogous to the marine prochlorophyte *Prochlorococcus*. Their emergent global distributions and physiological properties substantially correspond to observations. The attractive feature of such flexible representation of community structure is that it can be used to explore relations between ecosystems, biogeochemical cycles, and climate change (Follows *et al.*, 2007).

Importantly, simultaneous consistency was found between the geographical habitat, rank abundance, and physiological specialization of the emergent *Prochlorococcus* model ecotypes and their real-world counterparts. These parallels strongly suggest that the stochastic, self-organizing representation of marine ecosystems reflects real-world processes and is well suited for application in ecological and biogeochemical studies. This approach avoids some obstacles facing many current ocean ecosystem models, such as the a priori imposition of low diversity, the prescription of dominant functional types, and the problematic specification of the physiological rate coefficients that define them. This function-based approach may evolve to exploit the rapidly burgeoning genomic and metagenomic data mapping of the oceans in terms of genes and their encoded physiological functionality (DeLong *et al.*, 2006).

Microbial Life in the Deep Sea

The deep sea is a vast and strange expanse where many different creatures live. A great deal is now known, and much more still unknown, about deep sea habitats and their biodiversity (Koslow, 2007).

In the deep sea, food resources are very limited and include faeces, dead phytoplankton, crustacean moults, discarded mucus and noxious chemicals. The monotonous environmental conditions prevailing in the deep sea do not seem suited to high diversity but paradoxically, that diversity is really vast, and includes many fisheries, as many stony corals as on shallow reefs, and abyssal plains with as many species as tropical rainforests. The quantity and diversity of deep sea life often rival those documented for the rainforests.

The benthic communities of the sea bottom pose the strongest challenge of any ecosystem to our understanding of the Earth's biodiversity. The paradox is probably generated by a variety of factors, such as low productivity, which prevents any single species from multiplying enough to eliminate its competitors.

A highly diverse and abundant microbial flora and fauna flourishes in the greatest depths of the oceans. The rocks and sediments beneath the seafloor harbour most of Earth's microorganisms. Molecular approaches are now providing clues about the energy sources fueling their metabolic activity. Contemporary microbial surveys have considerably expanded the range of known habitats where microbial life is present. D'Hondt *et al.* (2004) found metabolically diverse, active microbial communities buried deep within marine sediments nearly 0.5 km below the seafloor. They inferred how subseafloor microbes eat and breathe; certain microbial activities appear to deviate substantially from standard models (Canfield *et al.*, 1993) of microbial metabolism in subseafloor sediments.

Scientific drilling into sediments and basaltic crust in the world ocean has revealed the omnipresence of microscopic life deep beneath the seafloor. Diverse communities of prokaryotic cells have been discovered in sediments and rock reaching a subsurface depth of 1 km. These microorganisms have few, if any cultured or known relatives in the surface world and are still only characterized by the genetic code of their DNA. Inagaki *et al.* (2006) and Biddle *et al.* (2006) have showed how they differ from microorganisms in the surface world and on the energy sources that sustain life in this buried ecosystem. The vast majority of sediment-associated microbes (about 97 per cent) occur in the upper 600 m of sediment. Microbial cell numbers range from 10^8 cells per gram of sediment just below the seafloor, to about 10^4 cells per gram of sediment 0.5 km deep in the subsurface. The subsurface microbial biomass is quite substantial. Studies conducted by D'Hondt *et al.* on deep sea sediment cores recovered from the equatorial Pacific Ocean off the coast of Peru led them to the following conclusions, some quite unexpected. D'Hondt *et al.* found substantial numbers of microbes in sediment samples, which decreased with increasing sediment depth. They also measured potential respiratory electron acceptors (oxidants), such as sulphate and nitrate, whose flux can serve as markers of specific microbial activities, because certain microbes use them to respire in the absence of oxygen. The presence or absence and distribution of such other microbial metabolic byproducts as carbon dioxide, ammonia, sulphide, methane, manganese, and iron also provide measures of microbial activity. Profiles of these biologically processed compounds serve as indicators of which metabolic pathways may be crucial (DeLong, 2004).

Whitman *et al.* (1998) had demonstrated that the "unseen majority" of microorganisms probably accounts for up to 85 per cent of Earth's prokaryotic biomass and about 30 per cent of the total living biomass.

Recent sampling work has focused on the eastern tropical Pacific, with sites ranging from the continental shelf to ocean depths of 5000 m, involving drilling through the seafloor and–at open-ocean sites–down to the basaltic crust, sediments with ages up to 35 million years old (Jorgensen *et al.*, 2006).

At all sites bacteria and archaea were found to be present below the seafloor. Their numbers dropped from more than 10^8 cm^{-3} at the sediment surface to less than 10^6 cm^{-3} just above the ocean crust, with an average density much greater than in the ocean above. Sometimes cell numbers up to 10^{10} cm^{-3} coincided with sediment horizons in which more energy was available from counterdiffusing methane and sulfate (Parkes *et al.*, 2005).

These large population sizes point to a great mystery of the deep biosphere. Although marine sediments harbour Earth's largest reactive carbon pool, the organic matter becomes more and more unreactive with depth and age and probably becomes inaccessible for microorganisms several million years after its burial. This prompts the question: how can there be sufficient energy for all these organisms to metabolize and grow?

Throughout their sediment cores, D'Hondt *et al.* detected such microbial processes as carbon oxidation, methane production and consumption, and reduction of sulphate, nitrate, and manganese. The existence of these processes deep within marine sediments is not so surprising as their location. Normally, electron acceptors (oxidants) diffuse into sediments from the overlying seawater and are then consumed sequentially in a predictable series of metabolic reactions, producing a microbially catalyzed oxidant-depletion profile in which oxygen is reduced first, then nitrogen, manganese, iron, sulphate, and finally carbon dioxide. These profiles probably reflect competitive processes that deplete available oxidants, with those yielding the greatest free energy being the first to be consumed (Canfield *et al.*, 1993; DeLong, 2004).

The metabolic activity of the subsurface populations can be calculated by transport-reaction modeling of pore water solutes that are used or excreted by the microorganisms (Jorgensen and D'Hondt, 2006). For the eastern Pacific seabed, the mean sulfate respiration is 10^{-18} mol per cell per year. Microbial cells need to metabolize a certain minimum amount of substrate before they can divide into two daughters. Calculating their minimum doubling time, the mean generation time of deep sea floor microorganisms is more than 1000 years (Jorgensen and D'Hondt, 2006). But such an extremely slow growth cannot be reconciled with our knowledge of the minimum energy requirements for life. All actively growing organisms have to operate their enzymatic machinery above a critical level to maintain vital cell functions such as replacement of degraded enzymes, repair of DNA damaged by high-energy radiation from natural radionuclides, and possibly the maintenance of an electrochemical gradient across the cell membrane (Price and Sowers, 2004).

One possible explanation for the low apparent rates of deep subsurface metabolism is that most subseafloor cells are either dormant or dead. However, a highly sensitive fluorescence technique used to detect the presence of ribosomes showed that many of the subsurface cells were alive (Schippers *et al.*, 2005).

The profiles of electron acceptors in deep sediments observed by D'Hondt *et al.* deviate substantially in some ways from the norm, pointing to unsuspected sources of microbial metabolites within subseafloor sediments. D'Hondt *et al.* reported that oxidants that normally diffuse downward from overlying seawater can also enter the sediments from subseafloor sources. In some cases sulphates had originated from brines below the sediment base; also nitrate and oxygen could enter from deep basaltic aquifers underneath the sediment column. This situation creates "upside-down" redox profiles, with atypical sources from beneath sediments providing oxidants such as sulphate and nitrate that enable microbes to respire anaerobically (DeLong, 2004). It appears that a physiologically diverse and active deep-sediment microbiota operates rather differently from model predictions (DeLong, 2004). Also that microorganisms in the deep subsurface (and their energetics) sometimes can differ substantially from well-studied model microorganisms in shallow near-surface sediments.

The identity and physiological state of the microbes of the deep subsurface are currently being determined with the effective toolbox of DNA-, RNA-, and biomarker-based techniques. DNA encoding 16S ribosomal RNA (a key gene for the phylogenetic identification of prokaryotic organisms) extracted from sediments provides thousands of genetic codes revealing novel lineages of microbial life. Most of the genetic types belong to groups that have no cultured relatives (Inagaki *et al.*, 2006).

In the context of the extremely low energy flux per cell in the deep subsurface, the search for additional energy sources has focused on molecular hydrogen (H_2), which is generated by chemical alterations in young basaltic crust along the mid-oceanic ridges. But most of the seabed lies on old, crack-permeable crust, in which the potential oxidants for H_2 (such as oxygen or nitrate) persist long enough to preclude a substantial H_2 supply. Another source of H_2 may be the decay of natural radionuclides of potassium, thorium, or uranium in the sediments; energy released by this decay dissociates water molecules into free radicals and H_2. Hence, this nuclear energy is not only destructive to microbial cells but may also support their metabolic activity (Jorgensen and D'Hondt, 2006). . According to Lin *et al.* (2005), water radiolysis could be the chief source of microbial energy in deep-sea sediments that are more depleted in organic matter than the eastern tropical Pacific sites discussed here. Such sediments with extremely low organic carbon flux cover large regions of the ocean floor, for example, in the central North and South Pacific Ocean.

This potential energy source is independent of biomass production by photosynthesis. In fact, it does not require an external oxidant. Water radiolysis produces not only H_2 but also oxidants such as H_2O_2 or O_2, which could be directly used for the energy-generating reoxidation of H_2. The rich communities at deep-sea hydrothermal vents also live on inorganic chemical energy, for example, from H_2 or H_2S, but they depend on O_2 produced from photosynthesis. An extreme low-energy subsurface biosphere driven by radioactivity would differ from all other ecosystems on Earth: it could proceed on a planet without surface life and solar energy (Jorgensen and D'Hondt, 2006).

Non-culture based methods of surveying microbial diversity suggest that a quite different suite of indigenous subsurface archaea and bacteria may predominate deep within sediments (Inagaki *et al.*, 2003; Newberry *et al.*, 2004), representing the indigenous, active members of deep-sea microbial communities. Subsurface microbes thriving deep in marine sediments ultimately rely on energy sources and oxidants produced from sunlight, rather than subsisting on geochemicals diffusing from Earth's interior.

Rare Bacteria in the Marine Biosphere

Introduction of molecular methods into microbial ecology in the 1980s (Pace *et al.*, 1986) yielded two surprises: (a) most DNA sequences found had no resemblance to organisms known from cell culture, and (b) the microbes that could be cultured were not found in molecular surveys. The very rare bacterium *Leeuwenhoekiella blandensis* isolated from the Mediterranean Sea has been cultured in the laboratory. It has not been retrieved by molecular methods. These surprises pointed to the probability that microbial diversity is really much greater than had been supposed. Further, some bacteria occur at such low abundance that molecular techniques are unable to detect them. In the past three decades, however, marine microbiology has become a very active and innovative field. Two recent reports by Markert *et al.* (2007) and Not *et al.* (2007) clearly demonstrate that today's molecular techniques can effectively reveal geographic distribution and characterize the symbiotic relationships which enable organisms to survive in extreme habitats. Also recent advances in sequencing technology have the potential to throw valuable light on the huge dimensions of the microbial diversity that has been hidden in nature (Pedros-Alio, 2006, 2007).

The marine environment constitutes the largest ecosystem on Earth that supports the richest wealth of rare species of microorganisms. Current estimates of the total number of bacterial species range from a few millions to hundreds of millions (Curtis *et al.*, 2002). A milliliter of seawater can contain 1 million bacterial cells, and quite a large number of species may be represented by just one cell each. However, because of the enormous volume of the oceans, even species represented by billions of cells would be rare!

One notable property of this rare biosphere (Sogin *et al.*, 2006) is that a single cell of a rare bacterial taxon can become abundant simply by clonal replication if conditions change and become suitable for its growth. This is in sharp contrast with sexual organisms, where finding a mate often becomes very improbable below a certain threshold abundance. Indeed, many small eukaryotes grow and divide without the need for a sexual partner and are not limited by being rare (Pedros-Alio, 2006, 2007).

A highly unlikely death is another special property of the rare bacteria (Pedros-Alio, 2006, 2007). The major causes of death for bacteria are attack by bacteriophages and predation by small heterotrophic protists that feed on bacteria. Viruses rely on encounters to find their prey, and if an organism is so rare, it can escape being attacked by its specific virus. Actively growing bacteria are usually bigger than starving bacteria, and protists selectively prey on the largest and most active bacteria (Pernthaler, 2005). Rare taxa are usually starved and grow slowly as compared to abundant taxa. So rare taxa are typically represented by smaller cells and tend to be overlooked by grazers–they persist longer in the environment. So, any habitat will have a very large biodiversity, formed by a few dozen abundant taxa plus a large collection of rare taxa that current molecular methods do not retrieve (Pedros-Alio, 2006, 2007).

Pure cultures are the first approach to explore this biodiversity. In principle any microorganism, regardless of its abundance, can be retrieved if the right selective conditions are created (Zengler *et al.*, 2005).

Another approach is based on using polymerase chain reaction (PCR) amplification of 16S ribosomal RNA genes found in a sample. These are then used to construct clone libraries. Typically, a certain number of clones is sequenced, and by applying statistical methods, the total number of different taxa in the library is estimated. This approach has helped to retrieve up to 500 taxa from marine samples, and estimates of the total diversity reach 2000 taxa in one library (Acinas *et al.*, 2005). Use of molecular tricks in PCR helps to find particular microorganisms–specific primers for certain groups of microorganisms can be used in the PCR reaction, or performing a nested PCR (Ward, 2002). These in turn can retrieve even rare sequences belonging to the target group. But this approach only finds the target group, not discover novel sequences. To critically probe the depths of the rare biosphere, the best way is to sequence a huge number of clones. Venter *et al.* (2004) adopted a brute-force shotgun sequencing approach to analyze samples from the Sargasso Sea. This method does not use PCR and so avoids bias against rare sequences. Its drawback is that the clone library contains every gene of all the microorganisms present in the sample. After sequencing 1 billion base pairs, Venter's group found 1800 taxa. Sogin *et al.* (2006) used a sequencing-by-synthesis approach to analyze some deep-water samples from the North Atlantic. To avoid the necessity of sequencing all genes present, they used one PCR step. Although all their sequences were relevant for diversity, they still suffered from the PCR bias. One of their samples revealed 5266 different taxa but statistical estimates pointed to a total richness of 18,191 different taxa (Pedros-Alio, 2006, 2007).

In the coming years, the number of sequencing reactions will increase to hundreds of millions, so we may anticipate the discovery of novel sequences to increase by at least another order of magnitude. The challenge is how to understand the bigger picture behind hose sequences. Can we exploit these genes in medicine or biotechnology?

Deep-Sea Microbes: *Photobacterium profundum*

Deep-sea life requires adaptation to high pressure, an extreme but common condition–oceans cover 70 per cent of Earth's surface and have an average depth of 3800 meters. Survival at such depths requires specific adaptation. *Photobacterium profundum* strain SS9 is a model for piezophily and shows high metabolic versatility.

Piezophiles have evolved in multiple lineages of the Bacteria and Archaea domains of life. Some high-pressure-adapted vertebrates and invertebrates have also been characterized. The genome of *P. profundum* has been sequenced. SS9 was previously isolated at a depth of 2500 m. It grows over a broad (90 MPa) pressure range and can be genetically manipulated. As a member of the family Vibrionaceae, it is related to a number of piezosensitive microbes for which whole-genome sequence information is available for comparison (Vezzi *et al.*, 2005).

The genome of SS9 shows a tripartite structure: a 4.1-Mbp major circular chromosome (chr. 1), a 2.2-Mbp minor circular chromosome (chr. 2), and an 80-kbp circular plasmid. The presence of two circular chromosomes is common to other Vibrionaceae, but the SS9 chromosomes are about 25 per cent larger than those of *Vibrio parthaemolyticus* and *Vibrio vulnificus* YJ016, the most closely related genomes to that of SS9 so far sequenced.

SS9 contains the maximal number of ribosomal RNA (rRNA) operons so far identified in a bacterial genome, 14 on chr. 1 and 1 on chr. 2. This probably reflects the ability to respond rapidly to favourable changes in growth conditions.

An overview of the genome reveals features probably related to the deep-sea environment. Notably missing from the SS9 genome are ORFs encoding light-activated photolyase genes; this is consistent with the absence of sunlight in the deep sea. A very uncommon trait is the presence of two complete operons for F_1F_0 ATP synthase, one on each chromosome. SS9 contains three complete sets of cbb3 cytochrome oxidase genes; the one on chr. 2 was possibly acquired from an α-proteobacterium, along with an unusual diheme cytochrome *c* gene. These findings support the idea that modified electron and proton transport are necessary for metabolic activity at high pressure. The reductases include those for both assimilatory and dissimilatory nitrate reduction, as well as those for tetrathionate, dimethylsulfoxide, fumarate, sulfite, and trimethylamine-N-oxide.

To obtain a comprehensive picture of high-pressure adaptation, Vezzi *et al.* used microarray technology to compare the transcriptional profile of SS9 shown at 45, 28, and 0.1 MPa. This led to the identification of 260 differentially expressed genes that could be grouped in 14 classes according to their Gene Ontology Biological Process (GOC, 2004). Comparison of the two different pressure conditions shows that most genes for amino acid transport, ion transport, protein folding, and glycolysis are up-regulated at 0.1 MPa.

Transcription is more active on chr. 1 than on chr. 2. Almost all of the highly expressed genes are on chr. 1. Complex carbohydrates are an important carbon source in oceanic abyssal environments as polymers sink down from shallower waters. The regulation of metabolic pathways for the degradation of different polymers, *e.g.*, chitin, pullulan, and cellulose is controlled by pressure, being activated at 28 MPa and turned off at 0.1 MPa (Vezzi *et al.*, 2005).

That SS9 is a true piezophile is confirmed by the observation that several stress-response genes are activated at atmospheric pressure. Indeed, four genes upregulated at 0.1 MPa are involved in protein folding and in response to stress conditions: *htpG*, *dnaK*, *dnaJ*, and *groEL*. In *Escherichia coli*, the abundance of the proteins encoded by these genes is known to increase after a high-pressure shock (Welch *et al.*, 1993).

Although the deep sea represents the commonest environment in the biosphere, adaptation to its prevailing conditions is still not well understood. Vezzi *et al.* have shown that SS9 has specific metabolic, regulatory, and structural adaptations to deep-sea conditions–biodegradation of relatively recalcitrant carbon sources is turned on at high pressure. Further, the sensitivity of SS9 to low pressure is borne out by the activation of different chaperones and DNA repair proteins at atmospheric pressure.

Symbiotic Bacteria in Deep Sea Hydrothermal Vents

Rich communities of animals living around deep-sea hydrothermal vents have been found in the eastern Pacific along with many other chemoautotrophic-based communities and animals. Such vents have an abundance of reduced chemicals (such as sulfide or methane) at the sea floor. Many such sites all over the world, *e.g.* mineral-rich hydrothermal vents and the oil- and gas-rich hydrocarbon seeps, are adversely affected by human activities. Assessing the impacts of these activities depends on a detailed understanding of the lifestyles of the creatures that thrive in this extreme habitat. Powerful molecular tools and the approach described by Markert *et al.* (2007) can enhance understanding of the abundant chemoautotrophic symbioses of the deep sea (Fisher and Girguis, 2007).

One conspicuous member of the hydrothermal vent community is the tubeworm *Riftia pachyptila*. It is a polychaete that has no mouth, gut, or anus and cannot feed by normal means. Instead, it depends on intracellular chemoautotrophic symbionts–which fill a large internal organ called the trophosome -for nutrition. The symbionts are γ-proteobacteria that functionally resemble plant chloroplasts in that they generate organic carbon as a food source for their worm host (using sulfide as an electron donor and oxygen as an electron acceptor). A clear understanding of the biology of their symbionts is crucial to understanding the biology of *Riftia* and other chemoautotrophic symbioses. Unfortunately, because no chemoautotrophic symbiont could so far be cultured in a laboratory, we know little about their metabolism. This lacuna has recently been partly filled by the work of Markert *et al.* (2007) who used a proteomic approach to examine protein expression in *Riftia* symbionts and to understand something of their biochemistry and metabolism.

Hydrothermal vents represent potentially dangerous environments. Temperatures can range from near freezing to over 300°C over centimeters. Within animal communities, temperatures can vary over more than 40°C within a few seconds. These vents are also ephemeral, with local sources of hydrothermal flow lasting only a few years. The habitat fluctuates, and vent fauna have to balance exposure to the hot and potentially toxic vent fluid with the need to obtain nutrition either directly from the fluid or from microbes found in the fluid. *Riftia* is well adapted for its symbiotic life style in these vents. These worms live with their vascularized gill-like plumes exposed to vent fluid and contain circulating hemoglobins that bind to both oxygen and sulfide reversibly and with high affinity and capacity (Arp and Childress, 1983). These conditions can enable *Riftia* to absorb and store these chemoautotrophic substrates, transport them through its tissues, and provide its symbionts with a plentiful supply of both. This makes the symbionts extremely efficient and productive–they fix carbon at high rates to support the host's growth (Girguis and Childress,

2006). Such adaptations also enable *Riftia* to grow fast and be fecund, while depending on its symbionts for nutrition.

Exploiting the power of genomic analyses, coupled with high-throughput protein profiling to obtain a snapshot of the proteins (or proteome) expressed by the *Riftia* symbiont, Markert *et al.* studied the extent to which *Riftia* symbionts achieve high rates of chemoautotrophic carbon fixation powered by sulfide oxidation. They found that 12 per cent of the total cytosolic proteome of these symbionts consists of three proteins involved in coupling energy production to sulfide oxidation–a notable departure from fast-growing, free-living heterotrophic bacteria that instead expend a substantial part of their energy synthesizing amino acids for their own cell division and growth. The dominance of these three proteins highlights the crucial importance of the symbionts to provide nutrition to the association by harnessing energy from sulfide (Fisher and Girguis, 2007).

Markert *et al.* observed quite high amounts of enzymes involved in the reductive tricarboxylic acid cycle in extracts of the *Riftia* symbiont. This might well be an important pathway of carbon fixation by the symbiont, with strong implications for some more energy-efficient carbon fixation.

As no more than 1 per cent of the microbes present in nature have been successfully cultured in the laboratory (and no chemoautotrophic symbiont has been cultured at all)–we may turn to genomic and proteomic approaches which provide valuable information on the symbiont's metabolic capabilities and evolutionary history. Quantitative proteomics also allows a researcher to use protein expression levels as a metric for assessing the importance of metabolic pathways used by these symbiotic microbes *in situ*.

Large Sulfur Bacteria and the Formation of Phosphorite

Phosphorite deposits in marine sediments are a long-term sink for the essential nutrient phosphorus which gets removed from the biosphere in this process. The first step in phosphorite formation is the precipitation of phosphate-containing minerals such as hydroxyapatite, followed by some other processes such as sediment transport, winnowing, and re-crystallization (Föllmi, 1996). A basic problem in explaining massive phosphorite deposits has related to identification of mechanisms that can concentrate pore water phosphate so much as to drive spontaneous precipitation of phosphorus minerals (Schulz and Schulz, 2005). These authors have suggested that apatite abundance in sediments on the Namibian shelf correlates with the abundance and activity of the giant sulfur bacterium *Thiomargarita namibiensis*, pointing to the possibility that sulfur bacteria drive phosphogenesis. Sediments populated by *Thiomargarita* showed sharp peaks of pore water phosphate (?300 micromolar) and massive pohsphorite accumulations (?50 grams of phosphorus per kilogram). Laboratory experiments revealed that under anoxic conditions, *Thiomargarita* released enough phosphate to account for the precipitation of hydroxyapatite observed in the environment. This abundant benthic bacterium (*Thiomargarita*) is specially adapted to survive under both oxic and anoxic conditions. It periodically contacts oxic bottom water to take up nitrate, and it can survive long periods of anoxia with nitrate stored internally (Schulz *et al.*, 1999). The phosphate uptake from different sources can take place either when *Thiomargarita* forms thick mats at the sediment surface or is suspended in the oxic water column (Schulz and Schulz, 2005).

While polyphosphate occurs in most living organisms, only some bacteria (such as *Thiomargarita* and *Beggiatoa*) and yeasts can accumulate large amounts. Bacterial phosphate accumulation has been most extensively investigated in waste-water treatment plants, where bacteria are used to remove phosphate. To induce luxury uptake of phosphate by bacteria in a waste-water treatment plant, an anaerobic phase needs to be introduced whereby phosphate is released and acetate is taken up and stored, *e.g.*, in the form of polyhydroxyalkanote (PHA). Uptake and storage of acetate need energy which the bacteria can derive from the breakdown of polyphosphate and consequent release of phosphate. In the following aerobic phase the polyphosphate-accumulating bacteria get energy by oxidizing the stored carbon using oxygen as the electron acceptor, and they take up an excess of phosphate, which they store as polyphosphate (Mino, 2000). The above reactions result in a sludge rich in bacterial polyphosphate, which can be eliminated from the system (Schulz and Schulz, 2005).

The mechanisms of phosphate uptake and release in *T. namibiensis* appear to be similar to that of polyphosphate-accumulating bacteria in wastewater, even though their main energy source is considered to be the oxidation of sulfide with nitrate or oxygen as the electron acceptor (Schulz et al., 1999).

T. namibiensis appears to have an unusual life mode–under anoxic conditions, it takes up sulfide and, perhaps acetate, which may be stored as glycogen. To address the problem of insufficient supply of a suitable external electron acceptor, some internally stored nitrate and polyphosphate are sacrificed and sulfide is oxidized to elemental sulfur to obtain energy. When conditions are oxic, the bacterium can derive energy from the oxidation of both sulfur and, presumably, glycogen. Simultaneously it also stores energy by accumulating polyphosphate and nitrate, the latter of which is stored in a central vacuole at concentrations of up to 0.8 M. Thus, *T. namibiensis* takes up these chemical compounds under conditions where they are readily available and use them under different redox conditions, when they are a valuable energy source that would otherwise be impossible to obtain at that time.

Marine Microbial Diversity and Drug Discovery

For long, culturable microorganisms have served as a rich reservoir of natural product drug discovery. They provide unique chemical compounds having direct application in curing diseases. These compounds also provide new chemical scaffolds from which highly efficacious molecules were derived. Largely because of some recent deceleration in natural product research in terrestrial habitats, interest in the exploration of marine microorganisms for novel metabolites increased following use of the biological and chemical techniques developed for such studies in terrestrial environments (Li and Qin, 2005).

According to Whitman *et al.* (1998), marine environments, including the subsurface, harbour ~3.67 x 10^{30} microorganisms which represent an extraordinary and highly dynamic gene pool of biodiversity. Most of these microbes have never been cultured, identified or classified, and their enormous chemical richness has not been tapped (DeLong, 1997). Fortunately, this situation has changed in the recent past as a result of the commendable progress made in phylogenetic studies based on small ribosomal RNA (rRNA) and genome-related technology independent of standard microbial cultivation (Pace, 1997; Hugenholtz, 2002).

The continuing rise of the global threat of drug-resistant pathogens warrants development of effective new strategies and resources to speed up and advance the drug discovery process. Rapid progress in metagenomics has heralded a new era in the study of marine microbial diversity that enables direct access to the genomes of uncultured microorganisms (Li and Qin, 2005).

Application of metagenomics has revealed more of the oceans' genetic diversity than has ever previously been seen. Sea-water samples collected by a yacht called Sorcerer II of the J. Craig Venter Institute, Rockville, MD, on a trip around the world contained gene fragments predicted to represent more than six million proteins–enough to almost double the number of proteins listed in online databases. The data set included new members of nearly all known protein families from bacteria and archaea, plus many viral genes, suggesting that viral diversity also has been undersampled so far (*PLoS Biol.* 5: e16, 2007).

Marine microbial communities, and their ecofunctionalities are best defined by the interaction between the biotic and abiotic components of marine ecosystems at both cellular and molecular levels. We lack a clear understanding of this interaction, partly due to the dominance of uncultured microbial inhabitants and the paucity of information on the metabolism of these uncultured organisms.

However, understanding of microbial communities has substantially increased over the past two decades from culture-independent phylogenetic studies, together with the recent development of oligotrophic culturing strategies (Stevenson *et al.*, 2004). Fifty three phylum-level lineages of prokaryotes, based on 16S rRNA sequence data (Handelsman, 2004), have been established and forms the foundation of modern bacterial taxonomy (Boone *et al.*, 2001). Importantly, some of these phyla comprise only marine uncultured microbes.

Much effort has gone into crossing the cultivation barrier by creating conditions that mimic the unique growth conditions and special nutrient composition of marine environments (Connon and Giovannoni, 2002; Kaeberlein *et al.*, 2002). Many organisms belonging to *Proteobacteria, Planctomycetes, Bacteroidetes, Acidobacteria* and *Verrucomicrobia* have already been cultivated.

The marine actinobacteria represent an interesting group whose population in the sea is greatly underestimated (Piza *et al.*, 2004). The actinobacterial population of sediments, from an uncontaminated marine site and a brackish water estuary, represents about 10 per cent of the total bacterial population of these sites, respectively. The uncontaminated marine site revealed a greater diversity although the total population was higher in the brackish estuary. However, the population of actinobacteria seems to be less than 5 per cent in oligotrophic seawater, a figure very close to that of the Firmicutes and slightly less than cyanobacteria (Venter *et al.*, 2004). Conceivably, the metabolic richness and functionalities of these marine actinobacteria could possibly brighten the prospects for the pharmaceutical industry if these organisms are successfully grown in culture. The genus *Salinospora* has been recently created to accommodate a widespread actinomycete taxon present in marine sediments (Mincer *et al.*, 2002). The novel compound, salinosporamide, isolated from some strains of *Salinospora*, shows strong anti-tumour activity (Feling *et al.*, 2003).

A useful approach at the genome level has been the massive marine metagenome sequencing project conducted at the Sargasso Sea near Bermuda by Venter *et al.* (2004). About 1800 new uncultured marine prokaryotic species were estimated from ~1500 litres of oligotrophic seawater, along with several species that have had their genome sequencing completed. Indeed, the work of Venter *et al.* represents the first metagenomic-sequencing project applied to the marine environment. It provides a very useful direction for future exploration of marine microbial diversity (Li and Qin, 2005).

Oceanic environments show a tremendous variety of pressure, salinity, temperature and nutrients. These environments enable marine microorganisms to develop unique capabilities for survival. This potentially offers an abundance of secondary metabolites that might differ from the metabolites produced by terrestrial microorganisms. Over 15,000 structurally diverse natural products with an amazing assortment of bioactivities have already been reported from marine microbes, algae and invertebrates (Salomon *et al.*, 2004). From such a large number of products, only three kinds of drugs originating from marine organisms are in the pharmaceutical market: cephalosporins as effective anti-microbials, cytarabine as an anti-cancer agent, and vidarabine as an anti-viral agent (Haefner, 2003). Happily, a few dozen drugs derived from marine microbes are currently either in clinical trials or are being clinically evaluated.

It is speculated that a large variety of biologically active substances originally isolated from marine invertebrates may in fact be produced by microorganisms that are associated with these invertebrates. Sponges and ascidians typically contain a variety of bacteria and archaea as symbionts (Proksch *et al.*, 2002). Indeed, some natural products derived from sponge show structural similarities with those produced by terrestrial insect symbionts (Piel *et al.*, 2004).

Marine chemodiversity constitutes a major target in the quest for natural drug products, and much effort is now devoted to this aspect to fulfil the ever-growing demand for new antibiotics and more effective medicines for cancer, heart and vascular diseases, and other diseases.

Marine microbial communities not only make up over 80 per cent of life on earth but also have a crucial role in primary energy and carbon recycling (Azam and Worden, 2004). Current assessments of this diversity based primarily on phylogenetic approaches have played an important role in the foundation of modern microbiology and taxonomy. Metagenomics enables direct access to the genome complex of marine ecosystems for heterogeneous gene expression and functional exploitation of those marine microorganisms which have not been cultured. Intensive research on metagenomics and marine microbial diversity has provided invaluable DNA sequencing data (Moreira *et al.*, 2004), and some few studies were involved in functional analysis of the gene clusters identified (Piel *et al.*, 2004). The accumulation of metagenomic sequencing data such as that gathered by Venter *et al.* at the Sargasso Sea near Bermuda, together with increased knowledge of the metabolism and functionality of uncultured microbes, may initiate a new era for the understanding and systematic investigation of the genetically-variable reservoir of bioactive products of marine microorganisms (Li and Qin, 2005).

Chapter 8
Microbial DNA, Genes and Genomes

Marine Microbial DNA

A world-wide hunt for marine microbial diversity has revealed an underexplored world of genes, proteins, and "species" in the sea. J.C. Venter and his associates reported over 7 million snippets of sequence, dubbed the Global Ocean Sampling, in three online papers published in *PLoS Biology* (2007). Venter *et al.* provided evidence of so many new microbial species that it warrants redrawing the tree of microbial life. What is more, the sequences have been translated into hypothetical proteins and some educated guesses have been made about their possible functions. The diversity of microbes uncovered appears enormous, so it will be very difficult to understand.

In fact a few years ago also the American millionaire genome sequencer had studied microbes living in the Sargasso Sea and had turned up the largest DNA sequencing of environmental samples ever accomplished (*Science*, 2 Apr. 2004). In 2007, he sailed from Halifax (Canada) through the Panama Canal and finished up 6500 km southwest of the Galapagos. He undertook this expedition in the *Sorcerer II*, his private yacht turned into a floating laboratory.

The researchers sampled some 40 locations, isolated and froze bacterium-sized cells, recorded the temperature, salinity, pH, oxygen concentration, and depth. Upon bringing the samples back to Venter's institute, the DNA was extracted and sequenced. Using a whole-genome shotgun approach, all the DNA in a sample was broken into fragments of specific sizes. These were sequenced, and assembled together by matching the ends of the DNA with a powerful overlap-hunting computer program. This approach can allow the reconstruction of entire genomes of the different organisms in any sample (Bohannon, 2007).

One conclusion that emerged from the study is that DNA in a typical community of marine microbes is so diverse that nothing close to a whole genome can be assembled, even with the best possible sequencing methods. Half of the 7.7 million DNA sequence fragments were so different that they could not be linked at all.

Nonetheless, the researchers did estimate the number of species in the samples based on slowly evolving marker genes, identifying over 400 microbial species new to science, and more than 100 of those were sufficiently different to define new taxonomic families.

It may seem strange but it is true that researchers are still unable to make a full accounting of the species contained in a drop of seawater. Indeed, it is this enormous genetic diversity that some have termed the "paradox of the plankton". As per traditional ecological theory when many species compete for the same resources–in the case of ocean microbes, light and dissolved nutrients–then only one or a few species should eventually outcompete the rest. If that were true, then many of the sequences gathered from the marine waters by Venter should map down onto a few dominant genomes. In contrast, however, the data obtained point to a countless mob. Most of the microbes that were trapped in Venter's filters were genetically unique, suggesting that there is actually an astoundingly extensive gene pool in the ocean.

It is possible that such a high variety of genes endows each species with sufficiently different metabolic tool kits to exploit slightly different combinations of resources, including the waste products of others, such that they can all coexist (Bohannon, 2007).

The newly detailed diversity also indicates that microbial taxonomy is in urgent need of thorough overhaul. The current taxonomy classifies microbes into different "ribotypes" by comparing the sequence of the highly conserved genes of the protein-synthesizing ribosome. However, in view of the fact that there is so high diversity within the DNA even after dividing them into ribotypes, Venter *et al.* propose to reject ribotyping altogether. Instead, they like to define groups of microbes based on the environment from which they were collected and how well their DNA matches a reference set of fully sequenced marine microbial genomes. This strategy has enabled Venter *et al.* to group sequence fragments into different "subtypes", with each of these subtypes representing a "distinct, closely related population" of microbes that fill a particular niche in their environment.

Another amazing outcome of work done on the samples collected during the expedition is that, based on the beginning and end of each gene teased out from the DNA sequences, Venter Institute researchers concluded that the DNA encodes 6.12 million hypothetical proteins–a finding that almost doubles the number of known proteins. Most of the predicted proteins are of unknown function, and a quarter of them show no similarity to any known proteins. Hopefully, some of these might be exploited to develop new synthetic materials, clean up pollution, or bioengineer fuel production (Bohannon, 2007).

The hypothetical proteins have opened up new windows to basic microbial biology. Comparison of predicted amino acid sequences with those of known proteins revealed a surprising abundance of signaling proteins thought to be used only by multicellular organisms! Among the hypothetical proteins from their marine samples, 28,000 of the so-called eukaryotic protein kinases, as well as another 19,000 of a group that are highly similar to these kinases have been found–three times the number previously known.

Indeed, the wealth of data produced by Venter's voyage constitutes a good starting point for protein, gene, and microbe discovery. In a way, Venter may be hailed as the Darwin of the 21st century!.

To enable researchers to deal with Venter's 100 gigabytes of sequence data and also other relevant information about a microbe's environment and location, Venter's team, in collaboration with some computer scientists in California, has built a metagenomics version of GenBank, the online genetic database that was curated by the National Center for Biotechnology Information in Bethesda, Maryland. Besides making the typical gene searches and genome comparisons, the new system, known as the Community Cyberinfrastructure for Advanced Marine Microbial Ecology Research and Analysis, is able to hunt for correlations between DNA sequence and environment for clues about co-occurring microbes.

Role of Extracellular DNA in Deep-Sea Ecosystem Functioning

The regeneration of phosphates from the organic P pool is crucial in the regulation of P availability in the oceans. What is not known, however, is why organic P is preferentially recycled in the deep ocean, where phosphates are not limiting. This undermines our understanding of the P cycle and ecosystem functioning on a global scale (Clark *et al.*, 1998).

Although DNA is a P-rich molecule, its role in P cycling has attracted only meagre interest–it is usually considered only as the genetic material associated with living biomass. Dell'Anno and Danovaro (2005) showed that DNA concentrations in deep-sea sediments worldwide are quite high (0.31 ± 0.18 g of DNA m^{-2} in the top cm), also that over 90 per cent of the DNA in these sediments is extracellular. The extracellular DNA fraction was estimated by calculating the difference between total DNA and the DNA within living biomass, determined by synoptic determinations on virus, prokaryote, protozoan, and small metazoan (meiofauna) abundances. Another, nuclease-based procedure that degrades only the extracellular DNA, supported this finding and revealed that 60 ± 4 per cent of the total DNA pool is enzymatically digestible.

Critical polymerase chain reaction and dot blot analyses showed that extracellular DNA estimates were not biased by the DNA released from living biomass during recovery. Worldwide estimates made by Dell'Anno and Danovaro indicate that the DNA content in the top 10 cm of the deep-sea sediments is 0.50 ± 0.22 Gt. This means that extracellular DNA in deep-sea sediments (0.30 to 0.45 Gt) constitutes the largest reservoir of DNA in the world ocean. This amount is six- to eightfold greater than that of DNA contained in all benthic prokaryotes inhabiting the uppermost 10 cm of the world marine sediments (Whitman *et al.*, 1998).

Certain pelagic-benthic coupling processes control the extracellular DNA distribution in global ocean sediments. This is borne out by the observed relationships between (i) DNA and phytopigment fluxes; (ii) DNA and phytopigment concentrations in surface sediments; and (iii) downward fluxes of DNA and sedimentary extracellular DNA content (Dell'Anno and Danovaro, 2005). The input of DNA from the upper (lighted) layer to the deep sea stimulates the production of benthic prokaryotes, which represent ~90 per cent of the total biomass (Danovaro *et al.*, 2001). Estimates indicate that the total DNA input to the sea floor is $1.26 \pm 0.18 \times 10^7$ metric tons year^{-1}, and extracellular DNA accounts for 13 per cent of the total organic P flux (9.91×10^6 metric tons year^{-1}, below 1000 m in depth) to the deep sea (Delaney, 1998). According to Dell'Anno and Danovaro (2005), the contribution of the P associated with extracellular DNA to the total organic P pool is ~3 per cent (= a global projection of 0.4 Gt of organic P) and the residence time, in the top cm, is 40.3 years for organic P and 9.5 years for DNA. The measurements of high deoxyribonuclease activities (0.50 ± 0.15 mg of DNA m^{-1} day^{-1}) directly prove that the extracellular DNA pool substantially contributes to P cycling. It was the application of a diagenetic model that enabled estimation of the extracellular DNA remineralization in the top 10 cm of the sediment also that it accounts for 17 per cent of the total organic P regeneration (range, 14 to 21 per cent). Thus it appears that the extracellular DNA in deep-sea sediments is selectively remineralized within the organic P pool and rapidly degraded (Dell'Anno and Danovaro, 2005).

Life in the deep-sea sediments is mostly dominated by prokaryotes (Turley, 2000). Extracellular DNA is a good C and N source, and so is critical for deep-sea benthic prokaryote metabolism. Dell'Anno and Danovaro calculated that the use of extracelular DNA alone supplies 4, 7, and 47 per cent of daily prokaryotic, C, N, and P demand, respectively. This is how the use of a labile C and N source found in extracellular DNA causes fairly fast P regeneration. The work of Dell'Anno and Danovaro suggests that the availability of extracellular DNA has strong implications for deep-sea ecosystem functioning, as it contributes substantially to P cycling and represents a key trophic resource.

Single-molecule DNA Sequencing

Harris *et al.* (2008) reported an amplification-free method for determining the nucleotide sequence of more than 280,000 individual DNA molecules simultaneously. A DNA polymerase adds labeled nucleotides to surface-immobilized primer-template duplexes in stepwise fashion, and the asynchronous growth of individual DNA molecules can be monitored by fluorescence imaging. Read lengths of >25 bases and equivalent phred software program quality scores approaching 30 have been achieved. The method has been used to sequence the M13 virus to an average depth of >150x and with 100 per cent coverage; thus Harris *et al.* could resequence the M13 genome with high-sensitivity mutation detection, demonstrating a method for high-throughput low-cost resequencing.

Although single-molecule sequencing has been known since 1989 (Jett *et al.*, 1989), more recent work of Braslavsky *et al.* (2003) points to the superiority and feasibility of single-molecule sequencing using DNA polymerase to sequence by synthesis, and a later study of single-RNA polymerase activity has showed that DNA sequence can be inferred from the serial observation of four identical single-molecule templates (Greenleaf and Block, 2006). What Harris *et al.* did is to use single-molecule DNA sequencing to resequence the M13 phage genome. Their library preparation process is quite fast, does not require the use of PCR, and results in single-stranded, poly(dA)-tailed templates. Poly(dT) oligonucleotides are covalently attached to glass cover slips at random positions. These oligomers are first used to capture the template strands, and then act as a primer for the template-directed primer extension that forms the basis of the sequence reading. Another alternative is to use them as a template replication step before sequencing. Up to 224 sequencing cycles may be performed, with each cycle consisting of adding the polymerase and labeled nucleotide mixture (containing one of the four bases), rinsing, imaging multiple positions, and cleaving the dye labels. Indeed, for the M13 work, this sequencing process could be performed simultaneously on over 280,000 primer-template duplexes (Harris *et al.*, 2008).

The Micro-Petri Dish for High-throughput Screening of Microorganisms

Although microorganisms can be grown on agarized medium in a Petri dish, the method suffers from weaknesses such as a poor facility for automation, the requirement for user skill, limited speed to result, the generation of waste (including the enrichment of pathogenic or genetically modified organisms), a short shelf life, and limited storage time of organisms *in situ* after culture. Further, agar is not suitable for the growth of all microbial species. Microbial culture on flexible porous membranes, on filters, or in chambers provides valuable alternatives by permitting manipulations such as staining *in situ* and changes in the nutrient environment of the cells, and culturing otherwise difficult species (Kaberlein *et al.*, 2002). But, these culture devices are not subdivided for creating enough growth compartments for genuine HTS (high throughput screening). This is because flexible membranes are very difficult to engineer at the micrometer scale and, it is usually assumed that micro-engineered mechanical systems (MEMS) devices cannot compete on cost with a cheap plastic disposable. Porous aluminium oxide (PAO) can serve as a good microbial culture support; it is highly porous (40 per cent by volume), inert, and stable with nano-range pores (20-200 nm diameter) and retains microorganisms on the rigid planar surface while allowing nutrients to pass (Ingham *et al.*, 2007).

These authors have fabricated a miniaturized, disposable microbial culture chip by microengineering a highly porous ceramic sheet with up to one million growth compartments. This highly versatile culture format, with discrete compartments as small as 7 x 7 µm, allowed the growth of segregated microbial samples at an unprecedented density. The chip has been used for four complementary applications in microbiology. (i) As a fast viable counting system that showed a dynamic range of over 10,000, a low degree of bias, and a high culturing efficiency. (ii) In HTC, with the recovery of 1 fluorescent microcolony in 10,000. (iii) In screening for an enzyme-based, nondominant phenotype by the targeted recovery of *Escherichia coli* transformed with the plasmid pUC18, based on expression of the *lacZ* reporter gene without antibiotic-resistance selection.

The rapid, successive changes in the environment of the organisms on the chip, needed for detection of β-galactosidase activity, point to a valuable property that has also been used to screen a metagenomic library for the same activity. (iv) In HTP screening of >200,000 isolates from Rhine water based on metabolism of a fluorogenic organophosphate compound, resulting in the recovery of 22 microcolonies with the desired phenotype. These isolates were predicted on the basis of rRNA sequence to include six new species. The above applications suggest that there is considerable potential for such simple, readily manufactured chips to impact microbial culture as it could facilitate the full automation and multiplexing of microbial culturing, screening, counting, and selection (Ingham *et al.*, 2007).

PAO can be microengineered by the greater stability, rigidity and intertness to temperature, wetting, or solvents of the starting material. A variety of disposable, surface culture, microbial growth chips, or 'micro-Petri dishes", have been designed by using a MEMS approach to engineer growth compartments on top of PAO, which acts as the surface on which a large number of microbial samples can be grown, assayed, and recovered.

Compartments on the micrometer scale could be successfully made with precise dimensions consistent and homogeneous with respect to microbial growth. Almost every compartment was available for culture, showing that selective removal of the laminate without degrading the functional properties of the PAO beneath was quite efficient.

Culture and containment of three very different microorganisms has been achieved. The culturability of these organisms on the chips was similar to that on unmodified PAO. If needed, improved containment was achieved by increasing wall height. However, because microcolonies of hundreds of cells can be contained in the standard format, the current height was sufficient for most detection and assay methods. Variant cells could be recovered manually or by means of a micromanipulator at a greater dilution than could have been possible by any conventional multiwell plate. An added merit is that in the micrometer-scale wells, the microcolonies are well supplied with nutrients from below and oxygen from above by diffusion (Ingham *et al.*, 2007).

The ready recovery of organisms from these chips, and the fact that 7-40 million compartments can be placed on a 96-well footprint, suggests that the chips are useful for detecting rare variants and mutations in HTS.

Good automation and throughput of the system required for HTS can be achieved by integrating additional elements including a chip holder with a microscope slide footprint which can be autoclaved, permits growth *in situ*, and facilitates positioning of the chip on an automated *XY* table. Appropriate software enables location of a specific growth area to aid recovery of organisms of interest.

Toxin-Antitoxin Gene Pairs

Microbes share genes between species so frequently that their evolutionary relationships become blurred. According to Edward Rubin of Joint Genome Institute in Walnut Creek, California, and his associates, there are some limits to this horizontal transfer.

Every time a microbe is sequenced, its DNA is transferred by researchers, using plasmids, into the model workhorse *Escherichia coli* in a large-scale experiment in horizontal transfer. The researchers take the genes that fail to transfer into *E. coli* from multiple species, and smuggle them into *E. coli* so that they would not be expressed until specifically induced. But when they were induced, the *E. coli* stopped growing, suggesting that the genes were toxic. These transfer-resistant genes could potentially be used as antibiotics in the future.

Bacterial horizontal gene pool contains an array of accessory mobile genetic elements which strongly affect genome plasticity, organization, and evolution (Thomas, 2000). Plasmids are autonomously replicating extrachromosomal elements of this pool and are important for bacterial adaptability and persistence because they provide certain functions not encoded by the circular chromosome (Hayes, 2003, 2003a). Plasmids can benefit host bacteria by enabling them to persist in otherwise hostile conditions. They also endow pathogenic properties. The clinical failure of antibiotics in recent years is partly connected to the rapid spread of plasmid-borne resistance genes in bacterial populations. Virulence determinants in bacterial pathogens are in some case borne on plasmids and are similarly dispersed. Antibiotic resistance, virulence, and other plasmids in bacteria make use of toxin-antitoxin gene pairs for ensuring their persistence during host replication. The toxin-antitoxin system eliminates plasmid-free cells that emerge as a result of segregation or replication defects. It contributes to intra- and interspecies plasmid dissemination. Chromosomal homologs of toxin-antitoxin genes are widely distributed among many pathogenic and nonpathogenic bacteria and can induce reversible cell cycle arrest or programmed cell death when exposed to starvation or other stresses. The elucidation of the interaction of toxins with intracellular targets and of the tertiary structures of toxin-antitoxin complexes have provided exciting insights into toxin-antitoxin behaviour.

Some examples of the contribution of plasmids to bacterial diversity are furnished by the foodborne pathogen *Bacillus cereus*, the anthrax agent *B. anthracis*, and the insect pathogen and commercial pesticide *B. thuringiensis*. Although these three are closely related species, from a human perspective, they show distinct biological properties that differ because of differences in the plasmid contents of the three species (Read *et al.*, 2003).

Considerable heterogeneity has been reported in the physical and genetic properties of plasmids, but large, low-copy-number plasmids are normally associated with antibiotic resistance and pathogenicity and include core regions needed for replication, segregational stability, and conjugative transfer. The segregational stability of these and other plasmids in bacterial populations emanates from the activity of plasmid-specified partioning proteins which direct plasmid copies to new daughter cells when a cell divides. In fact, Jaffe *et al.* (1985) identified plasmid-directed events resulting in selective killing or growth impairment of cells which had failed to acquire a plasmid copy. These mechanisms endow some advantage on plasmid-retaining cells by reducing the competitiveness of their plasmid-free counterparts;

this ensures the retention of the plasmid in the population. These bacterial cell-poisoning systems have attracted keen research interest (see Gerdes, 2000; Meinhart, 2003) because of enhanced understanding of these events and also because of much better appreciation of their widespread distribution both on plasmids of medical importance and on bacterial chromosomes. These toxin-antitoxin (TA) mechanisms are also known as postsegregational cell killing and addiction systems–cells are attacked from within. This differs from the action of colicins or antibiotics that are secreted by bacteria into their environment as inhibitors of neighboring microorganisms. The toxin component produced by TA cassettes maims bacterial cells; this points to the possibility that these factors might be used as novel antibacterial agents in the treatment of infectious diseases. Instead, restriction-modification enzyme pairs also can be either plasmid- or chromosomally encoded and are likewise being viewed as multifunctional TA systems that tend to promote segregational stability. They also protect against invading DNAs and directing genome rearrangements. The restriction enzyme may be analogous to the toxin, and the modification methylase is equivalent to the antitoxin (Kobayashi, 2001).

In TA cassettes the gene for the antitoxin component precedes the toxin gene. Quite often the two loci overlap, reflecting a common autoregulatory mechanism exerted by both components. Most TA modules conform to this arrangement, but in some TA cassettes either the gene order is reversed–the antitoxin alone exerts the regulatory effect or the product of the third gene is implicated (Gerdes, 2000).

The toxin gene codes for a stable protein. The antitoxin is either a labile protein or an untranslated, antisense RNA. The toxin gets neutralized by inhibition of toxin translation when the antitoxin is an RNA (type I), or by binding of the partner antitoxin when the latter happens to be a protein (type II). A plasmid-free variant can arise because of a replication error or some other defect in plasmid maintenance; in that case, the new cell still inherits the TA complex. The antitoxin component suffers more rapid degradation by host enzymes as compared with the toxin but it cannot be replenished in view of the absence of the plasmid. The TA complex releases the toxin which interacts with an essential host target to cause death or growth restriction of the plasmid-free cell (Hayes, 2003, 2003a).

Most TA loci identified on plasmids are of type II–the toxin protein is tightly sequestered by the antitoxin and there is no free toxin within the cell. But the antitoxin is more vulnerable to degradation by host proteases than the toxin. The implication is that if a plasmid-free cell appears, the toxin becomes available to target an essential intracellular host factor to kill the cell or at least retard growth (Gerdes, 2000).

DNA Metabolism Proteins as Targets

It is extremely difficult to analyze proteins that are inherently poisonous to their bacterial hosts. This is why few, if any, intracellular targets of plasmid-encoded toxin proteins have been characterized. DnaB is the primary replicative DNA helicase in *E. coli*.

DNA gyrase and DnaB are proteins required for DNA metabolism. These are the only known intracellular targets of plasmid-encoded protein toxins (Meinhart *et al.*,

2003). Although very few targets have been defined, it seems logical that a fundamental process such as DNA replication is disrupted by TA action, as this may impede proliferation of a plasmid-free cell most effectively. Translation is another basic macromolecular process and could be a target of chromosomal TA systems (Hayes, 2003).

Plasmid-based Toxin-Antitoxins

TAs of *E. coli* plasmids have been most thoroughly studied in view of their tractability. But databases and experimentation show that TA genes are widespread in various different bacteria, including pathogenic species (Meinhart *et al.*, 2003). *Shigella* causes acute dysentery and is responsible for many thousands of deaths annually worldwide. Isolates of one of its four main species, *S. flexneri*, typically have a large plasmid which specifies several virulence factors needed for intestinal epithelial cell invasion–an initial step in pathogenesis. This plasmid harbors a type II TA module, that has homologs in a variety of other disease-causing bacteria (Sayeed *et al.*, 2000).

Unrelated TA modules have been found on plasmids of other pathogens. *Streptococcus pyogenes* is a serious, reemerging human pathogen involved in inducing various conditions ranging from sore throat to toxic shock-like syndrome and acute rheumatic fever. Its plasmid pSM19035 harbors a type II TA cassette (Camacho *et al.*, 2002). The toxin protein of this plasmid can be up to three times larger than any other known toxin; it forms a heterotetrameric complex with an antitoxin. The formation of heteromultimeric species may be a characteristic feature of TA complexes (Meinhart *et al.*, 2003).

Enterococcus spp. are serious human pathogens involved in a variety of noscomial (hospital acquired) infections, including surgical wound, urinary tract, bloodstream and cardiovascular infections (Hayes, 2003). Enterococi often develop resistance to a wide spectrum of antibiotics, making treatment of enterococcal infections very difficult. Plasmid pAD1 of *Enterococcus faecalis* is a conjugative virulence plasmid that resembles many plasmids identified from clinical isolates of this bacterium. The *par* locus on pAD1 is a TA module that produces a 33-amino-acid toxin and an antitoxin that is a countertranscript RNA species (Weaver *et al.*, 2003).

Chromosomal Toxin-Antitoxin

TA modules represent a parasitic device. It ensures that those cells which have missegregated a plasmid, do not survive. Bacterial chromosomes also contain TA cassettes (cTAs) that show homology to those cassettes identified on plasmids but probably fulfill a different function to related elements (Hayes, 2003).

Programmed cell death (PCD, or apoptosis) occurs normally during development and tissue turnover in multicellular, eukaryotic organisms. However, several pathological conditions, *e.g.*, tumor formation, autoimmune and neurodegenerative disorders involve abnormal PCD mediated by either suppression or up-regulation of critical molecular components of the PCD apparatus. Being unicellular, bacteria were not thought to undergo PCD; however, in natural environments bacteria occur as multicellular colonies or as biofilms that carry out coordinated multicellular processes

(Shapiro, 1998). Colonial bacteria maintain discrete, ordered spatial structures. The idea that bacteria might also possess PCD mechanisms involved in regulating their mulitcellular organization was floated only in the last decade (Yarmolinsky, 1995). Some homologs of the eukaryotic PCD apparatus are found in bacteria (see Koonin and Aravind, 2002). Some parts of this eukaryotic machinery also can trigger bacterial cell death (Nanby-Wakao *et al.*, 2000). The *E. coli* genome, for example, contains several cTA modules that are either PCD genes or mediators of reversible cell cycle arrest. Conceivably, PCD in bacteria may enable surviving cells to scavenge nutrients from dead siblings or prevent the systemic spread of bacteriophages within a population. Another possibility is that cTA proteins arrest cell cycle, thereby allowing cells to enter a dormant or semidormant states as a protective measure against serious nutrient limitation and then to revive when environmental conditions become favourable (Pedersen *et al.*, 2002).

The *E. coli* K12 genome has some known type II cTA genes which either enhance segregational stability and/or display toxic-antitoxic behaviour when inserted ectopically into some multicopy plasmids.

Amino acid starvation and high levels of ppGpp activate the cTA locus *relBE*. RelE inhibits translation and induces a bacteriostatic response–a response that can be reversed if the RelB antitoxin is subsequently produced (Christensen *et al.*, 2001). RelE splits mRNAs within the ribosome but not free transcripts; *i.e.*, splitting depends on translation (Christensen and Gerdes, 2003). As the split mRNAs cannot be processed further, stalled ribosomes accumulate on damaged mRNAs and translation is inhibited, blocking cell cycle progression. Transfer-messenger RNA (tmRNA) sets free trapped ribosomes from defective mRNAs and tags polypeptides from these ribosomes for proteolysis. Replenishment of the pool of tmRNA might possibly allow recycling of amino acids from the RelE-stalled ribosomes and to induce translational restart. In case of RelBE, the negative effect of the RelE toxin can be reversed when its cognate antitoxin RelB is later produced. So it seems that RelBE is a modulator of the physiological response to poor nutritional conditions rather than bona fide PCD mechanism. But whether other cTA modules in E. coli evoke PCD or cell cycle arrest or have another role, is not clear.

What is clear, however, is that postsegregational cell killing by TA modules is a very effective strategy adopted by plasmids to ensure their persistence within bacterial populations. In contrast, chromosomally-located TA genes, homologous to those on plasmids, adjust the physiology of the cell in response to external cues, inducing either reversible bacteriostasis or cell death (Pedersen *et al.*, 2002). The mechanisms of action of type II plasmid-based TAs and cTAs are quite similar, yet the stimuli involved in antitoxin decay differ.

It is the metabolic state of the cell that triggers cTA activity. Since most bacteria seem to harbor multiple cTA genes, there is need to determine whether these are activated simultaneously in response to one or more inputs or separately in response to different signals (Hayes, 2003). The acquisition by bacteria of multiple cTA-based mechanisms is a widespread strategy for cell survival under stressful conditions.

The toxin constituents of TAs show activity not only in the hosts from which the elements are derived, but also in diverse bacteria (Grady and Hayes, 2003), and even in eukaryotes (Yamamoto *et al.*, 2002). This probably reflects an evolutionary relationship between TAs and PCD in eukaryotes. It seems that lateral transfer, probably mediated by plasmids or transposons, has contributed to the dissemination of TA genes (Gerdes, 2000).

Microbial Genomes

Fleischmann *et al.* (1995) were the first to publish the first complete sequence of a bacterial genome. This marvellous technical achievement was made possible by automatic DNA-sequencing machines. Today with over 60 microbial genome sequences determined and many more in progress, what is being learned? Have new concepts emerged about how cells work? Have some practical benefits accrued in the fields of medicine and agriculture? Is it feasible to determine the genomic sequence of every bacterial species on Earth? Although there have been some recent relevations, much of what is being found was quite expected.

Analyzing DNA sequences in unpurified or partially fractionated samples (*e.g.* in a drop of water from the Sargasso Sea or from an acid mine drain) can yield valuable information and has led to successful documentation of considerable organismal diversity. Extending this approach to terrestrial systems has revealed hitherto unsuspected and mixed communities of bacteria, archaea, and fungi. Vandenkoornhuyse *et al.* are studying the exchange of materials in one such niche by exposing pieces of turf from UK grassland (Scotland) or from French peatland (Normandy) to $^{13}CO_2$ and following the transfer of the isotope into ribosomal RNA of microbes associated with the plant roots. Preliminary work has revealed a broader-than-expected population diversity and a marked unevenness in the rate of uptake of photosynthetic products by the root-dwelling bacteria and fungi (see *PNAS* (USA) 104: 16970, 2007).

Much effort is being spent on comparisons of genomes. For medically important pathogens, comparisons between strains can reveal differences between the virulent and the avirulent. Comparisons between species can be informative about host or tissue specificity (Doolittle, 2002).

As regards size, the prokaryotic genomes from the Archaea and the Bacteria sequenced so far span more than an order of magnitude, from the mere 600,000 base pairs (bp) of some mycoplasmas (mollicutes) to almost 8 million for the nitrogen-fixing root-nodule bacterium, *Mesorhizobium loti*. By comparison, the genomes of eukaryotes sequenced so far range from less than 3 million bp for an intracellular microsporidian to the 4 billion found in the human genome (Doolittle, 2002). However, the genomes of eukaryotes typically contain high amounts of non-coding DNA, including both intergenic and intragenic (intron) sequences. In contrst, in most prokaryotes genes are tightly packed together with very little intergenic space, and they lack introns in the protein coding genes. If we count the genes themselves, the information content of prokaryotes and eukaryotes is not nearly so disparate as would seem from raw genome size. Indeed, the number of genes in the largest bacterial genomes actually exceeds the number in some eukaryotes–whereas *M. loti* has about 8,000 genes, *Saccharomyces cerevisiae* has only 6,200 and the microsporidian *Encephalitozoon cuniculi* only about 2,000 (Katlinka *et al.*, 2001).

Shotgum sequencing has been used for all the microbial genomes that have been sequenced. This method involves fragmenting of an informational polymer such as a protein or DNA molecule into large fragments determining partial sequences, and then putting the fragments in order by finding overlapping regions having identical sequences at opposite termini. The complete 1.83 megabase sequence of *Haemophilus influenzae* was determined by sequencing fewer than 20,000 fragments.

The overall DNA sequence is then reconstructed by the appropriate computer programs to find the genetic information. As a small portion of the genome codes for various RNA structures, including the ribosomal and transfer RNAs, these are easy to spot. Most of a bacterial genome is devoted to genes that code for proteins. These regions are identified by computer software that first translates the DNA sequence into amino acids according to the rules of the genetic code. If there is some uninterrupted set of amino-acid codons between an initiation and a termination codon, the DNA segment is presumed to encode a protein in that reading frame, and the region is referred to as an open reading frame (ORF).

After the ORFs have been tallied, the next step is to find out what they encode. This involves computer searches of large databases of known protein sequences. Significant matches are evaluated, and judgements are made about whether an ORF might encode exactly the same function as the protein retrieved from the database or some related function (see Doolittle, 2002). The judgement is influenced by two factors: the degree of similarity between the sequences (the new ORF and those in the database), and the closeness of the relationship of the organisms from which they were obtained.

Medically important bacteria account for only a small fraction of the Earth's diversity of prokaryotes; even so, three-quarters of the completely sequenced members of the Bacteria are of clinical significance.

Some of the newly completed genomes come from agriculturally important bacteria that were targeted because of their agricultural importance. Others have potential for bioremediation (the clean-up of polluted sites). Several members of the Archaea are methanogens that produce most of the world's atmospheric methane, a 'greenhouse gas'. There seem to be no known pathogens among the Archaea but the thermophiles among them are important as their thermostable enzymes might be useful to industry.

In each of the completely sequenced genomes so far there have turned up large numbers of putative genes for proteins of unknown function. This has been the biggest surprise in genome sequencing. The genome of the archaeon *Aeropyrum pernix*, for example, contains more than 1,500 ORFs–57 per cent of its total gene content–not recognizable by computer searching in any other organism (see Doolittle, 2002). More than 40 per cent of the approximately 4,000 ORFs found in *Mycobacterium tuberculosis* fall into the same category. In every genome examined so far, at least a quarter of the genes remain 'hypothetical', to which no function can be ascribed.

The hypothetical ORFs fall into two classes those found in various organisms, and which almost certainly encode functional proteins; and those unique to particular lineages. There are large numbers of unidentified genes in a variety of organisms that look conventional in every way. Where these unique sequences are coming from and what they do are not known.

What Makes a Pathogen?

Most medically important bacteria have both pathogenic and nonpathogenic strains. The 1944 classic experiment of Avery *et al.* involved the transformation of a non-capsulated, avirulent strain of pneumococcus (now called *Streptococcus pneumoniae*) with DNA from a capsulated, virulent strain. It was presumed that the capsule itself played a significant part in virulence, protecting the bacterium from the host's defence systems. The complete genome sequence has now revealed that some 12 genes in *S. pneumoniae* are needed for capsule synthesis (Tettelin *et al.*, 2001). The complete genome sequence of the avirulent strain has not yet been reported but microarray hybridization studies have shown the 12 genes to be absent; this confirms that the capsule is the primary virulence factor in these bacteria.

Some other genetically controlled characters that can confer virulence on bacteria include factors that enable the bacteria to attach to and disrupt host cells. Several pathogenic bacteria not only harbour a similar machinery for injecting proteins into the cytoplasm of the host's cells but also exchange the wherewithal to do so among them.

Pathogenicity islands are made of large gene clusters on bacterial chromosomes that are highly correlated with virulence. Pathogenic strains of *Helicobacterum pylori, e.g.*, have a 40,000-bp 'island' of DNA that includes a large number of genes involved in attacking host cells, but virulent strains lack this region. In *Escherichia coli* and *Salmonella*, clusters of genes that aid and abet the disruption of host cells are delineated in the chromosome by 'insertion sequences' or transposons. Among these are transfer RNA genes, which may serve as targets for special excision and integration enzymes. Occasionally the entire island is cut out and moved to a plasmid. Once on a plasmid, gene clusters may move to other bacteria where they are reintegrated in a new genome (see Doolittle, 2002).

Bacillus anthracis which (causes anthrax contains two large plasmids, one having a 44.5-kb pathogenicity island which contains genes for–among other things–the toxin lethal to humans; the island is flanked by inverted insertion sequences. The plasmid also contains some transposases and integrases.

The sequence of the main chromosome of *B. anthracis* is very similar to those of *B. cereus* and *B. thuringiensis,* neither of which carry the plasmids. *B. thuringiensis,* which produces a toxin lethal to lepidopteran caterpillars, is probably the most commonly used biological pesticide all over the world.

Adaptive Gene Losses

One highly fascinating phenomenon to emerge from the microbial genome studies is the extent to which parasitic bacteria have adapted to life in animal hosts.

Many of these organisms have lost numerous genes as they became dependent on their hosts. The very process of losing those genes is interesting. In the typhus bacterium *R. prowazekii,* almost 25 per cent of the genome is non-coding, in contrast to the 10 per cent non-coding DNA typical of most bacteria. Some of the non-coding DNA belongs to pseudogenes (segments that encoded proteins in the past, but which now contain stop codons and/or deletions that keep the gene from being expressed properly) (see Doolittle, 2002).

In the case of *Mycobacterium leprae,* most of the 1,100 pseudogenes appear to be fully functional ORFs in *M. tuberculosis.* Remarkably, the still intact genes of *M. leprae* have a very high sequence resemblance to those of *M. tuberculosis;* this implies that these two species have had a common ancestor quite recently.

Some parasites, *e.g., Mycoplasma* and *Borrelia,* depend on the host to provide key metabolites. Others, like *Rickettsia* and *Chlamydia,* are energy parasites that steal ATP from the host cell. These latter two bacteria are very distantly related and yet they do this with the same alien enzyme, an apparently unique and rare type of ADP/ATP translocase. Until recently, the only known relatives of the bacterial version were found in chloroplasts. How did these two widely divergent bacteria, both animal parasites, come to adopt the same strategy with the same enzyme? Compounding the mystery, the same protein was found to be encoded in the recently sequenced genome of the tiny eukaryote *Encephalitozoon cuniculi,* itself an obligate intra-cellular pathogen. Conceivably, the parasite uses the protein to scavenge ATP from its host cell, just like the small bacterial pathogens.

The streamlining associated with parasitism is not only associated with disease. In fact, the vast majority of instances in which bacteria live within eukaryotic cells are mutually beneficial to both host and guest.

Optimal Genome Size in Bacteria

Several hypotheses have been proposed to explain the fact that bacterial genome size is subject to natural selection, and that small genome sizes exemplify adaptations that promote competitiveness by reproductive efficiency (Mira *et al.*, 2001; Ochman *et al.*, 2000). The size and complexity of different prokaryotic genomes is estimated by measuring the number of open reading frames (ORFs); Ranea *et al.* (2005) defined complexity by the number of ORFs.

Genome size is not always related to phenotype or lineage in bacteria. This is evident from the observation that bacteria with a wide range of phenotypes and lifestyles often have similar genome sizes; also considerable diversity in genome size is observed for bacteria belonging to relatively narrow phylogenetic groups. Gene duplication and lineage-specific gene loss play a basic role in determining bacterial genome size, followed by horizontal gene transfer (Moran, 2002; Kurland *et al.*, 2003). The smallest genomes are usually derived from bacteria with larger genomes through what is aptly termed "evolution by reduction" (Dobrindt and Hacker, 2001). All these findings lead to the logical inference that gene duplication and loss and genome size variation have been interdependent events in evolution (Ranea *et al.*, 2004, 2005).

Bacteria are able to increase their survival by expanding their genetic repertoire but the tight packing of the genome coupled with the fact that the most evolved species do not necessarily have the biggest genomes point to certain other evolutionary factors as limiting their genome expansion. To understand these restrictions on size, Ranea *et al.* (2005) studied those protein families which contribute most significantly to bacterial-genome complexity. It emerged that all bacteria apply the same basic and ancestral 'molecular technology' to optimize their reproductive efficiency. Importantly, the very same microeconomics principles which govern the optimum size in a factory also explain the existence of a statistical optimum in bacterial genome size–an optimum that is reached when the bacterial genome obtains the maximum metabolic complexity (revenue) for minimal regulatory genes (logistic cost).

As per the analogy of an industrial factory, production capacity depends, *inter alia,* on the technology applied. Eukaryotes apply such additional technologies to reduce gene expression noise, as DNA methylation, nucleosomal chromatin or cellular compartmentalization (Bird, 1995). In contrast to eukaryotic cells, bacteria have adopted an evolutionary strategy based on speed in reproduction as a primary goal. Work on several size-dependent superfamilies clearly showed that universal molecular technology is shared by all prokaryotes to perform their metabolic and regulatory processes. It appears that all bacteria have probably used similar molecular technology to optimize reproductive efficiency. This efficiency could be achieved by striving for a balance between metabolic diversity and the associated regulatory cost that maximally exploits the environment while maintaining minimum cellular-doubling time (Ranea *et al.*, 2005).

Simple Sequence Repeats in Genomes

Simple sequence repeats (SSRs) in both prokaryotes and eukaryotes represent hypermutable loci subject to reversible changes in the SSR length (Rocha, 2003; Kashi and King, 2006). They are composed of in DNA sequences tandem repeats of short oligonucleotides and probably had functional and/or structural properties that distinguish them from general DNA sequences. Their variable length seems to affect local structure of the DNA molecule or the encoded proteins. Long SSRs (LSSRs) are common in eukaryotes but rare in most prokaryotes. In pathogens, SSRs can enhance antigenic variance of the pathogen population in a manner that counteracts the host immune response. On the basis of extensive analyses of representations of SSRs in >300 prokaryotic genomes, Mrazek *et al.* (2007) reported significant differences among different prokaryotes as well as among different types of SSRs. LSSRs composed of short oligonucleotides (1-4 bp length, designated $LSSR^{1-4}$) are often found in host-adapted pathogens with reduced genomes that do not comfortably survive in a natural environment outside the host. In contrast, LSSRs composed of longer oligonucleotides (5-11 bp length, designated $LSSR^{5-11}$) are found mostly in nonpathogens and opportunistic pathogens with large genomes. Comparisons among SSRs of different lengths suggest that $LSSR^{1-4}$ are probably maintained by selection. This is in accordance with the established role of some $LSSR^{1-4}$ in enhancing antigenic variance. By contrast, abundance of $LSSR^{5-11}$ in some genomes may reflect the SSRs' common tendency to expand rather than their specific role in the organisms' physiology.

There are also significant differences among genomes in terms of SSR representations (Mrazek *et al.*, 2007).

Some pathogens exploit SSRs to counteract the host immune response by increasing the antigenic variance of the pathogen population. In this case, SSRs located in protein coding regions or in upstream regulatory regions can reversibly deactivate or alter genes involved in interactions with the host (Rocha and Blanchard, 2002).

Long SSRs are generally less common in prokaryotes than in eukaryotes (Field and Wills, 1998). They are usually greatly overrepresented in eukaryotic genomes (Kashi and King, 2006).

Trait to Gene

Many genome databases and comparative genomic techniques have been developed and are allowing researchers to complement conventional gene identification strategies. The function of unknown genes can often be inferred by comparisons to well-characterized homologs. The successful use of phylogenetic profiles to predict gene function has led to identification of new members of developmental and biochemical pathways (McGuire *et al.*, 2000). Levesque *et al.* (2003) showed that, even if all of the homologs of a gene are unannotated, its function may be deduced through phylogenetic profiling. By designing algorithms that make functional predictions of genes based on orthology and set theory, they successfully identified 94 per cent of the clusters of orthologous groups that are known to be involved in flagella development or function in *Bacillus subtilis.* These algorithms make possible high-throughput functional prediction of genes beyond that provided by simple orthology-based annotation endeavors.

B. subtilis Knockout Experiments

Levesque *et al.* initially used the flagella trait to test the recall rate of their algorithms, but were surprised at the finding that the algorithm predicted that novel, highly conserved genes may be involved in this well-characterized system. They generated a short list of nine COGs (clusters of orthologous groups) that included four known flagellar COGs and five newly predicted flagellar COGs. They then knocked out the genes in the flagellate *B. subtillis* that corresponded to three of those newly predicted COGs. Further studies of the effects of knockout experiments led them to conclude that their algorithms were successful at recalling 29 out of the 31 known COGs involved in flagellar development and appeared to infer the function of at least 2 new genes in this well-studied bacterium. It is hope that, as more fully sequenced genomes become available, these algorithms may serve as a valuable resource for identifying novel genes in other traits of developmental, medical, or evolutionary significance (Levesque *et al.*, 2003).

Streamlined Genome of a Marine Bacterium

Very small, heterotrophic marine α-proteobacteria belonging to the SAR11 clade are found throughout the oceans where they account for about 25 per cent of all microbial cells. *Pelagibacter ubique*, the first cultured member of this clade, has the smallest genome and encodes the smallest number of predicted open reading frames known for a free-living microorganism. Among more than 200 complete published genomes from bacteria and archaea, *P. ubique* has the smallest number of genes (1354 open reading frames) for any free-living organism (Giovannoni *et al.*, 2005). In contrast to parasitic bacteria and archaea with small genomes, *P. ubique* has complete biosynthetic pathways for all essential amino acids and all but a few cofactors.

According to Morris *et al.* (2002), *Pelagibacter ubique*, strain HTCC1062, belongs to a most successful clade of organisms on Earth, but has the smallest genome (1,308,759 base pairs) of any cell known to replicate independently in nature. These bacteria occur as free-floating cells suspended in the water column and grow by assimilating organic compounds from the ocean's dissolved organic carbon (DOC) reservoir. They generate metabolic energy either by a light-driven proteorhodopsin proton pump (Giovannoni *et al.*, 2005) or by respiration. As the surface sea waters are poor in nutrients, the availability of N, P, and organic carbon often limits the productivity of microbial communities. *P. ubique* is one of the smallest free-living cells studied in a laboratory, but even its so small genome occupies (~30 per cent) of the cell volume. The small size of the SAR11 clade cells meets the requirements of the model proposed by Button (1991)–natural selection optimizes surface-to-volume ratios in oligotrophic cells, such that the capacity of the cytoplasm to process substrates is suitably matched to steady-state membrane transport rates.

Surprisingly, even this small genome probably codes for most of the basic functions of α-proteobacterial cells. The small genome size can be attributed to the near complete absence of nonfunctional or redundant DNA and the pruning of all except the most fundamental metabolic and regulatory functions. For example, *P. ubique* has no pseudogenes, introns, transposons, extrachromosomal elements, or inteins; few paralogs; and the shortest intergenic spacers yet observed for any cell.

The genome of *P. ubique* has the smallest number of paralogous genes observed in any free-living cell. Mira *et al.* (2001) postulated an interesting hypothesis that cells in some ecosystems are subject to powerful selection to minimize the material costs of cellular replication; this concept is known as *streamlining*.

The issue of genome reduction in prokaryotes, particularly in parasites, has aroused considerable debate–parasites have the smallest cellular genomes known. The relaxation of positive selection for genes used in the biosynthesis of compounds that can be appropriated from the host, plus some bias favoring deletions over insertions in most or all bacteria, seem to account for genome reduction in many parasites and organelles. The streamlining hypothesis rests on the assumption that selection acts to reduce genome size because of the metabolic burden of replicating DNA with no adaptive value. This hypothesis presumes that repetitive DNA arises when mechanisms that add DNA to genomes–for example, recombination and the propagation of self-replicating DNA (*e.g.*, introns, inteins, and transposons)–overwhelm the simple economics of metabolic costs (Giovannoni *et al.*, 2005). In contrast, evolutionary theory predicts that the probability that selection will act to eliminate DNA merely because of the metabolic cost of its synthesis will be greatest in very large populations of cells that do not suffer sharp periodic declines (Kimura, 1983).

The streamlining hypothesis has been invoked to explain genome reduction in *Prochlorococcus*, a photoautotroph that reaches population sizes in the oceans similar to those of *Pelagibacter* (Dufresne *et al.*, 2005). *Prochlorococcus* genomes range from 1.66 to 2.41 million base pairs (Mbp). Many organisms with reduced genomes, including some pathogens, also have very low G:C to A:T ratios (Ussery and Hallin, 2004). This may be due to biases in mutational frequencies, but alternatively might confer some selective advantage by lowering the nitrogen requirement for DNA synthesis, thereby reducing the cellular requirement for fixed forms of nitrogen. N and P are both critical constituents of DNA but are quite often limiting in seawater. The *P. ubique* genome is 29.7 per cent G+C. Of four complete *Prochlorococcus* genome sequences two, which lack the DNA repair enzyme 6-0-methylguanine-DNA methyltransferase, also have very low G:C to A:T ratios. The *P. ubique* genome does encode this enzyme, which suggests that other factors are the cause of its low G:C to A:T ratio.

The abundance of *Pelagibacter* populations indicates that they consume a large proportion of the labile DOC in the oceans. The global DOC pool is estimated to be 6.85×10^{17} g C (Hansell and Carlson, 1998) and roughly equals the mass of inorganic C in the atmosphere. It seems that in *P. ubique* efficiency is achieved in a low-nutrient system by dependence on transporters with broad substrate ranges (Button *et al.*, 2004) and several specialized substrate targets, in particular, nitrogenous compounds and osmolytes.

In respect of its simplicity the *P. ubique* genome is unique among other heterotrophic marine bacteria, such as *Vibrio* sp., *Shewanella*, and *Silicibacter*, which have larger genomes (4.0 to 5.3 Mbp) as well as global regulatory systems that enable them to implement various different metabolic strategies in response to environmental variation. According to Giovannoni *et al.*, conceivably *P. ubique* exploits the ambient DOC field whereas heterotrophic bacterioplankton with larger genomes are adapted to rapidly exploit pulses of nutrients at the expense of replication efficiency during the intervening periods (see Azam, 1998; Klappenbach *et al.*, 2000). This hypothesis is consistent with the finding that *P. ubique* contains only one ribosomal RNA (rRNA) operon and a low growth rate (0.40 to 0.58 cell divisions per day) that varies little in response to nutrient addition. In contrast, heterotrophic marine bacteria having larger genomes achieve some of the highest recorded growth rates and are highly responsive to nutrient concentration.

P. ubique has taken an unusual and distinctive path in evolution that is markedly different from that of other heterotrophic marine bacteria whose genome sequences are available. Evolution has divested it of all except the most fundamental cellular systems so that it replicates well under limiting nutrient resources; the outcome makes it the dominant clade in the ocean.

Phylogenetic and Bacterial Genomes

In most (except for the most reduced) bacterial genomes, distributions and compositional features of several genes indicate that they originated by lateral gene transfer (LGT). An ongoing process of gene acquisition in many bacterial genomes promotes adaptation and ecological diversification. Further evidence of LGT between distantly related organisms has come from phylogenetic studies involving large taxonomic samples (Ochman *et al.*, 2000; Brown and Doolittle, 1997). Incompatibility of phylogenies within and among bacterial phyla based on different genes has traditionally been attributed to LGT (Doolittle, 1999; Gogarten *et al.*, 2002). However, delineating molecular phylogenies for distantly related species is quite difficult and choice of phylogenetic methods, genes, or taxa can yield different results. Furthermore, distinguishing between orthologous genes (sequences that trace their divergence to the splitting of organismal lineages) and paralogous (duplicated) genes becomes more and more difficult when we consider more distantly related taxa (Daubin *et al.*, 2003).

The effects of LGT have been extrapolated from the deepest to the shallowest levels of bacterial relationships. Gogarten *et al.* (2002) proposed that the similarities in gene sequence and gene content which characterize widely accepted bacterial taxa reflect boundaries to gene transfer, rather than vertical transmission and common organismal ancestry.

According to Daubin *et al.*, it seems that LGT among prokaryotic phyla can be overestimated, even when applying a conservative statistical test, and phylogenetic studies based on small numbers of distantly related taxa sometime yield spurious results (Adachi and Hasegawa, 1996). As was reported by Nesbo *et al.* (2001), the addition of species to well-supported quartet phylogenies often decreases statistical support for LGT. Also, even when full genome sequences are available, the use of a poor criterion for orthology, such as the reciprocal best-hit method, can weaken ability to eliminate paralogous genes in distant species and may lead to false conclusions of LGT. The lesson is that evidence for LGT needs to be viewed very carefully when relying on poor taxonomic sampling.

Not all genes are equally subject to LGT. Some functional gene classes, *e.g.*, those involved in complex molecular interactions (termed "informational genes"), are less prone to LGT (Rivera *et al.*, 1998). Daubin *et al.* demonstrated a more pronounced dichotomy of genes, consisting of "acquired" and "ortholog" classes. These classes differ fundamentally: over 70 per cent of acquired genes code for proteins of uncharacterized functions, whereas ~80 per cent of genes in the ortholog class have functional annotations (of which only 10 to 15 per cent would be classified as "informational" genes). Because there is little evidence of LGT for orthologous genes, it was concluded that LGT is concentrated in a class of genes that are not candidates for phylogenetic analysis (Daubin *et al.*, 2003).

The fact that large amounts of alien genes have been discovered in bacteria has prompted the prediction that LGT causes phylogenetic disruption in bacteria. Invoking LGT is usually a last resort for explaining an observation of phylogenetic incongruence in animals or plants, but LGT is very often the first choice to explain incongruence in prokaryotes (Raymond *et al.*, 2002). Although LGT is an important driving force in prokaryotic evolution and adaptation, it cannot always explain instances of phylogenetic conflict. Nevertheless, if used critically, sequence data can be useful in reconstructing the evolutionary relationships of bacterial lineages (Daubin *et al.*, 2003).

Cyanobacterial Genome and the Origin of Photosynthesis

Cyanobacteria, an earliest branching group of organisms on Earth represent the only known prokaryotes to carry out oxygenic photosynthesis. They played a crucial role in the formation of atmospheric oxygen ?2.3 Gyr ago (Becker *et al.*, 2004). In recent years the availability of complete genome sequences of several freshwater and marine cyanobacteria has provided ample data for systematic analysis. The prokaryotic alga *Prochlorococcus*, often considered to be an anomalous green alga belonging to the Phylum Prochlorophyta is now thought to be a cyanobacterium. A comparison of the complete genomes from its three different strains has revealed a wide variety of gene complements within this genus due to strong genome reduction in some lineages (Hess, 2004; Dufresne *et al.*, 2005). Comparison of the genes shared by cyanobacteria and other photosynthetic organisms has enabled defining the "photosynthetic gene set" and demonstrated considerable lateral gene transfer (LGT) among phototrophic bacteria (Sato, 2002; Raymond *et al.*, 2003). It should, however, be noted that sequence data alone do not allow one to establish the direction of ancient lateral transfer of photosynthetic genes; that would require additional information from independent sources. Also, the tendency of photosynthetic genes to be laterally transferred between distantly related organisms makes it very difficult to identify the lineage that was the first to develop chlorophyll-based photosynthesis. A surprising outcome of this work was that genes for most proteins involved in photosynthesis were not represented in the photosynthetic gene set (Mulkidjanian *et al.*, 2006). These authors compared proteins encoded in 15 complete cyanobacterial genomes, including five genomes of *Prochlorococcus* spp., so as to look for the minimal set of genes shared by all cyanobacteria and to trace the status of conservation of these genes among other taxa. The fairly high level of sequence conservation among orthologous proteins from different cyanobacteria makes delineation of cyanobacterial gene clusters an easy task.

Their comparative analysis of the complete cyanobacterial genome sequences, including "near minimal" genomes of five strains of *Prochlorococcus* spp., revealed 1,054 protein families [core cyanobacterial clusters of orthologous groups of proteins (core CyOGs)] encoded in at least 14 of them. Most of the core CyOGs are involved in central cellular functions that are shared with other bacteria; 50 core CyOGs are specific for cyanobacteria, whereas 84 are exclusively shared by cyanobacteria and plants and/or other plastid-carrying eukaryotic algae. The latter group includes 35 families of uncharacterized proteins, which might also be possibly involved in photosynthesis. Only a few components of cyanobacterial photosynthetic machinery are represented in the genomes of the anoxygenic phototrophic bacteria *Chlorobium tepedium*, *Rhodopseudomonas palustris*, *Chloroflexus aurantiacus*, or *Heliobacillus mobilis*. These observations, coupled with recent geological data on the properties of the ancient phototrophs, point to the strong possibility that photosynthesis originated in the cyanobacterial lineage under the selective pressures of UV light and depletion of electron donors. According to Mulkidjanian *et al.*, the first phototrophs could have been anaerobic ancestors of cyanobacteria ("procyanobacteria") that performed anoxygenic photosynthesis by using a photosystem 1-like reaction center, rather similar to the heterocysts of modern filamentous cyanobacteria. From procyanobacteria, photosynthesis spread to other phyla through lateral gene transfer (Mulkidjanian *et al.*, 2006).

In view of the great complexity of the photosynthetic machinery, its origin and further evolution must have occurred in several steps under constant selective pressure. Such a selective pressure could come from at least two key factors: the need for the cells to acquire energy and to reduce the damaging effects of solar UV, which was much stronger in the absence of the ozone shield than it is now (Garcia-Pichel, 1998). Several researchers proposed that Reaction Centres 1 (RC1) probably evolved through several duplication events from simpler chlorophyll-binding membrane proteins, similar to the high light inducible proteins (HLIPs) of modern cyanobacteria (Garczarek *et al.*, 2003). Conceivably, such proteins might help in protecting DNA from the UV-induced damage (Mulkidjanian and Junge, 1997). The need for an alternative source of reducing power in the face of gradually decreasing atmospheric hydrogen content could possibly have led to the emergence of RC1. RC1 recycles NAD(P)H which has a redox potential similar to that of hydrogen and it can replace hydrogen in certain metabolic chains. The membrane hydrogenase and NADH-dehydrogenase are related enzymes which differ only in the substrate-binding module (Friedrich and Scheide, 2000).

As the atmosphere gradually became more oxidized, there was a need for adequate sources of redox equivalents and the need might have led to the formation of a small, more powerful RC2, possibly through fission/reshuffling of RC1 (Rutherford and Faller, 2003). Continuing depletion of electron donors upon oxidation of the available Fe(II) could possibly have stimulated the evolution of RC2 into the water-oxidizing PSII.

It is now generally agreed that modern cyanobacteria inherited their photosynthetic apparatus from ancestral phototrophs, whereas other phototrophic bacterial lineages acquired their photosynthetic machinery through LGT.

Genome of an Industrially Important Fungus

The filamentous fungus *Aspergillus niger* produces enormous amounts of several enzymes and specialty chemicals. Few other microbes can compete with this fungus in this respect. *A. niger* is the chief source of citric acid for food, beverages and pharmaceuticals and of glucoamylase, which is widely used for converting starch to food syrups and to fermentative feedstocks for ethanol production. Most of these fermentation processes are well established but the underlying genetics are not well understood. Pel *et al.* (2007) reported the genome sequence of *A. niger* strain CBS 513.88–knowledge of this sequence may lead to substantial improvement in the production of chemicals and enzymes by this fungus (Cullen, 2007).

What Pel *et al.* did was to sequence tiled bacterial artificial chromosomes representing the entire *A. niger* genome to produce a high-quality assembly of 19 supercontigs with a combined length of 33.9 Mb. About 5 per cent of the genome corresponding to rDNA, telomeric repeats and short gaps were not sequenced, but the orientation of supercontings on chromosomes could be deduced by high synteny within centromere regions to related *Aspergillus* species (*e.g.*, Machida *et al.*, 2005). Automated and manual curation enabled annotation of a total of 14,165 genes.

Citric acid fermentation by *A. niger* is quite efficient with routine yields of >200 g/l at >90 per cent conversion of feedstock sugar into citric acid. Enzyme production is also very efficient, and glucoamylase secretion occurs at levels >10 g/l. For total production of commercial enzymes, *A. niger* glucoamylase may be second only to *Bacillus* protease (Cullen, 2007).

Pel *et al.* integrated the annotated genome data into a model of central carbon metabolism in *A. niger* (see David *et al.*, 2003). The model covers the involvement of multiple malate dehydrogenase-like sequences, three of which encode cytoplasmic forms and one, a mitochondrial form. Genetic multiplicity was revealed for citrate synthase (three genes) and aconitase (four genes, two cytoplasmic and two mitochondrial).

Genetic redundancy was also observed among enzymes involved in the biogenesis of another commercially important organic acid, gluconic acid. (Genetic redundancy appears to influence the high production levels). Unlike citrate, gluconic acid is synthesized extracellularly, the first step being catalyzed by glucose oxidase. The resulting D-glucono-1,5-lactone and hydrogen peroxide are further metabolized by gluconolactonase and catalase, respectively. Genes encoding four glucose oxidases, four gluconalactonases and 11 catalases were detected, and several of these sequences featured N-terminal signals typical of secreted proteins. But whether or not such redundancy significantly influences organic acid production is not known.

A. niger has several genes encoding carbohydrate-active enzymes, many of which participate in the degradation of plant polysaccharides. Categorized on the basis of structurally related catalytic domains and as compared with those of other fungal genomes, *A. niger*'s glycoside hydrolase and polysaccharide lyase genes are distributed in a manner that reflects its ecological role.

As expected for a saprophyte isolated from soils and decaying vegetation, an array of cellulase-, xylanase-, amylase- and pectinase- encoding genes can be identified. Compared with related Aspergilli, no pectate lyase genes were predicted, but pectin lyases and polygalacturonases were abundant. *A. niger* acidifies the substratum, a condition under which pectate lyases are less effective.

Besides the organic acids and native enzymes, heterologous gene expression in *Aspergillus* has attracted much research (Archer and Turner, 2006). Several foreign proteins have been expressed in *A. niger*, but relative to native enzymes such as glucoamylase, yields have been low.

The current availability of genome data from related fungi enables comparative analyses of genome structure and evolution. Assembled or complete genomes of nine *Aspergillus* species include important opportunistic human pathogens (*e.g.*, *A fumigatus*), mycotoxin producers (*e.g.*, *A. flavus*) and essential experimental systems of genetic analysis (*e.g. A. nidulans*) (Nierman *et al.*, 2005).

The *A. niger* genome highlights both challenges and opportunities for future research. Impressive genetic diversity, especially among the carbohydrate-active enzymes, points toward novel enzymes of potential biotechnological significance. Identification and appropriate manipulation of components of the protein secretion machinery could potentially raise yields of native and foreign proteins, and even to directed metabolic engineering and increased production of specialty chemicals and pharmaceuticals. These areas will benefit greatly from detailed analysis of individual genes together with high-throughput characterization of the metabolome, proteome and transcriptome (Cullen, 2007).

Chapter 9
Microbial Cooperation, Mats and Biofilms

Microbial Mats

Microbial mats are probably the oldest form of life on earth, as witnessed by the oldest known microfossils being found in stromatolites which have been dated to over 3.5 billion years old. Stromatolites are the most dominant sedimentary structures in rocks of the Precambrian era.

Although the abundance of stromatolites declined during the Cambrian era and today, microbial mats can be found in a variety of different biotopes such as hypersaline lagoons, alkaline lakes and streams, and hydrothermal vents, both terrestrial and in the deep sea.

Microbial mats are stratified microbial communities that develop in the environmental microgradients established at the interfaces of water and solid substrates. They form a laminated multilayer of biofilms and strongly change the environmental microgradients in this interface as a result of their own community metabolism. The development of these microbial communities generates steep environmental microgradients and the establishment of a well-defined diffusion boundary layer immediately proximal to the multilayered biofilm.

A steep redox gradient at this interface leads to the depletion of oxygen at the mat surface and the buildup of sulfide (and often methane) with depth in sulfate-containing water bodies. Along with the establishment of the oxygen/sulfide interface, steep gradients of nutrients, fatty acids, and available metals can develop. Diffusion of small molecules along these steep microgradients occurs very fast, allowing rapid internal cycling of carbon, nitrogen, sulfur, and phosphate within the mat to support the high rate of metabolism of microbial mats.

The mats that develop in shallow waters are made up mainly of photosynthetic microorganisms, particularly oxygenic cyanobacteria. The metabolism of the cyanobacterial mats depends on available light for photosynthesis. Steep spectral light gradients are seen in these mats. Day-night fluctuations expose the mat communities to vast diurnal fluctuations in both their chemical and optical microenvironments. To cope with these fluctuations, the organisms must either migrate diurnally along with the changing microgradients or develop physiological flexibility to enable them to function under both oxidizing and reducing conditions.

The application of microelectrodes for measurements of oxygen, sulfide, and pH, together with development of micro-optic fibers for the study of spectral light distribution in microbial mats and careful electron microscopy has revealed the dynamics of microenvironmental conditions within mat communities. The development of microassays for the study of specific microbial metabolism, including oxygenic photosynthesis, aerobic respiration, and sulfate reduction, at microscale enhances the understanding of specific microbial activities and their interrelations within the mat communities.

Several mat-forming cyanobacteria carry out facultative anoxygenic photosynthesis using sulfide, hydrogen, and possibly reduced iron as alternative electron donors. In recent years, novel nonphotosynthetic sulfide-oxidizing bacteria have been discovered with diverse habitats and physiologies, many of which are highly versatile and flexible metabolically with respect to both electron donors and electron receptors.

Beggiatoa mats have been reported to be abundant in many marine coastal environments as well as deep-sea hydrothermal beds.

Cyanobacterial Hyperscums

A hyperscum may be defined as a crusted buoyant cyanobacterial mat that can be decimeters thick and in which the organisms are so densely packed that free water cannot be seen (Zohary, 1985). The hyperscum is formed as an undesirable consequence of excessive eutrophication of fresh waters, caused by the accumulation of surface cyanobacterial blooms in wind-protected sites on lee shores where they can persist for weeks to months, depending on the prevailing weather.

Hyperscums differ from scums or water blooms in the magnitude of their spatial and temporal dimensions and cell concentrations. Whereas relevant dimensions for *scums* are usually millimeters (for thickness), hours (for duration, and 10^2 to 10^2 mg of chlorophyll liter^{-1} as indicator of algal biomass, those for *hyperscums* would be decimeters, weeks or months, and 10^5 mg liter^{-1}, respectively. The term neuston refers to organisms adapted to life at the air/water interface. Neuston differs from hyperscum in the sense that in a hyperscum, most of the population is below the surface, and cells at the interface die of photo-oxidation and dehydration.

Surface scums of the dominant gas-vacuolated, bloom-forming genera *Anabaena, Aphanizomenon* and *Microcystis* occur throughout the world in a wide variety of freshwater bodies (Table 9.1). Similar phenomena in the marine environment are surface blooms of the cyanobacterium *Oscillatoria erithraea (Trichodesmium)*, which appear in windrows that can be several miles long (Wood, 1965).

Table 9.1: Selected Records of Occurrence of Hyperscums in Eutrophic Water Bodies where Cyanobacterial Blooms Occur Regularly (after Zohary, 1989)

Water Body	*Trophic Status*	*Dominant Bloom Genera*	*Mention of Hyperscums*	*Description of Scums or Hyperscums*
Fishponds, Israel	Eutrophic	*Microcystis, Oscillatoria, Spirulina*	No	Surface blooms in early mornings disperse in afternoons
Lake Brielle, The Netherlands	Hypertrophic	*Microcystis*	?	"Drijflagen" (floating layers)
Kremenchu Reservoir, USSR	Hypertrophic	*Microcystis, Anabaena*	Yes	"Dry crusts", "neustonic phase"
Lake Mendota, Wis.	Eutrophic	*Aphanizomenon*	No	"Massive blooms"
Potomac River, Washington, D.C.	Eutrophic	*Microcystis, Oscillatoria*	No	"Green paste"
Lake George, Uganda	Eutrophic	*Microcystis*	No	Surface accumulations in early mornings disperse in afternoons
Hartbeespoort Dam, South Africa	Hypertrophic	*Microcystis*	Yes	Crusted buoyant cyanobacterial mats of decimeter thickness
Lake Kasumigaura, Japan	Hypertrophic	*Microcystis, Anabaena*	?	"Heavy neustonic mats of decimeters thickness"
Queensland Coast, Pacific Ocean	Eutrophic	*Trichodesmium*	No	Bloom occupying up to 52,000 km^2

According to Reynolds and Walsby (1975), cyanobacterial surface scums form only when three preconditions coexist: a preexisting cyanobacterial population, positive buoyancy in a significant proportion of the organisms, and turbulence that is too weak to nullify the tendency of buoyant organisms to float. These preconditions seem to be essential but not sufficient for *hyperscum* formation. The additional requirements are as follows. (i) The pre-existing standing stock of positively buoyant cyanobacteria must be large (>100 mg of chlorophyll m^{-2} and 10^2 to 10^3 kg of chlorophyll per lake). This requirement limits hyperscum occurrence to hypertrophic water bodies. (ii) Calms (no winds) or low-speed winds (<3.7 m s^{-1}) must occur continuously over long periods (weeks) and be uninterrupted by storms. (iii) Shore morphometry must be such that there are wind-protected accumulation sites in the lee of the prevailing winds. (iv) High solar radiation levels must be present (Zohary, 1989). Of course, all the above-mentioned preconditions usually do not co-occur. This makes hyperscums an unusual sight. But, with increasing hypertrophy throughout the Third World (Robarts, 1985), the frequency and distribution of hyperscum occurrence are likely to increase.

The hyperscum community is usually a cyanobacterial-bacterial association in which *M. aeruginosa* tends to be the dominant organism, making up >95 per cent of the volume. Bacteria, can however, exceed *Microcystis* in cell numbers and show a wide range of sizes and morphological types.

Three structural zones may be distinguished in a mature hyperscum (with a well-developed crust). These include a crust about 2 mm thick; a compact layer, 5 to 10 mm thick, of partially dehydrated colonies below the crust; and the less compact bulk of the hyperscum, which can extend to a depth of several decimeters. Below the hyperscum can lie several meters of oxygenated, freely circulating lake water.

Scanning electron microscopy of the crust material reveals tightly packed, dehydrated *Microcystis* cells that are heavily colonized by bacteria.

Quite large gas bubbles, often 10 mm in diameter and containing end products of anaerobic decomposition can be seen trapped within this layer.

Under the aphotic, anaerobic conditions that prevail in the hyperscum, growth of obligate phototrophs such as *Microcystis aeruginosa* would be disadvantaged, and a prevalence of anaerobic decomposition processes is to be expected. Analysis by gas chromatography of the chemical composition of the trapped gas bubbles showed that they contained no oxygen, 28 per cent methane, 19 per cent, CO_2, 53 per cent N_2, and traces of H_2,–end products of bacterial anaerobic degradation of organic matter.

M. aeruginosa can survive for several months under dark, anaerobic conditions deeper in the hyperscum. Cyanobacteria are known to withstand long periods of dark, anaerobic conditions while maintaining their photosynthetic capacity (Stanier and Cohen-Bazire, 1977). *M. aeruginosa,* in particular, can overwinter on dark, hypolimnetic sediments of temperate lakes. Some filamentous nonplanktonic cyanobacteria are capable of heterotrophic growth in the dark, but *M. aeruginosa* is not a known heterotroph (Stanier *et al.*, 1971). The mechanism of how *M. aeruginosa* maintain itself for several months under dark, anaerobic conditions both in hyperscums and in hypolimnetic sediments is not known. The ecological role or implications of hyperscums are also not understood.

Methanogenesis in Hypersaline Environments

Whereas methanogenesis in certain ecosystems such as the rumen, sludge, and aquatic sediments has been well studied, environments representing biological extremes in terms of temperature, pH, and/or salinity have not received attention.

Biogenic methane found in some hypersaline environments shows much variability with respect to sulfate content, pH, salinity, and the abundance of methane within their anoxic waters and sediments.

Methane is present in all or most of the hypersaline environments examined to date. In all the environments except the Dead Sea, the methane appears to be of biogenic origin.

Methanogenic bacteria, as well as sulfate respirers, occupy a terminal niche in the degradation of organic matter in anaerobic environments. Complex organic compounds are degraded to simpler molecules by the various nonmethanogenic members of the microbial food web. Some substrates include hydrogen plus carbon dioxide, acetate, formate, methanol, methylated amines, and dimethyl sulfide. There is much diversity among methanogens with respect to substrate utilization. Most but not all species use hydrogen plus carbon dioxide, and a few can also metabolize formate. A few can grow by using the energy released from acetate cleavage, whereas some have a methylotrophic affinity in that they can be grown on methylated amines, dimethyl sulfide, or methanol (Oremland and King, 1989).

Microbial mats are centres of intense primary productivity, especially by a variety of oxygenic and, under some conditions, anoxygenic cyanobacteria. The availability of both dissolved and particulate organic carbon as a result of excretion from and lysis of these cyanobacteria results in the active growth of heterotrophic bacteria in the mats. The cyanobacteria and the heterotrophic bacteria interact; these interactions can range from beneficial (commensalism, protocooperation, and mutualism) to harmful (competition, amensalism, and parasitism). In particular, certain heterotrophic bacteria attach themselves to cyanobacterial heterocysts (Marshall, 1989).

Oxygen levels in microbial mats vary from supersaturation to depletion within the space of a few millimeters. Supersaturation of oxygen leads to CO_2 depletion and sensitization of cyanobacteria to photooxidative death. Under these conditions, intense heterotrophic bacterial activity creates O_2-deficient microenvironments around the cyanobacteria where CO_2 is replenished, thereby protecting the cyanobacteria from photooxidation. The close association of heterotrophic bacteria with cyanobacterial heterocysts also results in more effective N_2 fixation at higher oxygen levels.

Other benefits to cyanobacteria from associated heterotrophic bacteria can involve exchange of vitamins and other growth factors that enhance cyanobacterial growth.

Paerl and Keller (1978) proposed that the specific association of bacteria with cyanobacterial heterocysts stimulates nitrogen fixation by creating a more reduced microenvironment around the heterocyst, thereby protecting the oxygen-sensitive nitrogenase from the high oxygen levels found in blooms and algal mats.

In the upper sediment of many shallow water bodies, mainly in the marine coastal environment, phototrophic bacteria and other primary producers contribute to the first level of the food chain. In laminated sediments, beneath cyanobacterial or algal layers, they contribute effectively to the total photosynthetic production; this production may be considered to be a complement of oxygenic primary production (Caumette, 1989).

Oxygenic Photosynthesis and N_2 Fixation in Marine Microbial Mats

Paerl *et al.* (1981) showed that in spite of being widely distributed in nutrient poor oligotrophic marine waters, microbial mats usually show quite high rates of primary production and biomass accumulation. Localized productive mats bring to mind coral reefs, which also occur in nutrient-depleted waters. The availability of biologically utilizable nitrogen has been linked to levels of planktonic and benthic fertility in diverse marine environments (Carpenter and Capone, 1983; Paerl *et al.*, 1989). Unlike surface water planktonic communities, in which nitrogen demands are met mainly by regeneration and physical inputs (upwelling, vertical and horizontal transport, and terrigenous runoff), microbial mats and coral reefs are sites of significant biological N_2 fixation (Paerl *et al.*, 1981) In these, N_2 fixation acts as an internal source of nitrogen for supporting primary production.

Most microbial mats are often dominated by cyanobacteria, which are good initial (pioneer) colonizers and mat builders. Eukaryotic diatoms, anoxygenic phototrophic bacteria, and chemotrophic bacteria also constitute important subdominant microbial groups in older mats. Specific mat-building cyanobacterial, anoxyphotrophic, and chemotrophic bacterial genera can fix N_2 (Jones, 1974). Mat N_2 fixation is regulated by diverse environmental factors including ambient O_2 concentration, inorganic and organic nutrient availability and suppression by NH_3. The often dominant role of dissolved O_2 in determining both spatial and temporal N_2 fixation patterns and potentials in cyanobacteria-dominated mats is particularly important.

Paerl *et al.* (1989) described a paradoxical situation faced by dominant nonheterocystous cyanobacteria and associated heterotrophic eubacteria, thought to be the chief contributors to mat N_2 fixation rates. The paradox comes from the fact that resident microorganisms get most of their organic carbon requirements for growth and energy production from oxygenic photosynthetic primary production. However, associated O_2 evolution represents a potentially inhibitory step with respect to nitrogenase activity among these species. In contrast to heterocystous cyanobacteria, which have some structural and physiological protection from O_2 inhibition of nitrogenase in the form of thick-walled, O_2 deficient heterocysts, some organisms of the dominant mat cyanobacterial genera, *e.g. Lyngbya, Microcoleus, Oscillatoria*, and *Phormidium* spp., lack heterocysts. This prompts the question of how these nonheterocystous cyanobacteria and their associated eubacteria contemporaneously and contiguously optimize oxygenic photosynthesis and O_2-sensitive N_2 fixation within the confines of a highly productive mat microenvironment (microzone)?

Since mat N_2 fixation is strongly regulated by photosynthetic (O_2) histories, night-time acetylene reduction rates usually exceed day-time rates. To accurately assess mat N_2 fixation, both night-time and day-time measurements are essential. Diel N_2 fixation inputs based on daytime measurements alone may lead to underestimates on the order of 25 to 75 per cent (Paerl *et al.*, 1989).

Nonphototrophic Sulfide-Oxidizing Bacteria

The high diversity of the nonphototrophic sulfide-oxidizing or colorless sulfur bacteria is truly remarkable. Sixteen genera are known already and new taxa are being discovered rapidly (Hüber *et al.*, 1986; Kuenen, 1989). Further, well-known existing species have now been found to be able to use reduced-sulfur compounds as a sole, or supplementary, source of energy, *e.g., Paracoccus denitrificans*, a *Hydrogenobacter* and, among the phototrophic bacteria, some species of *Thiocapsa, Chromatium*, and *Ectothiorhodospira* (Kuenen, 1975).

Many of the colorless sulfur bacteria are taxonomically totally unrelated but share an ability to obtain metabolically useful energy from the oxidation of sulfur compounds, using widely different physiological types of metabolism and metabolic pathways.

The sulfur-oxidizing bacteria can be found whenever suitable sources of (in)organic sulfur compounds, besides a suitable electron acceptor, are present. In most cases oxygen or nitrate is used.

A sulfide/oxygen interface occurs in many sediments. At this interface, diverse metabolic types of sulfide oxidizers are found co-existing together at the highly dynamic interface because of, for example, diurnal or tidal changes. These changes lead to alterations in the fluxes of sulfide, oxygen, other nutrients such as organic compounds, and such alternative electron acceptors as nitrate.

To study the flexibility and reactivity of the organisms, pure- and mixed-culture studies were also carried out under fluctuating conditions (for example, with an alternating supply of autotrophic [thiosulfate plus CO_2] and heterotrophic [acetate] substrates) and under alternating starvation and nutrient-limiting conditions for different periods. The outcome of these studies has allowed the construction of an ecophysiological profile for specialist and versatile types. Not only is metabolic flexibility an important property for survival in nature, but also reactivity (*i.e.*, the ability to respond rapidly to a changing environment) is a crucial factor for survival of certain metabolic types (Kuenen, 1989). Table 9.2 gives a comparative ecophysiological profile of specialist and versatile taxa.

Table 9.2 Differences between a metabolically versatile and a specialized chemolithotrophic *Thiobacillus* sp. (after Kuenen, 1989)

Property	
Specialist Thiobacillus neapolitanus	*Versatile Thiobacillus strain A2*
Few energy substrates utilized (S^{2-}, S^0, $S_2O_3^{2-}$, $S_4O_6^{2-}$, etc.)	Many different inorganic and organic energy substrates utilized
High specific growth rate on single substrate (μ_{max}[a] on thiosulfate, 0.35 h^{-1})	Low specific growth rate on single substrate (μ_{max} on thiosulfate, 0.10 h^{-1}); relatively high specific growth rate on mixed substrates (μ_{max} on thiosulfate plus acetate, 0.22 h^{-1}).
High affinity for reduced-sulfur compounds	Relatively low affinity for reduced-sulfure compounds
Low flexibility with respect to energy generation and organic-carbon assimilation; constitutive enzymes	High flexibility; inducible enzymes for energy generation and carbon assimilation.
High reactivity toward few substrates	Low reactivity toward many substrates
Metabolic lesions	Many pathways, often overlapping
Low endogenous respiration	High endogenous respiration in autotrophically and heterotrophically grown cells.
Very resistant to starvation	Less resistant to starvation
Ecological niche: environments with continuous or fluctuating supply of reduced sulfur compounds and a low turnover of organic compounds	Ecological niche: environments with simultaneous presence of both inorganic and organic substrates (mixotrophic conditions)

[a]μ_{max} Maximum specific growth rate.

The role of organic compounds in the selection of different organisms is only one of the many selective forces exerted on the organisms living at the interface. Even under the artificially strong selective forces in a chemostat, stable coexistence of different physiological types can be predicted and observed. Mathematical models further predict that the number of coexisting species can never exceed the number of variables in a certain environment. In most microbial biofilms or mats, the number of variables is very large indeed, and, therefore, the number of coexisting species is also quite large (Kuenen, 1989).

Ecophysiology of *Beggiatoa*

A colorless sulfur bacterium typical of microbial mats, *Beggiatoa* sp., is one of the few well-studied mat-building chemolithotrophs.

Nelson and Jannasch (1983) and Nelson *et al.* (1986) grew several pure cultures of autotrophic marine strains of *Beggiatoa* in artificial sulfide-oxygen gradients. Test tubes with a sulfide-containing agar were inoculated with *Beggiatoa* cells and overlaid with a soft agar overlay. The organisms rapidly moved to the sulfide/oxygen interface and started to multiply. Autotrophic growth could be established.

Nelson *et al.* (1986) studied the sulfide and oxygen microprofiles in artificial gradients. In the uninoculated control tube, the oxygen and sulfide coexisted at fairly high concentrations. Fluxes of oxygen and sulfide entering the interface were calculated from the slopes of the gradient by using Fick's first diffusion law. The half-life of oxygen and sulfide in the control was more than 50 min. *Beggiatoa* species apparently outcompete the chemical oxidation by lowering the concentrations of the oxygen and sulfide at the sulfide/oxide interface (Kuenen, 1989). By their efficient sulfide oxidation at the sulfide/oxygen interface, the sulfur bacteria have the potential to protect the aerobic environment from toxic sulfide and the anaerobic environment from toxic oxygen (Kuenen, 1989).

Under continuous darkness the organisms tend to form large, dense mats on sulfide-rich sediments overlaid by aerobic water bodies. When these sediments become exposed to (day)light, a complex diurnal pattern of competing and interacting sulfide-oxidizing bacteria develops (Larkin and Strohl, 1983). When the balance of sulfide and photosynthetic oxygen production in the top layer is such that the oxygen reaches below the photic zone, *Beggiatoa* species grow at the sulfide/oxygen interface deep in the sediment during daylight hours. At night, these organisms, move up to the top of the sediment and continue to oxidize the sulfur compounds there. However, if light penetrates to the sulfide layer, a situation can develop whereby the phototrophs use the sulfide both in the light and in the dark. In this situation *Beggiatoa* species can be outcompeted (Jorgensen and DesMarais, 1986).

In some situations *e.g.* a sulfide-rich sediment, *Beggiatoa* species can be found during the day under an oxygen-producing *Oscillatoria* bloom. Under the *Beggiatoa* layer a red layer of the motile *Chromatium* species was found by Jørgensen, (1982). After a transient situation occurring after sunset, the sulfide production by this particular mat was such that the sulfide entered the first few cm of the aqueous phase above the sediment. Being unable to move into the water, *Beggiatoa* spp. remained at the top of the sediment, but the motile *Chromatium* cells could move into the water and take over the role of the *Beggiatoa* spp. as a chemolithotroph (Kuenen, 1989). It appears that *Beggiatoa* spp. not only grow at the interface but also can survive substantial anaerobic periods. Intracellular sulfur formed during aerobic periods probably serves as an electron acceptor for anaerobic heterotrophic metabolism in these species. Indeed, it may be imagined that, by moving up and down through the oxygen/sulfide interface, the *Beggiatoa* spp. can regenerate the electron acceptor (sulfur) to benefit optimally from substrates in the anaerobic zone. Such a strategy might be particularly important for the many heterotrophic *Beggiatoa* strains.

Beggiatoa spp. are typical interface organisms specialized on the one hand to microaerophilic life in a solid substrate and on the other hand to life at a moving interface (Kuenen, 1989).

Medical Biofilms

Microbial biofilms have a significant role in clinical medicine. Serious infections can be caused by biofilms formed on implanted medical devices. The determination of biofilm architecture, particularly the spatial arrangement of the colonies (clusters of cells) relative to one another, affects the functioning of these complex communities. Standard antimicrobial treatments fail to eradicate biofilms because the organisms develop adaptive resistance to antibiotics. The need for the development of effective antimicrobial molecules or combination techniques to counter biofilm infections is urgent.

Pathogenic microorganisms often attach to living and non-living surfaces such as medical devices including urinary, venous, and arterial catheters, shunts, heart valves and tubes. The attached microbes aggregate and multiply into a mushroom like shape that is held together by glycocalyx, an extracellular "polysaccharide matrix" termed biofilm.

Microbial biofilms are often formed by antimicrobial-resistant organisms and are responsible for over 60 per cent of infections treated in the developed world (Costerton *et al.*, 1999). About 24 per cent of adults have lost at least 4 mm of periodontal attachment, and 40-60 per cent of young people as well as adults suffer from some form of gingival (biofilm) infection due to biofilm related infection. Many infections are related to urinary catheters and intravascular devices. In fact, about 95 per cent of urinary tract infections are associated with a urinary catheter, 85 per cent of pneumonias are associated with mechanical ventilation, and 85 per cent of bloodstream infections are associated with an intravascular device (Richards *et al.*, 1999).

The percentage of human infection is usually so high because the microbes present in the biofilm are strongly resistant to killing, and to treatment with microbial agents (Costerton, 1987). Shedding of the bacteria from the biofilms into the surrounding tissue and the circulation system also causes acute illness. Table 9.3 lists some common microorganisms that cause human infection. Most biofilm related infections are caused by *E. coli*, *Staphylococcus aureus*, *S. epidermidis*, and *Pseudomonas* (Prasanna and Doble, 2008).

Use of modern imaging techniques has identified the structural and developmental complexity, and mechanism of biofilm formation on surfaces. One such technique is confocal laser microscopy. Sessile bacteria growing in heterogeneous matrix and enclosed by micro colonies interspersed with open water channels facilitate efficient uptake of nutrients from the bulk phase into the biofilm. Such exchange of optimum nutrients and waste products provides the first link between formation and its function (Stoodley *et al.*, 1994).

Table 9.3: Some Common Microorganisms that Cause Human Infections by Forming Biofilms (After Prasanna and Doble, 2008)

Body Site	*Implant or Device*	*Causal Bacteria*
Urinary tract	UT catheters CV catheters	*Escherichia coli*, *Staphylococcus epidermidis*, *S. aureus*
	Temporary pacemaker	*S. epidermidis*, *S. aureus*
Percutaneous	Peritoneal dialysis catheters	*S. epidermidis*
	Orthopedic pins	*S. aureus*
Subcutaneous	Cardiac pacemaker	*S. epidermidis*
	Mammary prosthesis	*S. aureus*
Soft tissue	Intraocular lenses	*Pseudomonas*, *S. epidermidis*
	Prosthetic heart valve	*S. aureus*, *S. epidermidis*
Circulatory system	Vascular graft	*S. aureus*, Gram negative bacteria
Bones	Prosthetic hip	*S. epidermidis*, *S. aureus*
	Total knee	*S. epidermidis*, *S. aureus*

Five stages are usually involved in biofilm formation on medical devices (Figure 9.1). The identification and association is mediated through long (*e.g.*, gravitational, van der Waals, and electrostatic) and short (*e.g.*, hydrogen bonding, dipole-dipole, ionic, and hydrophobic) interactions.

Stages 3 and 4 involve the aggregation of cells into micro colonies followed by growth and maturation of the biofilm. These stages take about 2-3 h and are characterized by stronger adhesion between the bacteria and the foreign material. Specific chemical reactions between cell surface compounds and substrate surfaces lead to irreversible molecular bridging. Polysaccharides on the bacterial membrane surface and adhesion proteins within facilitate attachment to substrate surfaces. The biofilm can be flat or mushroom shaped, depending on the nutrient source. In stage five, termed "transient motility", the biofilms are sloughed or shed and daughter cells are formed (Sauer *et al.*, 2002). These daughters then move downstream to form new attachment sites.

The extracellular polysaccharide (EPS) produced by bacteria is involved in their aggregation and adhesion to the surface. This matrix houses the microorganisms. It not only allows the formation of stable communities ("micro consortia") of synergistic strains but also enables them to degrade recalcitrant substances. By retaining water,

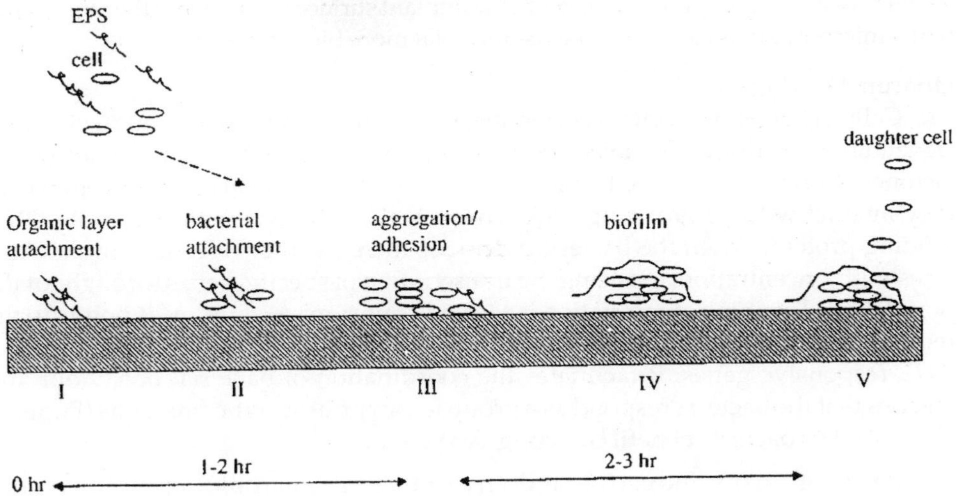

Figure 9.1: Mechanism of Formation of Biofilm (after Prasanna and Doble, 2008).

EPS prevents desiccation. EPS can be exopolysaccharides or capsular polysaccharides (when the bacteria grown in a liquid culture are centrifuged, the extracellular polysaccharide that remains associated with the cell is known as capsule, whereas those that remain in the supernatant are known as exopolysaccharides). It is this latter type that plays a critical role in determining the architecture of the biofilm. In most cases the polysaccharides cannot be distinguished as they are not easily separable from the biofilm. The exopolysaccharides synthesized by microbial cells vary greatly in their composition, chemical and physical properties. Some are neutral macromolecules, but most are polyanionic due to the presence of either uronic acids (D-glucuronic acid being the most common, although D-galacturonic and D-mannuronic acids are also found) or ketal-linked pyruvate. Phosphate, and rarely sulphate, may also confer polyanionic status. The composition and structure of the polysaccharides determine their primary conformation.

Microorganisms that form biofilms occupy some enclosed surface space and again stability in the growth environment. Biofilms also protect from such environmental challenges as metal toxicity, dehydration, salinity, UV, antibiotics or antimicrobial agents (Hall-Stoodley *et al.*, 2004).

Some important properties conducive to good biofilm formation are surface area, the type of surface (rough or smooth), porosity, charge on the surface and surface hydrophobicity. Rough surface is preferred by the microbes as it is better for stronger adhesion. Crevices also favour attachment. Hydrophobicity in case of polymeric materials enhances biofilm formation. Hydrophobic organisms prefer hydrophobic surfaces. Divalent cations such as Mg^{2+} and Ca^{2+} influence biofilm formation through their effect on electrostatic forces. Functional groups present on the surface of the material influence both hydrophobicity and surface charge which in turn affect bacterial adhesion and proliferation. Several physical, chemical and biological techniques may be employed to modify the implant surface to minimize the attachment of the microorganism and to make the material more biocompatible.

Quorum Sensing

Cells in bacterial biofilms communicate among themselves through chemical signaling. Small, diffusible molecules belonging to the class of N-acylated homoserine lactones (AHLs) are released by biofilm bacteria into their local environment where they interact with neighboring cells. These AHLs associate with a cognate DNA binding protein. When the bacterial densities increase, the AHLs accumulate to a threshold concentration, inducing the transcription of specific genes throughout the population. This type of regulation is termed "quorum sensing", as it points to the requirement for a sufficient population of bacterial cells needed for the activation of AHL-responsive genes. It facilitates the coordination of bacterial behaviour and ensures that the bacteria respond as a group to carry out specific functions (Fuqua *et al.*, 1994). The bacteria benefit by acting as a group.

For long, many industrial and environmental microbiologists have been concerned in understanding the formation of biofilms and biofouling. In recent years, the contribution of biofilms to human infections has been widely recognized. The medical fraternity is deeply interested in finding ways of inhibiting the formation of biofilms on medical devices. Some of these methods exploit the use of small molecules.

Antibiotics and Anti-microbial Molecules

The MIC (minimum inhibitory concentration) has long been the widely accepted standard for testing the susceptibility of microorganisms to an antibiotic. It measures the efficacy of antibiotics against planktonic organisms and serves as an important reference in the treatment of many acute infections. A common practice is to treat a bacterial infection with several antibiotics such as gentamicin, carbenicillin, co-trimoxazole, tetracycline, and ceftizoxime. The chief problem here is that the biofilm becomes resistant to the antibiotics. Rastegar *et al.* (1998) reported that *P. aeruginosa*, a common cause of wound infections showed 95 per cent resistance to the above mentioned antibiotics. Some antibiotics *e.g.* ampicillin, ceftiofur, cloxacillin, oxytetracycline, penicillin G, streptomycin, tetracycline, enrofloxacin, erythromycin, gentamicin, tilmicosin and trimethoprim-sulphadoxine act on cultures like *Actinomyces pyogenes, Corynebacterium renale, C. pseudotuberculosis, Staphylococcus aureus, S. hyicus* and *Streptococcus agalactiae*. But the biofilms formed by all these organisms turned out to be quite resistant (Olson *et al.*, 2002). Chlorine as sodium hypochlorite is a most effective antimicrobial agent; a 600 fold increase in its concentration is required to treat the biofilm formed by *S. aureus* when compared to the planktonic cells of the same species (Luppens *et al.*, 2002).

The mechanisms for the resistance of biofilms to biocidal agents are of three main types: (1) adaptive phenotype, (2) restricted entry of the drug molecules, and (3) reduced metabolism and slow growth. Of these, the second is quite important. The EPS slimy matrix bars the penetration of drugs and so protects the organisms against various treatments. The antimicrobial agent becomes adsorbed onto the EPS, effectively diluting its concentration before it reaches individual bacterial cells in the biofilm

(Dibdin *et al.*, 1996). The negatively charged EPS restricts penetration of the positively charged molecules of antibiotics, such as amino-glycosides, by chemical interaction or molecular bindings (Donlan and Costerton, 2001; Stewart, 2002). In case the antibiotic becomes inactivated or is attached ionically to the surface layers, its penetration to the depths of the biofilm is retarded (Lewis, 2001).

Regardless of the mechanism involved, the final result is the development of resistance towards antimicrobial agents or antibiotics. Some other approaches to addressing the problem include modifying the material properties of the medical devices, *e.g.*, adsorption or coating of antimicrobials onto the surface, antimicrobial impregnated matrices and physico-chemical surface modification (Figure 9.2).

Polymer catheters or metal stents can be coated with antimicrobial organic molecules with a view to inhibiting the bacterial adhesion. These "active" coatings release high initial fluxes of antibacterial agents during the critical few hours of post-implantation period to block the initial adhesion of the bacteria. This lead to a reversible, weak interaction between the microbe and the coated material (An *et al.*, 2000).

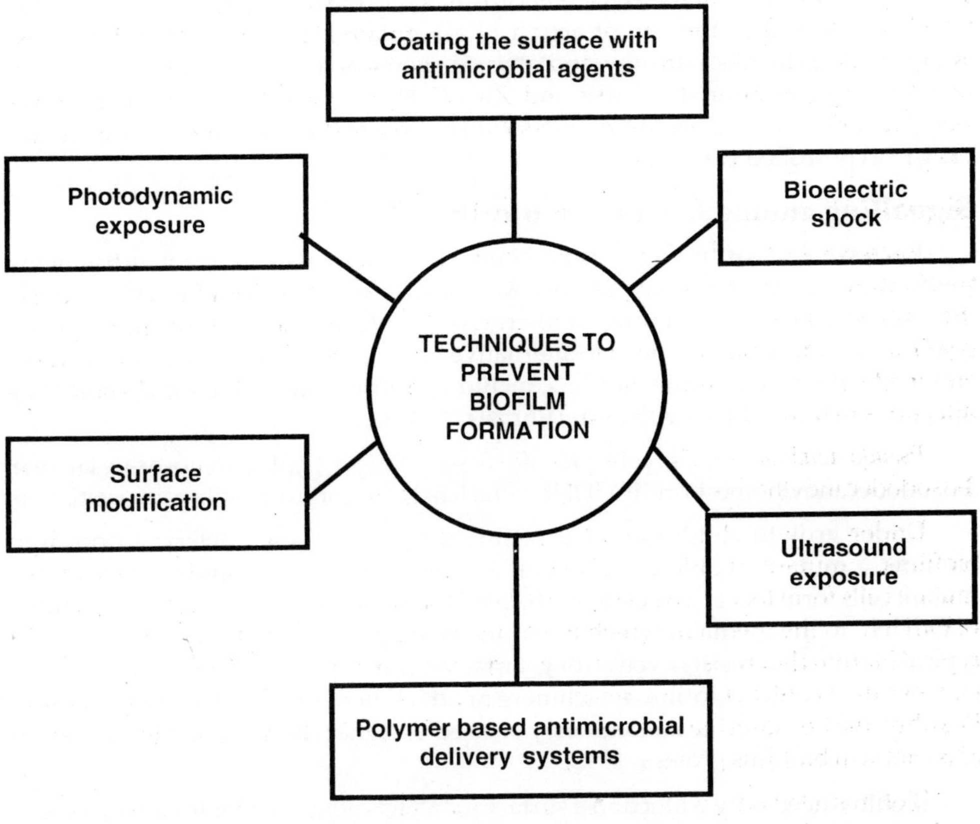

Figure 9.2: Techniques Commonly Used to Prevent the Formation of Biofilms (After Prasanna and Doble, 2008).

The bioelectric application not only prevents biofilm formation but also enhances the activity of the antimicrobials against established biofilms. Antibiotics and a weak electric field are applied simultaneously. When direct current electric fields between 1.5 and 20 V/cm are used, the effective concentrations of antibiotics effective against biofilm bacteria fall from approximately 5000 to 4 times greater than that needed for free living bacteria in the absence of electric current (Costerton *et al.*, 1994).

Ultrasound enhances antibiotic transport through the biofilm. Low-frequency ultrasound (70 kHz) of low acoustic intensity increased the growth rate of *S. epidermidis*, *P. aeruginosa* and *E. coli* on polyethylene surfaces, presumably by increasing the transport of oxygen and nutrients to the cells (Pitt and Ross, 2003).

Photodynamic technique is effective against pathogens associated with the skin and in the oral cavity. Its application is limited to parts where light can reach. Photosensitizing drugs produce reactive oxygen species against which the microorganisms cannot defend themselves, leading to a breakdown of the biofilm (see Soukos *et al.*, 2000).

Biodegradable microparticulate polyester hydrogels, micelles and fibrous scaffolds are used as effective drug carriers (Varde and Pack, 2004; Freiburg and Zhu, 2004). The antimicrobial molecule is encapsulated in the microspores of the polymer which is released at the site of action in a regulated manner, but stability and encapsulation efficiency are low, and physico-chemical compatibility of the polymer and the drug is poor. Freiburg and Zhu (2004) reported that biodegradable poly(lactide) (PLA) and its co-polymers with glycoside (PLGA) release encapsulated drug in a controlled way.

Signalling among Bacteria in Biofilms

Bacteria in biofilms exchange chemical signals as a way of influencing multicellular developmental patterns. Microbial growth in biofilms is notoriously difficult to control. Therefore, deliberately interfering with these intraspecies communication signals can lead to alternative treatments for controlling industrially and medically troublesome biofilms, including those responsible for chronic lung infections in individuals with cystic fibrosis (CF).

Pseudomonas aeruginosa cells in biofilms secrete a particular homoserine lactone, 3-oxododecanoylhomoserine (OdDHL) which helps to control biofilm differentiation.

Under growth conditions that ordinarily permit these bacteria to produce biofilms, a mutant unable to make OdDHL does not form a biofilm. Instead, the mutant cells form looser associations that weak surfactants readily disrupt. Addition of OdDHL to the media in which the mutants are growing restores formation of a typical biofilm that resists even strong surfactants. It appears that other homoserine lactones may control biofilm detachment or attachment in other bacterial biofilms. Possibly, the homoserine lactones may be used as signals between different species of bacteria in biofilms (Potera, 1998).

Biofilms need only water and a surface for attachment in order to cause havoc in medical and industrial settings. Bacteria forming biofilms produce mushroom shaped, polysacchride coated pillars that are surrounded by water channels. This bizarre internal structure of biofilms protects the organisms from destruction by antibiotics and biocides. Bacteria in biofilms can be 1,500-fold more resistant to antibiotics than are their free-floating counterparts.

P. aeruginosa is also the leading cause of morbidity and mortality in CF patients. Its biofilms coat the lung surfaces of CF patients, clogging their airways by mucus and other debris. Use of OdDHL (or some of its analogues) may help block the development of such biofilms.

Besides medical settings, biofilms foul industrial water systems. Paints used to prevent biofilm fouling on boats contain toxic heavy metals. If homoserine lactone-blocking analogs could be used instead of heavy metals to prevent biofilm formation, paint producers and boating enthusiasts would welcome the development (Potera, 1998).

Bacterial Flagellar Motors for Biofilms

Biofilms are slimy carpets less than a millimeter thick that appear on rocks, leaves, pipes, teeth, or any other place that has some nutrients and water. These biofilms are formed by many species of bacteria whose cells must first attach to a surface, which in many species requires swimming propelled by rotating helical flagella. Cells then stop expressing genes that encode components of the flagellum, and they secrete a sticky matrix of polysaccharides that holds them together on the surface (Branda *et al.*, 2005). At a surface, swimming becomes a hindrance and an inverse relationship between swimming and attachment is often seen in diverse species (Kolter and Greenberg, 2006).

The flagellar motor in bacteria consists of a rotor ~45 nm in diameter that spins at several hundred revolutions per second, surrounded and pushed by a ring of protein complexes anchored to the cell wall and powered by an ionic current across the cytoplasmic membrane (Berry and Armitage, 2008). The motor is connected to a propeller composed of a helical filament which extends many micrometers outside the cell. In *Bacillus subtilis*, motility and matrix formation are linked by a protein SinR that upregulates expression of flagellar genes and downregulates expression of matrix-forming genes. Deletion of the *sinR* gene causes cells to form biofilms (Kearns *et al.*, 2005).

Blair *et al.* (2008) showed that flagellar filaments do not rotate in these aggregates. And these bacteria do not swim even in the absence of a gene needed for matrix formation. A search for mutations that allow swimming without *sinR* revealed the off-switch for rotation: *EpsE*, probably glycosyltransferase protein involved in building biofilms. The *epsE* gene is found in an operon that encodes the proteins that make the matrix and is repressed by SinR. Induced expression of EpsE in *B. subtilis* stopped cell motility.

To determine whether EpsE acts as a brake that locks the motor, or a clutch that leaves the rotor freely spinning, Blair *et al.* tethered bacteria to a substrate by their filaments and noticed rotation of the cell bodies around single flagellar motors. Under the influence of EpsE, cells stopped spinning but continued to perform free rotational Brownian motion, indicating a clutch mechanism.

The direct inhibition of motor rotation by EpsE seems to be a new control mechanism for bacterial swimming. Bacterial flagella are large protein complexes that require about 40 to 50 genes to assemble (Macnab, 2003). The chief advantage of the EpsE mechanism over transcriptional control of flagellar genes is speed. In B. subtilis, only one protein, EpsE, needs to be expressed to stop the motor. This may be important if cells are to stay put in the early stages of biofilm formation. But the advantages of a clutch over a brake mechanism are not so clear. Perhaps free rotation of flagella–or, alternatively, reduced motility during the transition to the EpsE-inhibited state–is important for the formation of well-structured biofilms (Wood *et al.*, 2006).

Social Evolution in Microbial Biofilms

The large majority of bacterial life in nature occurs in surface-bound communities called biofilms (see Characklis and Marshall, 1990) rather than in isolated planktonic cells (Kolter and Greenberg, 2006). In biofilms, cells attach to surfaces and secrete slime (polymeric substances). The central importance of biofilms to microbial life is underlined by their crucial role in many chronic diseases, antibiotic resistance, biofouling, and wastewater treatment (see Fux *et al.*, 2005). Biofilms seem to require high levels of cooperation because extracellular polymeric substances are a shared resource produced by one cell that can be used by others. Biofilms represent densely packed systems in which cells share many secreted molecules such as enzymes, iron-scavenging siderophores, and extracellular polymers. Such a sharing of resources indicates that they are open to the evolution of "cheater" strains that use the resources of others without contributing anything to the group pool (West *et al.*, 2006; Fiegna *et al.*, 2006). When many unrelated strains occur together in biofilms, strong evolutionary conflicts are predicted over contribution to shared group resources (Foster, 2005).

The matrix or slime that surrounds the cells is a characteristic feature of many biofilms. These extracellular polymeric substances (EPS) embed bacteria. Polymers are a shared resource that benefits the biofilm by maintaining its structure (Hughes *et al.*, 1998) and protects it from dehydration, ultraviolet radiation, or predator grazing (Palmer *et al.*, 2001; Kolenbrander *et al.*, 2005). Polymers also facilitate extracellular enzymatic activity and signaling. Kreft (2004) analyzed the success of strains with different rates of substrate uptake in mixed biofilms. He reported that slow-growing, high-yield strains could be exploited by high-growth, low-yield strains, and concluded that cooperation and altruism were required for optimal biofilm formation. Like slow substrate uptake, polymer production reduces the available energy for growth. This prompts the question: what maintains polymer production in the face of selection for competition and rapid growth?

Strong cooperation in biofilms is predicted if they contain a single strain (high genetic relatedness). Genetically identical cells, as seen in most multicellular organisms, do not have evolutionary conflicts of interest, and they are expected to behave simply as it is optimal for the group (Frank, 1998; Foster, 2004).

Very little is known about within-species genetic diversity in biofilms. However, considering the tremendous diversity of species that can occur within biofilms (Palmer *et al.*, 2001; Kolenbrander *et al.*, 2005), it is not likely that all or most biofilms have the clonal structure required to generate perfect cooperation throughout. This prompted Xavier and Foster (2007) to examine whether cooperation was necessary for polymer production in an individual-based simulation. They examined this hypothesis by using a detailed individual-based simulation of a biofilm to investigate the outcome of evolutionary competitions between strains differing in their level of polymer production. Their model included a biochemical description of the carbon fluxes for growth and polymer production, and it explicitly calculated diffusion-reaction effects and the resulting solute gradients in the biofilm. An emergent property of these simple assumptions is a strong evolutionary advantage to extracellular polymer production. Polymer secretion is altruistic to cells above a focal cell: it propels later generations into better oxygen conditions, but it harms others. Also polymer production suffocates neighboring nonpolymer producers.

The advantage of polymer production in the focal cell comes from differential effects on cells lying above versus laterally to the founding focal cell. For those cells which occur above the focal, polymer production is altruistic and favored by kin selection–the focal EPS cell incurs a cost by dividing more slowly than the EPS–cells at the beginning of the simulation, but it gains a fitness benefit through relatives because the polymer pushes its descendents into oxygen-rich conditions where there is also less competition for space. That this behaviour is altruistic is borne out by the fact that by producing polymer, the focal cell slows down its division to enable other cells to divide. This form of altruism operates through cells helping descendent cells in the same lineage–it is altruism only to the extent that actions that reduce an individual's personal reproduction to help existing offspring reproduce are also considered altruistic (Foster and Ratnicks, 2005).

Strong local competition is known to prevent altruism among relatives (West *et al.*, 2002). But the upward expansion of the biofilm reduces local vertical and horizontal competition for space within a lineage, and it favors their cooperation. There is strong lateral competition at biofilm initiation, however, as evidenced by the reduced fitness of cells lying horizontal to the polymer-producing lineage. For these cells, polymer production appears to be spiteful: the focal cell reduces its reproductive rate to produce polymer which in turn reduces the fitness of the lateral neighboring cells (Lehmann *et al.*, 2006). The extracellular polymer, therefore, gives a competitive push to a lineage by allowing its cells to rise up and over other cells and suffocate them. It is this suffocation effect that accounts for the frequency-dependent nature of the advantage, whereby polymer producers invade most easily in those biofilms which contain many nonproducers. When the frequency of polymer producers increases in a biofilm, they have less and less chance to overgrow other lineages (increased competition for oxygen and space), which reduces the benefit of polymer production (Xavier and Foster, 2007). This property is somewhat analogous to vertical growth in plants and suggests that polymer secretion confers great competitive advantage to cell lineages within mixed-genotype biofilms–global cooperation is not required.

The model proposed by Xavier and Foster radically changes the perception of how biofilms might respond to changing social conditions. The presence of multiple strains in a biofilm should promote rather than inhibit polymer secretion (Xavier and Foster, 2007).

Yeast Biofilm

Standing alone, fungal and bacterial pathogens are relatively easy prey for antimicrobial drugs. But many of these germs cling together in resilient sheets and globs called biofilms which resist traditional chemical attack. A knowledge of what causes bacterial microfilms to form can help provide potential new targets for infection-fighting drugs. Fungal biofilms frequently contaminate medical devices, cause chronic vaginal infections, and lead to life-threatening systemic infections in people with weakened immune systems.

Reynolds and Fink (2001) have now reported that they've coaxed a harmless fungus, bakers' yeast (*Saccharomyces cerevisiae*) to form a biofilm. As the yeast is so well studied–its entire genome has already been sequenced–its use as a new biofilm model will expose vulnerabilities that can be targeted in other, pathogenic fungi.

Bakers' yeast occasionally forms a film on the surface of sherry but it does not naturally congregate into a typical biofilm: simply, a film that sticks to plastic. To induce bakers' yeast to do this, the researchers tested several strains and tweaked the yeast's nutrients until they hit on a combination that does produce a robust biofilm. The bakers' yeast built the largest biofilms and stuck most stubbornly to plastic when it was given low concentrations of glucose, suggesting that lean times induce the yeast to change form.

When grown in low-glucose medium, the yeast cells adhered to a variety of plastic surfaces. On semi-solid (0.3 per cent agar) medium they formed 'mats': complex multicellular structures composed of yeast-form cells. Both attachment to plastic and mat formation require Flo 11p, a member of a large family of fungal cell surface glycoproteins involved in adherence (Reynolds and Fink, 2001). These authors found that while the adherence of yeast to plastic was enhanced as the glucose concentration was lowered, it was reduced in the complete absence of glucose; this points to a requirement for active metabolism. Examination of the attached cells by microscopy revealed that they were round yeast-form cells.

Quorum-sensing: Microbial Communication

Bacterial cells show several density-dependent phenotypes–a pheno-menon referred to as quorum-sensing. It occurs through the production of an autoinducer by the organism which, when it reaches a threshold concentration (a direct reflection of population density), activates certain genes and produces the concerned phenotype. This cell-to-cell communication has played an important role in many diverse community-based functions and also helps in the establishment of a population in a changing environment according to the requirement. This sensing can be exploited for controlling plant and animal diseases and regulation of the production of useful/toxic fermentation products.

A wide range of microorganisms are able to perceive and respond to the presence of neighbouring populations. This 'quorum-sensing' is accomplished by the extracellular accumulation of small, self-generated chemical signalling moieties that induce a concerted effort on behalf of a population to produce the desired phenotypic effect (De Kievit and Iglweski, 2000).

Fuqua *et al.* (1994) first used the term quorum-sensing to reflect the minimum threshold level of individual cell mass required to initiate a concerted population response. They termed the signal molecule used for communication as 'autoinducer', owing to its origin from inside the bacterial cell. The desired response materializes by attainment of quorum employing the autoinducer and the process is called 'autoinduction'. The whole circuit relies on the intracellular production and export of a low-molecular-mass signalling molecule, whose extracellular concentration grows with the population density of the producing organism. The signalling molecule can be sensed and reimported into these cells, allowing the whole population to respond to changing environment/requirement once a critical concentration (corresponding to a particular cell density) is achieved.

Till now, several classes of microbially-derived signalling molecules have been identified. Broadly, they can be divided into two main categories: (i) amino acids and short peptide derivatives, commonly utilized by Gram-positive bacteria, and (ii) fatty acid derivatives, called homoserine lactones (HSLs), frequently utilized by Gram-negative members. Regardless of the nature of the signal molecule, the whole network functions by its reentry into the cell either via diffusion or through active transport (Dunny and Winans, 1999; Whitehead *et al.*, 2001). The signalling mechanism involves subsequent interaction of the signal with an intracellular effector that induces the pathway for the concerned phenotype (Gera and Srivastava, 2006).

Bioluminescence: The *Lux* System

Certain marine species such as *Vibrio fischeri* and *V. harveyi* free-living in sea water (at low cell density) are non-luminescent. But when grown to high cell densities in the laboratory, a *V. fischeri* culture becomes bioluminescent with a blue-green light. This bacterium forms symbiotic relationships with some fishes (such as the Japanese pinecone fish *Monocentris japonica*) and squid species (such as *Euprymna scolopes*). These marine animals carry a specialized light organ which houses *V. fischeri*. *E. scolopes* shows luminescence in darkness when there is a high-density of *V. fischeri* population (10^{10}-10^{11} cells ml^{-1}) in the light organ. The squid exploits this bioluminescence to perform a behavioural phenomenon called counter-illumination. At night, the squid camouflages itself from predators found below it by controlling the intensity of light that it beams downwards, so eliminating a visible shadow created by moonlight. In return, *E. scolopes* provides the *V. fischeri* population with nutrients. The increase in luminescence as a function of the cell density is attributed to the transcriptional regulation of the enzyme luciferase which in turn corresponds to a threshold density of cells. This whole circuit is based on the bacterial assessment of its population density through release of autoinducers. The autoinducer establishes a communication between the cells, as reflected in the expression of the luciferase gene (*lux*). Bioluminescence expressed by *V. fischeri* has served as a model system to study density-dependent expression of gene function (Hastings and Greenberg, 1999).

Many diverse microorganisms, both Gram-negative and Gram-positive, employ quorum-sensing systems to regulate phenotypes ranging from mating to virulence against the host, antibiotics and production of other metabolites, and many others (Dunny and Winans, 1999). One common feature between these diverse phenotypes is that a microbial function can be performed successfully, based on appropriate population size, cross-communication to ascertain its own community, or to even differentiate self from non-self. For instance, *Streptococcus pneumoniae* uses a peptide-mediated quorum-sensing system to acquire competence for genetic transformation. In *Bacillus subtilis*, on the other hand, high cell density contributes to the regulation of at least two different developmental processes–development of genetic competence and sporulation. The mating system in *Enterococcus faecalis* has a peptide pheromone-based quorum-sensing system (Lazazzera and Grossman, 1998).

Virulence in Gram-negative Bacteria

Besides the well-known example of bioluminescence by *Vibrio* species, virulence of several Gram-negative members is regulated by quorum-sensing. The opportunistic human pathogen *Pseudomonas aeruginosa* produces and secretes multiple extracellular virulence factors which cause much host tissue damage. Production of virulence factors is under the hierarchical control of two pairs of LuxI/LuxR homologues, LasI/LasR and RhlI/RHlR. Both LasI and RhlI are autoinducer synthases that catalyse the formation of HSLs, N-(3-oxododecanoyl)-homoserine lactone (OdDHL) and N-(butanoyl)-homoserine lactone (BHL) respectively.

At high cell density, LasR binds to its cognate HSL autoinducer, and together they bind to promoter elements preceding the genes encoding some secreted virulence factors–elastase (*lasB*), a protease (*lasA*), endotoxin A (*toxA*) and alkaline phosphatase (*aprA*). LasR bound to autoinducer activates *las*I expression, which generates a positive feedback loop. Some plant-associated bacteria also employ quorum-sensing for their pathogenic as well as symbiotic life-styles.

Quorum-sensing in Gram-positive Bacteria

Some Gram-positive bacteria also use quorum-sensing systems. The nature of the signal molecules used in these systems differs from those of Gram-negative organisms (Dunny and Leonard, 1997). Quorum-sensing is used to regulate the development of bacterial competence in *Bacillus subtilis* and *Streptococcus pneumoniae*, conjugation in *Enterococcus faecalis*, and virulence in *Staphylococcus aureus*. In pneumococci, five genes have been implicated in the peptide-mediated regulatory circuit, *viz.*, *com ABCDE* for competence development. The peptide signal required for development of the competent state is called CSP (competence stimulating peptide). CSP, a 17-amino acid peptide is produced from a 41-amino acid precursor peptide called ComC (see Gera and Srivastava, 2006).

Microbial Cross Communication

The occurrence of cell-cell signalling is widespread among different bacterial species. Many organisms utilize the same species of molecule to regulate different phenotypes. It is conceivable that some form of interspecies communication prevails in environments where different autoinducer-producing bacterial species inhabit the same habitat. Various LuxR homologues can interact with non-cognate acylHSL molecules. Depending on the LuxR homologue and the acylHSLs in question, such interactions can result in the activation of the specific transcriptional regulator. Conversely, when assayed in the presence of the cognate acylHSL, other species of acylHSL essentially block activation of the LuxR homologue, presumably by competing for the ligand-binding site on the protein (Gera and Srivastava, 2006).

Many of the known acylHSL-producing microorganisms associate with higher organisms either in a pathogenic or symbiotic relationship. Higher organisms have evolved mechanisms that enable them to detect and respond to acylHSL messaging systems so as to prevent or limit infection (Teplitski *et al.*, 2000). For example, the macroalga *Delisea pulchra* produces furanones that can specifically interfere with acylHSL-mediated quorum-sensing systems.

While serving as prokaryotic cell-to-cell signals, some acylHSLs also act as virulence factors. In particular, one of the *P. aeruginosa*-produced molecules, OdDHL, is a potential modulatory agent of the mammalian immune systems. Interspecies communication through the use of autoinducers has been invoked as a possible mechanism by which the pathogenicity of certain virulent bacteria such as *Burkholderia cepacia* is enhanced (McKenny *et al.*, 1995).

Since many bacteria make use of quorum-sensing for regulation of various phenotypes as a part of their pathogenic or symbiotic lifestyles, the ability to block or promote these systems can be a powerful tool to solve many problems and enhance productivity. Genetically modified plants could be used to produce AHLs to manipulate plant-bacterium associations. Quorum-sensing may also be a desirable strategy to control multiple-drug-resistant strains of bacteria that employ quorum-sensing for virulence and pathogenicity and for control of trigger of host-cell response, for novel antimicrobial targets and immunomodulatory agents.

Cooperation in Bacteria

Cooperation can occur at many levels of biological organization. Within a cell, genes cooperate to replicate in a coordinated manner; in multicellular organisms cells act together to build a functioning body. Within social groups animals often cooperate to forage and reproduce. Many cooperative actions happen to be evolutionarily derived characters. Rainey and Rainey (2003) and Velicer and Yu (2003) have investigated how a cooperator produced by spontaneous mutation may spread in an ancestral population of non-cooperating individuals.

The defining aspect of cooperation being the investment of resources toward a common good, a costly strategy tends to be adopted that can benefit other individuals, irrespective of whether those others contribute to the common good or not (Velicer, 2003). Any newly emerging cooperator would thus improve the status of the community, but carry the costs alone and thus be unable to thrive. And yet, there exists considerably cooperation in nature, suggesting that the above dilemma can be solved. It may be that rare cooperators, such as the progeny of a single cooperative mutant, interact more often with one another than with non-cooperating residents. This would confine benefits from cooperation to the initially small group of cooperators, and help this group to prosper (Ackermann and Chao, 2004).

It is conceivable that individuals do not recognize one another, but relatives remain spatially close and so interact preferentially. This requires the physical environment to be such that the dispersal of newborn individuals is limited. We do not know how often natural environments fulfill this condition. The papers by Rainey and Rainey (2003) and Velicer and Yu (2003) on cooperation in microbes, suggest a new and simpler solution. Rainey and Rainey studied the colonization of a new niche by cooperative groups of bacteria. They evolved experimental populations of the plant pathogen *Pseudomoans fluorescens* in static broth microcosms–unshaken glass beakers containing growth medium. The experiments started with wild-type cells that inhabited the liquid phase of the medium and thus suffered from oxygen limitation. Two days after the start of the experiment, cells with a new phenotype emerged–the so-called 'wrinkly spreaders'–which could overcome oxygen limitation by colonizing the air-broth interface.

This colonization was facilitated by the over-production of an adhesive polymer, causing the cells to adhere to one another and to surfaces through the formation of a self-supporting mat at the surface of the broth, providing the cells with access to both oxygen and nutrients. The evolved trait can be interpreted as a cooperative act because the mat is a common good that allows all its inhabitants to have access to oxygen, regardless of whether they contribute to the mat or not.

Velicer and Yu (2003) worked on cooperative motility in the bacterium *Myxococcus xanthus* whose wild type strains swarm in groups across soft surfaces by a mechanism known as 'S-motility'. Velicer and Yu at first incapacitated S-motility by knocking out the *PilA* gene, leading to the loss of the type IV pili. The mutant populations were then allowed to evolve under conditions that imposed strong selection for motility; only cells that could move to the edge of a growing colony had a chance of surviving to the next generation. The question that arose was: would the cells regain some of the motility that they lost through the knockout mutation? It was found that they did: after only 32 rounds of selection, all eight populations showed increases in motility (see Ackermann and Chao, 2004).

What happened was that the two strains had acquired another motility mechanism that is normally not suited for soft surfaces, the 'A-motility', in which propulsion seems to occur through the extrusion of slime. Modification of the A-motility alone, however, failed to explain all of the regained motility. Both strains showed enhanced production of fibril, forming an extracellular, matrix consisting of carbohydrates and proteins¾matrix that was essential for the regained motility, perhaps by increasing the efficiency of the A-motility on the soft surface. Interestingly, the presence of the matrix increased motility of all cells in a colony, regardless of whether they contributed to it or not. Fibril production therefore constituted an act of cooperation.

In both the new studies the cellular change that generated the cooperative act also increased cohesion. Cohesion promotes the attachment of daughter cells to each other and acts as a mechanism to create clusters of cells in an otherwise well-mixed population, such as in the liquid cultures raised by Rainey and Rainey. Although *Myxococcus* cells live on surfaces and not in a liquid habitat, their motility in the absence of cohesion would also produce random distributions in a two-dimensional world. With cohesion, a single trait not only confers cooperation but also enhances the probability of preferential interactions. In effect, with cohesion, we see two adaptations for one trait (Ackermann and Chao, 2004).

There are at least two kinds of cooperation, minimal and synergistic. Minimal cooperation is simply a contribution to the public good and has been shown directly in both studies. Synergism is a special case of cooperation–the situation where a group of individuals derive some benefit together that they could not obtain alone (see Velicer, 2003). With synergism, groups can acquire previously unattainable functions and possibilities. Both the studies (of Rainey and Rainey, and Velicer and Yu) point to the possibility that the microbes may interact synergistically. The pertinent question is : can the surface of the microcosm used by Rainey and Rainey only be colonized by groups, but not by individual wrinkly spreaders? And, likewise, is the modified motility system of *M. xanthus* only effective in groups?

Bilingual Bacteria

Several bacteria make use of chemical signals to coordinate group behaviour. Many bacteria show a spectacular cooperative activity. The marine *Vibrio fischeri*, for example, produces bioluminescence in the light organs of deep-sea fish. Sometimes the bacterial group behaviour is deadly–many bacteria become virulent only when they reach a certain local concentration (Parsek, 2007). Such coordinated actions are performed when bacteria communicate with each other by transmitting chemical signals; this process is termed quorum sensing.

Vibrio cholerae has been known to be "bilingual": it uses two distinct signaling molecules to suppress its virulence (see Miller *et al.*, 2002; Zhu *et al.*, 2002). One of these signals is a molecule that many bacterial species use for quorum sensing. The identity of the second signal has been revealed only recently by Higgins *et al.* (2007). Their discovery represents a new structural class of quorum-sensing signal that could make it a possible lead for drug discovery.

In quorum sensing, bacteria release signals into the surrounding environment; if the signals reach a critical concentration, they can be sensed by bacteria in the vicinity and they stimulate a response. In *V. cholerae*, quorum sensing operates through two parallel systems, either of which can independently initiate a response. The first of these involves the AI-2 molecule. This signal is also used by many other species. Al-2 is detected by the sensory proteins LuxP and LuxQ found in the bacterium's cell membrane.

The second signal is probably detected by a membrane-associated sensor called CqsS. Both systems in V. cholerae funnel information into the same signaling cascade (through the transducing proteins LuxU and LuxO) so that an analogous functional response is produced in each case: LuxO is inactivated, resulting in increased activity of HapR, a negative regulator that represses expression of virulence genes (Parsek, 2007).

Higgins et al. introduced the *cqsA* gene into *Escherichia coli*, producing recombinant strain that generate a much stronger signal than the parent *V. cholerae* species. By extracting the culture medium of the *E. coli*, a mixture of compounds was obtained that could be separated into its constituent parts. The purified compounds were then tested on a strain of *V. cholerae* that had been engineered to emit light in response to the unknown signal. Certain fractions of the mixture were thousands of times more active than controls.

Higgins *et al.* identified the active compound as 3-hydroxytridecan-4-one (Figure 9.3) and confirmed this by synthesizing the compound chemically, and testing the synthetic version in their activity assay. They proved conclusively that 3-hydroxytridecan-4-one is the second signal for this bacterium.

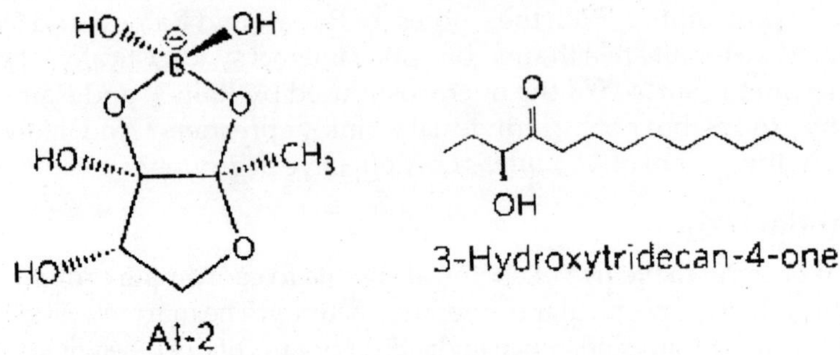

Figure 9.3: Structures of Two Quorum Sensing Signals (After Parsek, 2007). (Reprinted by permission from Macmillan Publishers Ltd. (*Nature* 450: 805-806, 2007).

According to Higgins *et al.* the Cqs signal might be exploited therapeutically to dampen *V. cholerae* virulence. Quorum-sensing systems that trigger virulence have already served as targets for therapies against *Pseudomonas aeruginosa*–an opportunistic pathogen that causes infections in people with weakened immunity. Usually the aim is to block quorum sensing with a small-molecule inhibitor– challenging task because such inhibitors must be specific, stable, easily deliverable to the infection site and should outcompete the natural quorum-sensing signal for the target receptor.

Chapter 10
Microalgae and Protists

Introduction

The recent upsurge of worldwide interest in several aspects of environmental phycology–phycology deals with the study of algae–not only attests to their profound importance in local, regional and global environmental problems but has also contributed to a much better appreciation of the algal services to aquatic and atmospheric environments besides, of course, the well-known role of the blue green algae (also called cyanobacteria) in the nitrogen economy of paddy fields (see Kumar, 1999). The most important of these services relates to the photosynthetic carbon fixation and oxygen evolution by oceanic phytoplankton. Besides the terrestrial forests, the marine phytoplankton is the major sink for the greenhouse gas carbon dioxide. Any decrease or loss of phytoplankton caused, say, by the increased incidence of ultraviolet-B radiation resulting from stratospheric ozone depletion, may have adverse impacts not only on the world's fish harvests (a food chain effect) but also global warming. Global warming might produce some drastic rearrangements of flora, fauna and fishing grounds in some sea areas but the overall oceanic productivity may not change significantly.

A clear connection between marine algae and climate has become apparent in recent years. It occurs through the production of a compound (called dimethylsulphoniopropionate, DMSP) which breaks down to release volatile dimethyl sulphide (DMS) and acrylic acid. DMS is a key substance in the global sulphur cycle. Its oxidation products influence atmospheric acidity, cloud formation, and the Earth's temperature. The DMSP is synthesized by algae to help maintain their osmotic balance with sea water. The chief sources of DMSP in the natural environment are marine phytoplankton blooms, where high levels of primary production and biomass occur on a timescale of weeks.

The Kingdoms of Eukaryotes

In the five kingdom system of classification of Whittaker, the kingdom Protista represents a rag-bag of convenience for all eukaryotes that are not animals, plants or fungi. Most taxonomists agree that only monophyletic groups–an ancestor and all of its descendents–should be formally classified as higher taxa, disregarding any notions of convenience or communication. Recent advances in molecular and cell biology have improved our understanding of evolutionary relationships. Eukaryotes may be divided into just a few major and probably monophyletic groups.

1. Opisthokonta

This group contains animals and true fungi, as well as several unicellular groups, including the free-living choanoflagellates, a diverse range of parasitic forms called Ichthyosporea or Mesomycetozoea, and a group of free-living amoebae called the nucleariids. Two significant groups of spore-forming parasites, myxozoa and microsporidia, usually considered as protists, are really animals and fungi, respectively (Simpson and Roger, 2004).

Choanoflagellates, most animal sperm and the zoospores of chytrids (the only fungi with flagella) all swim with their single flagellum emerging from their posterior end; this unique arrangement may have been inherited from the common ancestor of opisthokonts. Molecular phylogenies demonstrated over a decade ago that animals, chaonoflagellates and fungi are specifically related. Some other groups have been added since then, through analyses of ribosomal (r)RNA gene sequences.

Animals are virtually the only multicellular predators on Earth. Although small animals overlap ecologically with some larger unicellular eukaryotes, they monopolize large heterotroph niches in all environments. Fungi are dominant osmotrophs, especially in terrestrial systems, and play crucial roles as decomposers and as symbionts or parasites of plants. Microsporidia infect a wide range of animals and insects, and are probably amongst the most biodiverse of parasite groups (Simpson and Roger, 2004).

2. Amoebozoa

Includes most cells moving and feeding by means of broad or finger-like pseudopodia. Included are classical amoebae with broad pseudopodia (*e.g., Amoeba*) and the major groups of slime moulds and some mitochondrion-lacking organisms–pelobionts and entamoebae. Most Amoebozoa are free-living heterotrophs which engulf other cells by their pseudopodia (some large amoebae even eat small animals). Such amoebae play significant role in benthic ecosystems. There are several facultative or obligate parasites (*e.g. Entamoeba histolytica*).

The slime moulds are amoebae that periodically form a stalked spore-producing fruiting body. In cellular slime moulds (*e.g., Dictyostelium*), many individuals aggregate to form a superorganism, in which some cells are sacrificed to create the stalk of the fruiting body–a form of multicellularity and differentiation, derived independently of animals, plants and fungi. Myxogastrid slime moulds (*e.g. Physarum*) form giant multinucleate super cells (plasmodia), common in terrestrial ecosystems.

Traditional classifications grouped diverse amoeboid organisms together, uniting most or all Amoebozoa, but grouping them also with the unrelated Radiolaria, Foraminifera and 'heliozoa'. Recent studies of rRNA sequences and other analyses and sampling of more amoebae suggest that these organisms are related, as do analyses of individual protein sequences (especially actin) and sophisticated studies of multiple proteins (Simpson and Roger, 2004).

3. Plantae

'Primary endosymbiosis' refers to the origin of a eukaryotic organelle by the engulfment, enslavement and genomic reduction of a prokaryotic cell. Three photosynthetic groups have plastids that originated by primary endosymbiosis: (i) land plants (embryophytes) and green algae such as *Chlamydomonas*; (ii) red algae, and (iii) an obscure group called the glaucophytes. The plastids of these groups form a single lineage specifically related to cyanobacteria; primary plastid endosymbiosis seems to have happened just once in eukaryotic evolution, with the host being a common ancestor of these three groups. Some phylogenetic analyses of the gene for elongation factor 2 indicate that 'reds' and 'greens' are closely related, with glaucophytes perhaps being their sister group. This whole group is treated as 'Plantae', but it should be noted that most botanists use 'Plantae' for subsets of this group, such as green algae plus land plants.

The incorporation of the primary plastid had a pronounced impact on the genetic potential of the host organisms. Most Plantae are specialist phototrophs; a few are non-phototsynthetic parasites, but even these still contain reduced plastids. Plantae appear to lack entirely the ability to engulf particulate food.

Multicellularity evolved on several occasions within Plantae: probably once in red algae, but repeatedly within green algae. The multicellular assemblage, 'charophytes', gave rise to the embryophytes that dominate land habitats, but are of very minor importance in the ocean (where extremely small unicellular green algae are significant).

4. Chromalveolata

In 'secondary endosymbiosis' a eukaryote already harbouring a primary plastid is engulfed by another host eukaryote, and in course of time becomes reduced to an organelle. The new plastid-containing host has been termed a 'secondary alga' (Simpson and Roger, 2004). Although secondary endosymbiosis happened more than once in eukaryotes, most groups of secondary algae seem to have descended from one particular endosymbiosis that involved a red algal symbiont. These organisms, plus their many non-photosynthetic relatives, make up the group Chromalveolata. The group unites four major groups of eukaryotic algae: dinoflagellates, cryptophytes, haptophytes and stramenopiles (~heterokonts), plus many non-photosynthetic forms. The first three groups are unicellular, with a few colonial forms. Stramenopiles range from tiny unicells, through to elaborate unicells and colonies (*e.g.* diatoms), and multicellular and massive life forms (*e.g.* kelps). Dinoflagellates and diatoms comprise the dominant 'large' phytoplankton in the ocean. Dinoflagellates and stramenopiles also contain many heterotrophic forms (and mixotrophs–organisms that subsist by both photosynthesis and heterotrophy). Heterotrophic stramenopiles are important consumers of bacteria in aquatic environments, but also include some animal parasites/commensals, and a diversity of fungal-like forms (The Irish potato famine pathogen, *Phytophthora infestans*, is an oomycete stramenopile). Heterotrophic and mixotrophic dinoflagellates are important micro-predators in the plankton, and there are also many parasitic forms.

Notwithstanding its apparent algal origins, Chromalveolata includes two best known groups of non-photosynthetic microbial eukaryotes. Ciliates, the dominant micro-predators in many habitats, include such important laboratory models as *Paramecium* and *Tetrahymena*. Apicomplexa (= Apicocomplexa) may well be the most successful group of specialist parasites on earth, and include the agents of toxoplasmosis, cryptosporidiosis, coccidiosis and, of course, malaria (*Plasmodium*). Apicomplexa live inside their hosts, usually in total darkness and yet most have retained a non-photosynthetic plastid; this betrays their algal ancestry and provides an attractive target for novel drug therapies (Simpson and Roger, 2004).

5. Rhizaria

Over the last decade, this novel major grouping of eukaryotes has been delimited mainly thanks to improvements in the sampling of rRNA genes. Rhizaria brings together great diversity of free-living unicellular organisms, many of which feed by means of fine 'filose' pseudopodia, plus some fungi-like plant parasites, such as plasmodiophorids, and some animal parasites such as Haplosporidia. Foraminifera and Radiolaria are the best-known free-living Rhizaria.

Foraminifera are planktonic or benthic marine amoebae, many of which are large; they build external mineralized shells that fossilize easily. Radiolaria are large planktonic marine cells, generally with beautiful rapidly symmetrical spiked internal skeletons. Most other free-living Rhizaria are heterotrophic flagellates or amoebae that consume other microbes associated with surfaces; they may have extremely important roles in benthic and soil habitats. The chlorarachniophytes, are mixotrophic secondary algae that harbour a plastid of green algal origin, stemming from a completely separate secondary endosymbiosis from that of chromalveolates. Many Foraminifera and Radiolaria contain algal symbionts (Simpson and Roger, 2004).

6. Excavata

These are unicellular, mostly heterotrophic flagellates. Several groups such as trypanosomatids, diplomonads aand parabasalids cause disease–sleeping sickness, giardiasis and trichomoniasis, respectively. Each parasitic group also has some free-living relatives which consume other microbes, often capturing them out of suspension by means of a distinctive feeding groove. Many excavates possess greatly modified mitochondria, which are not used for oxidative phosphorylation. These cells are quite common in low-oxygen habitats, such as animals guts. Members of one line, Heterolobosea, have evolved as broad-pseudopod-forming amoebae independently of Amoebozoa, and even include their own group of slime moulds, the acrasids. Another group is the secondary algae, the euglenids.

Both protein phylogenies and a complex fusion of pyrimidine synthesis genes suggest that Opisthokonta is most closely related to Amoebozoa. This grouping has been referred to as 'unikonts'. It appears that the root of eukaryotes falls between unikonts and everything else (see Stechmann and Cavalier-Smith, 2003).

Ancestral Eukaryotic Cells

Eukaryotic cells are very different from their presumably prokaryotic ancestors. The last common ancestor of living eukaryotes appears to have been a 'complete' eukaryotic cell that had a nucleus, endoplasmic reticulum and Golgi apparatus, and that underwent mitosis and meiosis. Its mitochondria were capable of oxidative phosphorylation, amongst other functions. It had a complex eukaryotic cytoskeleton including eukaryotic flagella (most likely a pair of them), and was heterotrophic, consuming food particles by phagocytosis. The only major eukaryotic features that seem to be of later origin are plastids. What we still do not understand is: did living eukaryotes diverge soon after the rapid evolution of the eukaryotic cell, or was this cell assembled gradually, but with modern eukaryotes then replacing all intermediate forms (Simpson and Roger, 2004, 2004a).

Opisthokonta

Two of the major multicellular kingdoms, Metazoa and Fungi, constitute a monophyletic clade, the opisthokonts, which comprises some unicellular lineages, such as Choanoflagellata, Nucleariidae, Ichthyosporea (also known as the DRIPs or Mesomycetozoea), *Ministeria* and *Corallochytrium* (Steenkamp and Baldauf, 2004). It appears that the closest living relatives of the multicellular fungi and metazoans may be sought among these unicellular opisthokonts. According to Lang *et al.* (2002), choanoflagellates are closely related to metazoans with the Ichthyosporea being the next closest outgroup. By contrast, the status of the nucleariids is rather controversial. *Nuclearia* sp., was recently renamed as *Capsaspora owczarzaki* as a filose amoeboid symbiont of a pulmonate snail that has never been definitely assigned to any opisthokont lineage (Ruiz-Trillo *et al.*, 2004). These authors have reported large subunit ribosomal and actin sequences of C. *owczarzaki* and an actin sequence of the novel ichthyosporean species *Sphaeroforma arctica.* Their researches have indicated that *Capsaspora owczarzaki* is not a nucleariid but an independent opisthokont lineage that is more closely related to choanoflagellates and ichthyosporeans. *Nuclearia* branches as a sistergroup to Fungi. Interestingly, the clade containing *Capsaspora,* Choanoflagellata and Ichthyosporea, seems to be most closely related to Fungi and Nuclearia, rather than to animals. Since *C. owezarzaki* occupies an important position near the base of the Metazoa, *Capsaspora* is a key candidate for future comparative genomic studies aimed at understanding the origin of metazoans.

Biodiversity of Eukaryotic Microbes

Reconstruction of eukaryotic phylogeny and history has been adversely affected by paucity of knowledge of extant microbial eukaryotic diversity. Some models of deep eukaryotic evolution have depended on comparisons among only a few microbes, primarily cultivated model organisms and pathogens (Philippe *et al.*, 2000). Sparse taxon representation can strongly influence the accuracy of phylogenetic reconstructions. For the super-kingdoms of Bacteria and Archaea, cultivation-independent identification of environmental organisms by rRNA gene sequences has opened up a highly diverse microbial world that is not represented by cultured organisms (Hugenholtz *et al.*, 1998), and it appears that may also be the case for eukaryotes. Recent rRNA gene-based surveys of eukaryotes in oxic planktonic environments have revealed novel rRNA gene sequences (see Dawson and Pace, 2002; Lopez-Garcia *et al.* 2001).

The most deeply divergent of known eukaryotic lineages in phylogenetic trees are some anaerobic or aerotolerant organisms of anoxic environments which harbour a highly adverse assemblage of eukaryotic microbes (Fenchel and Finlay, 1995). Most microbes cannot be cultured, however; also microscopic descriptions are biased toward morphologically conspicuous organisms. Consequently, organisms detected by cultivation or morphology may not always represent the abundance and diversity of naturally occurring ones.

Anaerobic microbes have had little representation in phylogenetic studies, are the least known of eukaryotes, and may well be the most phylogenetically diverse. To provide fresh perspective on the natural diversity of eukaryotes in anoxic environments and to discover novel sequences for evolutionary studies, Dawson and Pace (2002) conducted a cultivation-independent, molecular phylogenetic survey of three anoxic sediments, including both freshwater and marine samples. Many previously unrecognized eukaryotes were identified, including representatives of seven lineages that are not specifically related to any known organisms at the kingdom-level and branch below the eukaryotic "crown" radiation of animals, plants, fungi, stramenopiles, etc. The survey further identified new sequences characteristic of known ecologically important eukaryotic groups with anaerobic members. These phylogenetic analyses with the new sequences have enhanced our understanding of the diversity and pattern of eukaryotic evolution (Dawson and Pace, 2002).

Dawson and Pace purified environmental DNAs from two marine and one freshwater anoxic sediments and used them as templates for amplification of rRNA genes by PCR, with degenerate primers selective for eukaryotic rDNA. PCR products were cloned and screened by RFLP to identify the unique types, which were then sequenced. Over 3,000 clones were screened by RFLP, and 125 unique eukaryotic SSU rDNA sequences from the three anoxic sediments were successfully determined. Sequences were then aligned with a database of ~5,000 other eukaryal rDNA sequences, and phylogenetic analyses were conducted to compare the sequences with those of known organisms. Eight independent clades were identified in the three environments and were found not to be specifically affiliated with known, kingdom-level eukaryotic lineages. In one typical phylogenetic tree that includes a broad representation of eukaryal rRNA sequences, seven of these novel lineages consistently branch outside the eukaryote "crown" radiation of animals, plants, fungi, etc. (Knoll, 1992). Several of the new kingdom-level lineages branch at intermediate levels in the overall eukaryotic radiation. The new lines punctuate the long lines that previously separated the basal radiation from more peripheral branches in most phylogenetic analyses (Dawson and Pace, 2002).

The new environmental sequences detected by Dawson and Pace enlarge the known extent of eukaryal rRNA diversity. Whereas only a few thousand rRNA sequences of eukaryal microbes are currently known, more than 200,000 "morphospecies" have been described on morphological or other grounds (see Levine *et al.*, 1980; Dawson and Pace, 2002). This discrepancy between the number of identifying gene sequences and the number of classically described organisms underscores the need for further molecular surveys, both in the environment and in culture collections, so as to better understand the diversity and evolution of eukaryotes.

Some of the new sequences are not specifically affiliated with any molecularly described taxonomic group and so indicate novel, kingdom level relatedness groups. The new sequences add considerable variation to the sequence collection, by enhancing the accuracy of the phylogenetic calculations. The general results of these and other phylogenetic studies based on rRNA sequence comparisons are shown in Figure 10.1.

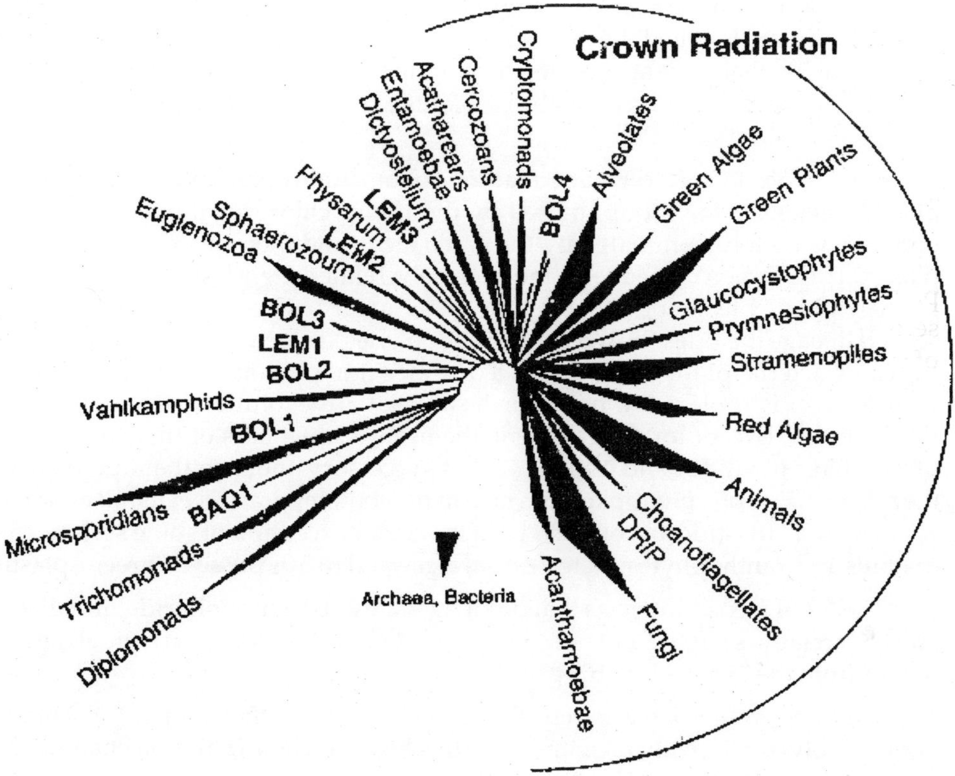

Figure 10.1: Schematic Sketch of the Evolution of Eukarya. Summary tree of global SSU rRNA phylogeny including the novel lineages from the three environments surveyed. The areas of the wedges reflect the number of SSU rRNA sequences of these groups in GenBank. The DRIP lineage is a recently defined protistan clade near the animal-fungal divergences (after Dawson and Pace, 2002). (Copyright (2002) National Academy of Sciences, U.S.A.).

Modern Eukaryotic Phytoplankton

The term 'phytoplankton" describes a diverse, polyphyletic group of mostly single-celled photosynthetic organisms that drift with the currents in marine and fresh waters. Although accounting for less than 1 per cent of Earth's photosynthetic biomass, this phytoplankton carries out more than 45 per cent of Earth's annual net primary production (Field *et al.*, 1998). Whereas on land, photosynthesis is mainly dominated by the Embryophyta containing nearly 275,000 species, there are less than ~25,000 morphologically defined forms of phytoplankton distributed among several major divisions or phyla.

Numerically, most of the phytoplankton in the contemporary oceans is composed of cyanobacteria, the only extant prokaryotic group of oxygenic photoautotrophs. Oxygenic photosynthesis (PS II and PS I) may have evolved only once, but it subsequently spread via endosymbiosis to a wide variety of eukaryotic clades. The earliest oxygenic photosynthetic eukaryotes probably arose from the wholesale engulfment of a coccoid cyanobacterium by a eukaryotic host cell that already contained a mitochondrion. The engulfed cyanobacterium became a membrane-bounded organelle called the 'plastid". The mitochondrion and plastid are the only two organelles that were appropriated via endosymbioses. Gene loss in the course of evolution essentially reduced both symbionts to metabolic slaves within their host cells (Falkowski *et al.*, 2004).

Early evolution of oxygenic eukaryotic photoautotrophs gave rise to two major plastid lineages. One group, united by the use of chlorophyll b as an accessory pigment, is largely dominated by the green algae and their descendants, the land plants. This "green" lineage is more closely related by plastid phylogeny and photosynthetic physiology than by the evolutionary history of the host cells. The second lineage includes the red algae (rhodophytes), which retain the most features of cyanobacterial pigmentation, and a diverse set of phytoplankton and seaweeds whose plastids (but again, not their host cells) are evolutionarily derived from rhodophytes. Except for the red algae themselves, members of this "red" lineage utilize chlorophyllide c and its derivatives as accessory photosynthetic pigments. Of the major eukaryotic phytoplankton taxa in the contemporary oceans, all except one contain "red" plastids. In contrast, with the minor exception of some soil-dwelling diatoms and xanthophytes, all terrestrial algae and plants possess "green" plastids.

The fossil record suggests that dinoflagellates, coccolithophorids, and diatoms and their relatives largely, but not completely, displaced other eukaryotic algae in the oceans from the Mesozoic to the present.

The three groups of marine eukaryotic phytoplankton that dominate the modern oceans evolved relatively recently in Earth's history. Their radiation changed both ecosystem structure and biogeochemical cycles in the oceans. In the contemporary oceans, higher trophic levels (*e.g.* fish and marine mammals) depend on short food chains driven by bloom-forming phytoplankton, usually based on diatoms. The intermediaries in this energy transfer are metazoan zooplankton, such as copepods. It appears that the rise of armoured phytoplankton through the Mesozoic to the present represents an expansion in diversity of secondary and higher order food web dynamics. The long-term increase in metazoan diversity in marine ecosystems in the Cenozoic partially reflects adaptive evolution associated with a long-term rise to prominence of eukaryotic algae in pelagic ecosystems.

Plankton Fluorescence and Ocean Plant Health

Phytoplankton are microscopic algae that grow abundantly in oceans all over the world. They are at the base of the marine food chain. As these algae have certain growth requirements, they can indicate change in their environment. Like the green plants growing on land, phytoplankton require sunlight, water, and mineral nutrients for their growth. Sunlight being most abundant at and near the sea surface, phytoplankton are found at or near the surface. Because they have a global-scale influence on climate, phytoplankton strongly interest oceanographers and Earth scientists. A unique signal detected by NASA's *Aqua* satellite helps researchers to check the health and productivity of ocean plants. Fluorescent red light emitted by marine planktonic algae and detected by *Aqua* shows how efficiently the microalgae are converting sunlight and nutrients into food through photosynthesis.

Phytoplankton account for about one half of all photosynthetic activity on Earth. The health of these marine plants affects commercial fisheries, the amount of CO_2 the ocean can absorb, and how the ocean responds to climate change. In recent years, scientists have used satellite sensors to estimate the amount and distribution of the green pigment chlorophyll, which is an indicator of the amount of plant life in the ocean. But it is the Moderate Resolution Imaging Spectroradiometer (MODIS) on NASA's *Aqua* satellite, that enabled researchers to observe 'red-light fluorescence' over the open ocean.

Most plants absorb more energy from the Sun than they can actually consume through photosynthesis. The extra energy is mostly released as heat but a small fraction is re-emitted as fluorescent light in red wavelengths. MODIS is the only instrument of its kind that can record this signal on a global scale.

Red-light fluorescence throws valuable light on the physiology of marine plants as well as the efficiency of photosynthesis. The fluorescence is greater when phytoplankton suffer iron deficiency–iron is a critical nutrient in seawater. The iron which the algae need for their growth is carried to the sea surface by wind blown dust from deserts or arid areas, and from upwelling currents in the ocean near river plumes and islands. Fluorescence data from MODIS has enabled researchers to study these dynamics on time-scales of weeks to months, to track plankton responses to iron inputs from dust storm and the transport of iron-rich water from islands and continents. Gathering and analyzing such data over extended periods of a few decades may help in detecting long-term trends in climate change and other human-caused disturbances to the ocean.

Oceanic Productivity and Pollution

Biological productivity in the world's oceans is mostly controlled by the supply of nutrients to surface waters. The balance between supply and removal of nutrients–including nitrogen, iron and phosphorus–determines which nutrient limits phytoplankton growth. Nitrogen limits productivity in much of the ocean (Tyrrell,

1999; Corredor *et al.*, 1999) but large portions of the tropics and subtropics suffer from extreme nitrogen depletion. In these regions, microbial denitrification removes biologically available forms of nitrogen from the water column and produces substantial deficits relative to other nutrients (Tyrrell and Law, 1998; Deutsch *et al.*, 2001). Beman *et al.* (2005) showed that nitrogen-deficient areas of the tropical and subtropical oceans are acutely vulnerable to nitrogen pollution. Despite naturally high nutrient concentrations and productivity, nitrogen-rich agricultural runoff fuels large (54-577 km^2) phytoplankton blooms in the Gulf of California. Runoff has a strong, consistent effect on biological processes, in many cases stimulating blooms within days of fertilization and irrigation of agricultural fields. According to Beman *et al.*, by the year 2050, 27-59 per cent of all nitrogen fertilizer will be applied in developing regions located upstream of nitrogen-deficient marine ecosystems. Their views highlight the vulnerability of these ecosystems to agricultural runoff (Beman *et al.*, 2005).

Diversity of Small Eukaryotes in Deep-Sea Plankton

Picoplankton is composed of cells with a diameter of less than 3 mm which are the major contributors to both primary production and biomass in open oceanic regions (van der Staay *et al.*, 2001). However, as compared to the prokaryotes, the eukaryotic component of picoplankton is poorly known. Recent discoveries of new eukaryotic algal taxa based on picoplankton cultures point to the existence of many undiscovered taxa. Traditional approaches based on phenotypic criteria have limitations in depicting picoplankton composition due to their tiny size and lack of distinctive taxonomic characters. van der Staay *et al.* analyzed 35 full sequences of the small-subunit (18S) ribosomal RNA gene derived from a picoplanktonic assemblage collected at a depth of 75 m in the equatorial Pacific Ocean, and showed that there is a high diversity of picoeukaryotes. Most of the sequences were previously unknown but could still be assigned to important marine phyla including prasinophytes, haptophytes, dinoflagellates, stramenopiles, choanoflagellates and acantharians. They also found a novel lineage, closely related to dinoflagellates and not previously described.

Phylogenetic information from ribosomal RNA genes directly amplified from the environment has changed our view of the biosphere by revealing an extraordinary diversity of previously undetected prokaryotic lineages. Using ribosomal RNA genes from marine picoplankton, several new groups of bacteria and archaea have been identified, some of which are quite abundant. Little is known, however, about the diversity of the smallest planktonic eukaryotes, and available information in general concerns the phytoplankton of the euphotic region.

Most previous work on oceanic eukaryotic picoplankton focused on its photosynthetic component because chlorophyll fluorescence makes it easy to detect by flow cytometry, and pigment signatures permit inferences to be made about its taxonomic composition. Among autotrophs, haptophytes constitute one of the major picoplanktonic lineages, as suggested by the dominance of the diagnostic carotenoid 19'-hexanoyloxyfucoxanthin in most oceanic waters. Another key group in the icoplanktonic autotrophs is the prasinophytes, primitive green algae that have been epeatedly isolated from marine waters.

Stramenopiles or Heterokonta contain such important oceanic algal classes as he ubiquitous diatoms, but also heterotrophic groups such as the bicosoecids.

Dinoflagellates, like the stramenopiles, contain both autotrophic and eterotrophic taxa.

The most intriguing discovery from this work was that of an environmental ineage consisting of six clones that could not be assigned to any known eukaryotic axonomic group. This seems to be a close relative of both the extant dinoflagellate ineage and the Perkinsozoa, a newly established phylum regrouping the marine parasites *Perkinsus* and *Parvilucifera*, of which the latter infects marine dinoflagellates.

Among purely heterotrophic protists, acantharians are typical blue-water organisms abundant in temperate and subtropical regions of the world oceans that are important in the global oceanic budget of elements such as strontium and barium. The vegetative cells of acantharians are usually much larger than picoplankton.

Two sets of sequences clustered with the heterotrophic choanoflagellates which are the closest known relatives to sponges. One of these is related to the naked odosigidae (*Monosiga*) and the other the marine Acanthocidae (*Acanthocorbis*), which isplay delicate silicified lorica.

Lopez-Garcia *et al.* (2001) amplified 18S rRNA genes and studied them from amples taken at 250, 500, 2,000 and 3,000m deep at the Antarctic polar front limit in transect along the Drake passage (59°,19' 48"S, 55°45'11"W, sea floor at 3,671 m). his sampling site is a region of water-mass mixing from the Atlantic and Southern ceans. It corresponds to cold and oligotrophic waters where microbial biomass, specially at 3,000m deep, reached minimal values in the area as deduced from DNA ields. 18S rRNA environmental gene libraries were constructed from the 0.2-5 mm lanktonic fraction and, for comparison, also from the microbial fraction> 5 mm at ,000m deep. From this could be recovered two new groups of alveolate sequences elated to dinoflagellates that were found at all studied depths. These may be important omponents of the microbial community in the deep ocean.

It appears that many new eukaryotic lineages remain to be uncovered from the marine environment, especially within the smallest size class. The genuineness of the new lineages found by van der Staay *et al.* such as that related to both dinoflagellates and Perkinsozoa, is shown by the fact that they are represented by several different sequences.

From the ecological viewpoint the above finding suggests that a large fraction of eukaryotic picoplankton is heterotrophic because 24 of the 35 sequences retrieved belong to heterotrophic lineages, some of which are probably parasitic. This importance of parasitism might require a revision of our current views on the structure of marine microbial food webs, which consider oceanic heterotrophs mainly as phagotrophic predators. When more eukaryotic sequences from a variety of environments become available, it should be possible to determine the morphology and ecological role of these unknown protists (van der Staay *et al.*, 2001).

Gliding Motility

Cell locomotion in some eukaryotic protists through gliding motility represents substrate-dependent motility that takes place without the aid of cilia or flagella, and also without the kind of cell deformation seen in amoeboid crawling. Several different protistan cell types engage in gliding motility and the process has been intensively studied in diatoms and the apicocomplexan parasite *Plasmodium* (Menard, 2001). Both diatoms and the Apicomplexa use an actomyosin-based motility apparatus, while the substrate-dependent gliding motility of *Chlamydomonas* (a biflagellate green alga) seems to be driven by a microtubule-based motility system located in the flagella but is independent of that used for its alternative axoneme-powered swimming behaviour (see Heintzelman, 2003).

Plasmodium and *Toxoplasma* cause malaria and toxoplasmosis, respectively; both are obligately dependent upon gliding motility for their active invasion of host cells (Menard, 2001). Gliding motility in the Apicocomplexa shows actin-dependence, and is associated with certain accompanying motor molecules, specifically myosins A novel family of myosins, has been identified in *Toxoplasma* (Heintzelman and Schwartzman, 1997), and its homologues have been reported in *Plasmodium* and other related parasites (Heintzelman, 2003). *Toxoplasma gondii* myosin A–or its *Plasmodium* homologues–appear to be the most likely candidate for being the actin associated motor that drives gliding motility (Kappe *et al.*, 1999; Heintzelman, 2003).

Diatoms

Diatoms are single-celled algae characterized by having silica (glass) cell walls. They are increasingly used in environmental and other studies. Much of the groundwork on their identification and classification was carried out in the 19th and early 20th centuries using light microscopy. Since the 1970s there has been a proliferation of new generic names, based on electron microscope studies.

The intricate patterns and huge variety of shapes of diatoms have fascinated naturalists ever since microscopes were invented; today, unravelling their true biodiversity is a major challenge. What makes diatoms so fascinating is that their skeletons–the cell walls–are fashioned into beautifully ornate and precise shapes and patterns. These structures form the basis of diatom classification and identification.

Diatoms are typically a golden-brown colour because of brown pigments that mask their green photosynthetic pigments, the chlorophylls.

Diatoms are found almost everywhere where there is water, including the oceans, fresh waters and damp soil. Their freshwater habitats range from Antarctic lakes to thermal pools, from pristine to polluted rivers, from stagnant puddles to unmodified wetlands. There are both planktonic and benthic forms and they occur as single cells or in colonies (for example, as filaments). Many benthic diatoms are motile and can glide along surfaces or through mud.

The silica cell wall of diatoms consists of two almost equal parts (called valves) which fit together like the lid and bottom of a box of chocolates. There may be girdle bands between the two valves. The complete rigid cell wall is called the frustule. The cell wall structure means that many diatoms look completely different according to whether they are viewed from the side or the top (Kilroy and Bergey, 1999).

Diatoms are divided broadly into two types:

Centric, with round valves, so that the frustule is like a Petri dish. The tops of each valve may have regular or irregular holes or depressions (punctae), perhaps arranged in radial lines (striae).

Pennate, with elongated valves that have striae on the frustule surface often lying more or less at right angles to the long axis.

Within the second group, many diatoms possess a 'raphe' system. The raphe is a slit in the valve surface (Figure 10.2). There is usually a pair on each valve face. In one group, the frustule is twisted so that the raphe lies along the edge of the valve rather than in the centre.

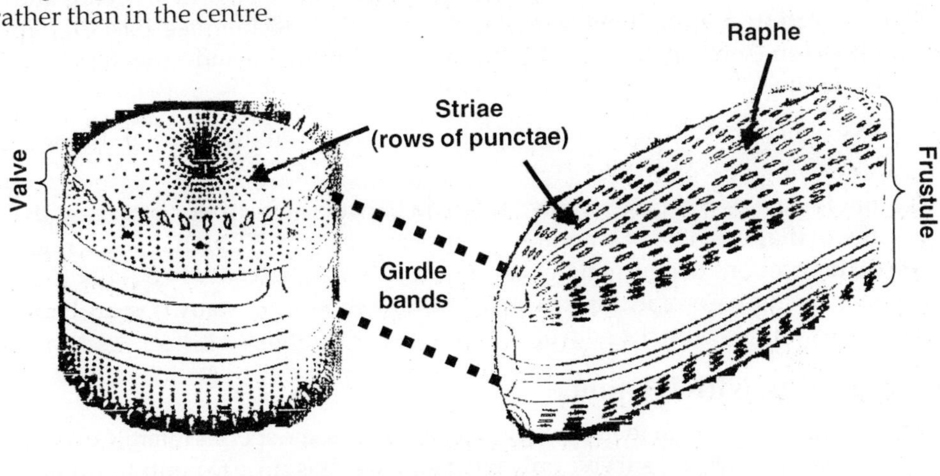

Figure 10.2: Centric Diatom (Left); Pennate Diatom (Right)
(*Source*: NIWA (New Zealand) *Water and Atmosphere* 7(3): 1999)..

Species may be distinguished by looking at : valve shape and size; arrangement and density of the striae; form and numbers of the punctae and other openings and structures on the frustule; and length, shape and arrangement of the raphe and raphe endings.

Other non-frustule features that are also important include the form of the chloroplasts and reproductive mechanisms (Kilroy and Bergey, 1999).

Some species vary greatly in their size because cell replication takes place by the frustule splitting into its two valves, each of which then generates another valve to make two complete new cells. The valves fit one inside the other, and the newly generated valves are always the ones inside–the smaller one. Thus the average size of cells in a dividing population gradually gets smaller. At a certain minimum size (which varies with species), the cells will not divide further. Sexual reproduction takes place and at this stage the maximum size of the cell is restored.

Diatoms typically range in size from 6 to 150 µm (1000 µm = 1 mm), though some marine genera can be up to 4 mm long (*e.g., Thalassiothrix*). Although tiny, they multiply rapidly and are extremely abundant. It is estimated that they are responsible for around 25 per cent of the Earth's primary productivity.

In aquatic habitats diatoms are one of the main components of the food web. Their high oil content makes them an energy-rich food source for single-celled animals and other small creatures. Indeed, the distinctive smell of fish probably originates from diatom oil, even though two or three levels of the food web may separate the two types of organism.

Diatoms are extremely useful as environmental indicators of both past and present conditions. They are used for monitoring the health of lakes and rivers. After death their ornate silica skeletons can persist intact virtually indefinitely. For example, they can be preserved in the sediments that accumulate at the bottom of oceans and lakes. By collecting deep sediment cores and using dating techniques, past diatom communities can be reconstructed and it can be determined if and how they have changed over time.

An essential prerequisite for the use of diatoms in such environmental studies is accurate identification of diatom species.

Technological developments have led to major revisions of our views on the systematics of diatoms. Use of scanning electron microscopy (SEM) has accelerated the rate of descriptions of new genera and species. New genera have been split from larger genera on the basis of features visible only under the SEM. Many taxa are being critically re-examined with the help of SEM.

Biomonitoring with Diatoms

Diatom species differ in their ecological tolerances. Some species tolerate or even require acidic water. Others survive only where the water is close to neutral. Aquatic habitats contain numerous diatom species and knowledge of their ecological tolerances enables algologists to infer the ecological conditions of the habitat. This forms the basis for biomonitoring using diatoms.

In biomonitoring, sites are sampled and examined for differences or changes in the diatom flora that may signal an impact.

A simple example is an assessment to check for leaking septic tanks next to a lake. In one case, diatom samples were collected from three sites along a line extending perpendicular from the shore. *Gomphonema parvulum* is a diatom that grows well in high-nutrient conditions, such as where sanitary wastes are present. *G. parvulum* was extremely abundant near the shore (92 per cent of all individuals) and less abundant at the other sites (41 per cent and 52 per cent), indicating that, indeed, enough septic waste was entering the lake to affect diatom populations.

Some advantages of biomonitoring over chemical means of testing water are:

1. Organisms 'sample' water continuously, whereas chemical testing is intermittent and often misses discrete events;

2. Organisms respond to a variety of conditions, whereas chemical tests are specific;
3. Biomonitoring is cost-effective relative to chemical testing; and
4. Biomonitoring reveals biological effects.

Despite its potential, diatom biomonitoring has been little used in many countries, use being limited not only by inadequate identification guides but also by a limited knowledge of ecological tolerances of diatoms within a country.

Adaptive Response to Toxic Agents

Adaptive Response

Organisms are vulnerable to the negative effects of various different physical and chemical genotoxic agents, both natural (*e.g.* solar ultraviolet light, ionizing radiation) and anthropogenic environmental pollutants released into the environment by human activities. Many defense mechanisms enable organisms to minimize genotoxic damage. One of these is induced radioresistance or adaptive response (AR). It means that a fairly small "conditioning" radiation dose induces increased radioresistance when the cells are irradiated with higher doses sometime later (Hillova and Drasil, 1967). Radioadaptive response induction expresses the ability of low dose radiation to induce cellular changes that change the level of subsequent radiation-induced or spontaneous damage. The AR is probably a nonspecific phenomenon–the exposure to minimal stress inducing a very low level of damage can trigger an AR resulting in increased resistance to higher levels of the same or of other types of stress (Wolff, 1998; Joiner, 1994; Asad *et al.*, 2004; Girigoswami and Ghosh, 2005; Yan *et al.*, 2006).

The AR has been reported in organisms ranging from bacteria, yeasts, higher plants, insect cells, mammalian cells, human cells *in vitro*, and in animal models *in vivo* during a protracted (low dose-rate) exposure prior to an acute dose treatment (Bryant, 1975, 1979; Howard and Cowie, 1978; Santier *et al.*, 1985; Wolff *et al.*, 1988; Boreham and Mitchel, 1993; Salone *et al.*, 1996; Asad *et al.*, 1997; Wolff, 1998; Wang and Cai, 2000; Tiku and Kale, 2004; Venkat *et al.*, 2001; Chankova and Bryant, 2002; Gajendiran and Jeevanram, 2002; Sedgwick and Lindahl, 2002; Jovtchev and Stergios, 2003; Savina *et al.*, 2003; Ulsh *et al.*, 2004; Zhou *et al.*, 2004; Chankova *et al.*, 2007; Rohankhedkar *et al.*, 2006).

Some different endpoints used to demonstrate an AR are cell survival, gene mutations, repetitive DNA loci mutations, chromosome aberrations and micronuclei induction, neoplastic transformation *in vitro*, microarrays showing gene expression changes, DNA single- and double-strand breaks and biochemical analyses of enzymatic/non-enzymatic antioxidant defence system (Dimova *et al.*, 2008).

According to one plausible hypothesis, AR might be induced by reactive oxygen species (ROS) (Feinendegen, 2005; Feinendegen and Neumann, 2006; Feinendegen *et al.*, 1999; Jones *et al.*, 1999; Shankar *et al.*, 2006). ROS are either formed in living organisms during their metabolism or after exposure to various biotic and abiotic stimuli such as UV, ionizing radiation, ozone exposure, heavy metals. ROS damage some cell constituents and cause oxidative stress (Bolwell *et al.*, 2002; Varnova *et al.*., 2002; Babu *et al.*, 2003). Ionizing radiation (IR) can damage DNA both by direct ionization and by indirect processes in which DNA is harmed by certain radiolytic reactive products. Free radicals can attack biomolecules such as DNA, proteins and lipids; they can cause lipid peroxidation and generate some intermediates which react with DNA (Halliwell and Gutteridge, 1989). ROS can also induce multiple local lesions in nucleotides, single- and double-strand breaks (SSBs and DSBs), DNA-DNA cross-links and DNA-protein cross-links (Asad *et al.*, 2004). For example, many bacteria show adaptive responses that protect them against the toxicity and mutagenicity of DNA alkylating agents (Sedgwick and Lindahl, 2002).

Different stress conditions can activate similar defense mechanisms in various biological systems (Joiner *et al.*, 1999; Babu *et al.*, 2003). The AR seems to involve the transcription of many genes and the activation of several signalling pathways that trigger cell defenses: more efficient detoxification of free radicals, DNA repair systems, induction of novel proteins in irradiated cells with a conditioning dose. AR also enhances antioxidant production (Wolff, 1998; Sedgwick and Lindahl, 2002; Girigoswami and Ghosh, 2005).

Experiments with restriction enzymes have indicated that DNA DSBs with blunt or cohesive ends can induce an AR (Wolff, 1998). Some radiosensitive DSB repair-deficient mutants show no induced radioresistance, suggesting the involvement of DSB rejoining. Changes in chromatin conformation may sometime reduce sensitivity of chromatin to damage by indirect effect of a test dose, or increase accessibility of damaged sites to repair enzymes (Belyaev *et al.*, 1996). According to Wolff (1998), poly-ADP-Ribose Polymerase-1 (PARP) may also be involved in the AR and probably interferes in the cell cycle control. In fact, Vaganay-Juery *et al.* (2000) reported that AR might be interpreted in terms of increased non-homologous end-joining of DSB or increased homologous recombination. On the other hand, according to Guruprasad *et al.* (2002), PARP does not seem to be involved in the induction of AR in mouse bone marrow cells after treatment with alkylating agents. .

DNA damage caused by hydroxyl radicals derived from the radiolytic decomposition of H_2O produces lesions that strongly induce DNA repair mechanisms (Boreham and Mitchel, 1991). Experiments in several biological systems to test whether DNA repair upregulation might be involved in the AR, showed that DNA repair underlies the AR induced by low radiation doses in human and plant cells (see Dimova *et al.*, 2008) by increasing the amount and rate of DNA repair (Joiner *et al.*, 1999).

We know nothing about the proteins and genes involved in adaptive responses in cells. Gene transcriptional levels change after exposure to ionizing radiation with low doses that induce AR (Wolff, 1998; Coleman *et al.*, 2005). One clue to the nature of the underlying process has come from work showing a dependence on de novo protein synthesis. Treatment of *Oedogonium*, *Chlamydomonas* and *Closterium* cells with protein synthesis inhibitors (cycloheximide and chloramphenicol) after the first 'conditioning' dose prevented the induced repair responses in these algae (Horsley and Lazlo, 1971; Bryant, 1975; Howard and Cowie, 1978).

Conceivably, cellular ability to induce AR could be affected by changing cellular oxidative stress levels. Enzymes form a crucial component of cell defense mechanisms that protect organisms from the harmful action of ROS damaging DNA and other biomolecules. Presumably, enzymatic, non-enzymatic and indirect antioxidant defense systems are involved in the formation of AR to oxidative stress (Yan *et al.*, 2006). Joksic *et al.* (2000) suggested that oxidative stress triggers an antioxidant response which includes changes in the activity of enzymatic defense system, mainly SOD.

Another possibility is that ROS (*e.g.* hydrogen peroxide and nitric oxide) might act as signal transducers in plant and animal cells (Varnova *et al.*, 2002). As signalling molecules, ROS could possibly affect the development of AR through participation in the damage-sensing process after conditioning dose exposure. It is conceivable that in some cases the increased scavenging of ROS by the antioxidant system reduces the induced damage resulting in AR.

Adaptive protection usually develops after a lag phase of a few hours and may last for days to months. It decreases steadily at doses above 100 mGy to 200 mGy and is not observed after acute exposures to more than 500 mGy (Feinendegen, 2005). Adaptive response depends on the experimental design. It is both dose and time-dependent with a maximal effect occurring several hours later, for example between four to six hours after exposure for the unicellular green alga *Chlamydomonas reinhardtii* (Chankova *et al.*, 2007). A dose-dependence of the radiation-induced AR was also observed in *C. reinhardtii* when DSB DNA rejoining was used as an indicator. A small conditioning dose of gamma rays exposure increased the rate of DSB rejoining but when the magnitude of the conditioning dose was progressively increased, there was a corresponding decrease in the fraction of damage remaining.

Besides being quite short, the AR to irradiation can also be modulated–Tiku and Kaler (2004) reported that AR was induced more effectively when a conditioning dose was given in small fractions. Conceivably, such fractionated irradiation of human fibroblasts might increase survival through repair of sublethal damage (Raaphorst *et al.*, 2000). The efficiency of activating cellular defense reactions varies depending on the level and type of the impact. In some cases the conditioning dose acts synergistically, thereby increasing the frequency of chromosomal aberrations observed following the test dose (Zhou *et al.*, 2004). Also different agents may have different impacts which could result in adaptive reactions of variable efficiency.

Relevance of Adaptive Response

Low-dose radiation sometimes exerts beneficial effects, *e.g.* protection causing DNA damage prevention and repair and immune stimulation (Upton, 2001). Low to intermediate doses of ionizing radiation enhance growth and survival, augment the immune response, and increase resistance to the mutagenic effects of further irradiation in plants, bacteria, insects and mammals (Upton, 2001). Immunity is stimulated in human populations after long-term exposure to high level natural background radiation. On the other hand, some have argued that the AR may not be a relevant mechanism for radiation protection because the *worst* outcome for the cell (cell death) is probably the *best* outcome for the organism as a whole since the low (conditioning) dose could also generate a risk of cellular transformation (Hofseth, 2004). Chronic low-dose occupational exposure to ionizing radiation also causes adverse health effects and increases DNA damage (Zakeri and Assaei, 2004; Sari-Minodier *et al.*, 2007). Likewise, greater DNA damage has been found in chromium platers, offset printing workers, welders (Iarmarcovai *et al.*, 2005, 2007) and pathologists exposed to formaldehyde.

According to Dimova *et al.*, improved understanding of AR can potentially open up new approaches for protection of cells. The AR and the ability to modify it may have important roles in fractionated radiotherapy and could facilitate verification of whether this phenomenon influences the estimation of the risk of low level radiation exposure. To better understand the AR, at least two hypotheses should be tested: (a) that radiation-induced AR may differ depending on the cell type and genotype; and (b) that modifying cellular oxidative stress levels could have an impact on the ability of the cells to initiate the AR.

Horizontal Gene Transfer in Eukaryotic Algae

Over the millennia natural selection has directed the evolution of once primitive, maladapted biological structures to the remarkably diverse molecular machines found within extant organisms. Charles Darwin's evolutionary synthesis was in fact based on this concept of natural selection as a slow-and-steady workhorse–He firmly believed that nature makes no jumps. The subsequent discovery of genes as the medium of inheritance and random mutation as the propagator of change reinforced Darwin's prescient views.

The more recent discovery of horizontal gene transfer (HGT) as a significant evolutionary driver, on the contrary, adds something to the Darwinian synthesis. Many organisms, particularly prokaryotes, do make evolutionary jumps by sharing genes with one another; in this way they open a back door to an adaptation or ability that was already evolved in another organism. The growing wealth of whole-genome data has made HGT a central force in the evolution of many different prokaryotes (Ochman *et al.*, 2000). In contrast, the impact of HGT on eukaryote genomes has been rather limited. Eukaryote taxonomy has leaned heavily on the species concept of genetically segregated germ lines. The barriers against HGT in bacteria are magnified in eukaryotes by strong complexities in transcription and translation, *e.g.*, the need for correct splicing of RNA transcripts studded with introns. It may also be thought that sexual reproduction affords many eukaryotes the same advantage that is gained through HGT in bacteria. Archibald *et al.* (2003) showed that HGT has played an important role in the evolution of the eukaryotic alga (see Douglas and Raven, 2003; Raymond and Blankenship, 2003). Archibald *et al.* (2003) analyzed 78 plastid-targeted proteins rom *Bigelowiella natans*, a member of the enigmatic algae chlorarachniophytes, and showed that at least about 20 per cent of these genes were probably acquired by HGT. This finding expands the number of examples of so-called transdomain HGT occurring between prokaryotes and eukaryotes. All plastid-containing eukaryotes acquired their photosynthesis when perhaps » 2 billion years ago, a primitive eukaryote engulfed a photosynthetic cyanboacterium. This primary endosymbiotic event conferred upon early eukaryotes a powerful metabolic ability that till then was seen only among photosynthetic bacteria; also led to, along with the enslavement of a proteobacterium that would become the mitochondrion, a marked horizontal transfer of genes into a primitive eukaryote. The modern progeny of this primitive photosynthetic eukaryote suggest a single primary endosymbiotic event. There is some evidence for the idea that endosymbiosis occurred multiple times, but many organisms became extinct later due to global catastrophes (see Douglas and Raven, 2003; Raymond and Blankenship, 2003). Some eukaryotes also developed photosynthesis through the same mechanism, though not by engulfing a cyanobacterium but rather a eukaryotic alga. This latter process is termed secondary endosymbiosis, and could have given rise to multiple independent groups of photosynthetic organisms, all of which characteristically contain plastids with three or more bounding membranes. The multiple origins of secondary endosymbiosis is borne out by the fact that it occurred in various lineages after the radiation of the three major primary groups of algae (red, green, and glaucocystophyte), leading to secondary algae with quite varying plastid phenotypes. Both primary and secondary endosymbiosis involved substantial loss of genes from the endosymbiont genome, many of which transferred into the host genome, with each host-encoded plasmid-targeted gene now having a transit peptide sequence that directs it back to the plastid (McFadden, 1999). Cryptomonads and chlorarachniophytes (including *B. natans*) are two groups of algae that evolved through secondary endosymbiosis. They still have a relict nucleus called nucleomorph, much reduced in size, from the originally engulfed algae (see McFadden, 2001).

Some transfer of genes has also occurred from the nucleomorph into the host genome, although despite these complex transfer events these genes in the host genome would be expected to have a phylogenetic signal consistent with the engulfed algae and hence should be grouped more broadly with cyanobacteria. Genes that were present in the host before endosymbiosis should cluster with other eukaryotic genes and so can be used to classify the original host. Most nuclear-encoded algal genes indeed show these stratifications. In *B. natans* there is clear evidence of a green algal endosymbiont origin. However, Archibald *et al.* (2003) showed that many of the plastid-targeted genes from *B. natans* diverge from this expectation. These horizontally transferred genes span varied functions, including chlorophyll biosynthesis, carbon fixation, and ribosome structure; they cluster with a similarly broad range of taxa other than green algae. There is overwhelming evidence in support of HGT having played a significant role in the evolution of this organism (Raymond and Blankenship, 2003).

Since all the genes studied by Archibald *et al.* are encoded within the nucleus of *B. natans* but actually function within the plastid, the signal and transit peptides are essential. It seems that this necessity would strongly favour HGT in and among algae, where nuclear-encoded genes targeted to plastids must navigate a similar network of endomembranes through the direction of transit peptides, and additional signal peptides in the case of other secondary algae. Many of the genes studied by Archibald *et al.* are consistent with this expectation, favouring phylogenies consistent with HGT from streptophytes or red algae into the *B. natans* genome. But, intriguingly, two of the analyzed genes indicate HGT from different bacteria. This exemplifies prokaryote-to-eukaryote gene transfer, also because these acquired genes initially would have lacked the proper leader sequence of entry into the plastid. Whether the appropriate targeting sequence was incorporated *de novo* through gene conversion or some other mechanism of replacement is not known.

Phenotypes that favour promiscuous plasmid sharing among bacteria have been known for long, but no plasmid analog exists in eukaryotes. According to Archibald *et al.* (2003), HGT in *B. natans* may occur in the same way it has for the several endosymbiotic events over the past 2 billion years, *i.e.*, by engulfing other organisms. It is clear that many eukaryotes and prokaryotes can indeed mimic evolutionary jumps through HGT (Raymond and Blankenship, 2003).

Algae as Ecosystem Engineers

There are many organisms, including algae, that play a role in the creation, modification and maintenance of habitats–activities that do not involve direct trophic interactions between species but are nevertheless important and quite common. These organisms are called ecosystem engineers. They directly or indirectly modulate the availability of resources to other species by causing physical state changes in biotic or abiotic materials.

Blooms of marine phytoplankton scatter and absorb light in upper layers of the ocean. This enhances warming of surface waters that can initiate the development of a thermocline. Freshwater phytoplankton in lakes also intercepts light in upper water column, smaller algae being more effective as compared to larger species. Light interception leads to shallower mixing depth, lower metalimnetic temperatures and lower heat content of water column.

Cyanobacterial communities in deserts and arid lands also act as engineers by exuding mucilaginous organic compounds. These compounds glue the cells, organic matter and soil particles together to form microphytic crumbs. The slimy exudates change infiltration, percolation, retention and evaporation of water and reduce soil erosion. Indeed, the blue-green algae have been called the creatures of earth, wind and fire, a reference to their ability to thrive in harsh and varied environments.

Crustose coralline algae such as *Lithophyllum* found in coral reefs overgrow and cement together detritus on outer algal ridge of barrier reef. This tends to break force of water and protect corals against strong wave action.

An open ocean ecosystem experiment has been conducted to answer a puzzling question: Why do vast regions of the oceans have high concentrations of N and P and yet support low phytoplankton biomass? One hypothesis is that phytoplankton of these regions is limited by iron. Bottle experiments are not suitable for testing this hypothesis because corrections for zooplankton grazing cannot be made properly and also there is a high probability of contaminating the bottles with traces of iron. With a view to overcoming these limitations, the first open ocean fertilization experiment, called IRONEX, was carried out in the equatorial Pacific in 1995. Dissolved iron was added to an unenclosed 64 km^2 patch of water so as to increase the ambient concentration 100-fold. It was found that photosynthetic efficiency increased within the first 24 hr, and by the third day, chlorophyll concentration as well as primary productivity had increased three times. Increased photosynthesis caused a small but significant drawdown of CO_2 in the surface waters but was not sufficient enough to produce detectable changes (decline) in nutrients. This short term experiment did not indicate whether the modest productivity increase and negligible impact on N and P levels resulted from the disappearance of available Fe, increased zooplankton grazing, or the short duration of the experiment. In the follow-up IRONEX II, multiple additions of iron, over several days, were used to simulate natural iron input, and both zooplankton grazing rates and concentrations of other potentially limiting nutrients were monitored. The huge phytoplankton bloom formed by the relief of iron limitation was not checked by either grazing or secondary nutrient limitation, thus strongly supporting the view that the high-nitrate low-chlorophyll condition of these waters is caused by iron-limitation of algal growth. IRONEX II showed increases in all phytoplankton groups, with diatoms showing the greatest increase in biomass over ambient (about 85 times). The smaller algae were less affected, with *Synechococcus* (1 micron) and red fluorescing picoplankton (less than 2 microns) only doubling in biomass. The principal results of IRONEX II were the following.

1. Specific growth rate (cell division rate) of the phytoplankton doubled, phytoplankton abundance increased by over 20 times, and nitrate concentration declined by half. Diatoms dominated.
2. Indicators of phytoplankton photosynthetic capacity (*e.g.*, light absorption) shoed immediate and sustained increases.
3. As the bloom developed, the partial pressure of CO_2 in the centre of the patch decreased rapidly, reducing the ocean-to atmosphere CO_2 flux by 60 percent.

Mesoscale iron fertilization experiments bring out not only the feasibility but also utility of engineering efforts in the open ocean.

Some have suggested that fertilization of the entire Southern ocean with Fe could sequester substantial quantities of atmospheric CO_2, delaying predicted greenhouse warming of the earth to a modest degree. Debates about the feasibility, wisdom, and ethics of such climate engineering are greatly affected by our knowledge of the response of oceans to nutrient enrichment. Experiments like IRONEX can contribute something towards this effort.

Algal Blooms in Lakes, Ponds and Estuaries

High nutrient concentrations cause phytoplankton blooms in many shallow water bodies. Indeed, the occurrence of blooms has been recorded since prehistoric times; the biblical description of the waters of the river Nile "turning to blood" (Exodus 7) may well be the earliest recorded example of an algal bloom akin to a "red tide". However, as compared to those times, it appears that we are now experiencing a global epidemic of blooms of previously unimportant species of algae. This represents yet another manifestation of the environmental degradation associated with modern societies. The loads of both nitrogen and phosphorus to many lakes, ponds and shallow coastal ecosystems have been increasing, and such nutrient enrichment stimulates phytoplankton growth.

High nutrient concentrations in water bodies result in rapid algal growth, producing high densities of algal biomass, called phytoplankton blooms.

Phytoplankton blooms cause several problems for water supplies and aquaculture. In estuarine and coastal waters the commonest bloom-related problems include fish kills; contamination of shellfish with plant products that are toxic to humans; oxygen depletion caused by the eventual decay of the biomass; and the production of surface scums and loss in water clarity.

Many village ponds in India are infested with perennial blooms of *Microcystis aeruginosa* which has become infamous as a "nuisance" organism. Most workers have overlooked the valuable services of this organism to the environment: it is a strong accumulator of heavy metals such as nickel, cadmium and copper, and other pollutants from the ambient water. Parker *et al.* (1997, 1998) showed that potassium ions strongly inhibit the growth of *Microcystis* and so offer a new possibility for the control or prevention of its blooms.

Chapter 11
Environmental Microbiology

Introduction

For billions of years, virtually everything in the physical environments of Earth has been transformed to some extent by the activities of microorganisms.

Microorganisms tend to concentrate at interfaces, where nutrients and energy fluctuate. At the interfaces between the lithosphere, hydrosphere, and atmosphere, life can be turned to stone with the recycling of biological matter into physical matter (Newman and Banfield, 2002). The huge biomasses of microorganisms that grow at these interfaces can exert a strong impact on regional geochemistry; for example, in mineral and hydrocarbon deposition, and the nutrient and waste of plankton blooms. Indeed, planktonic organisms may have strongly influenced the evolution of Earth's atmosphere (Kasting and Siefert, 2002).

Along with the recent realization that microbial life is ubiquitous, novel forms of metabolism have been discovered in extremophilic organisms that live in extreme physico-chemical conditions (Reysenbach and Shock, 2002). Despite these insights into the scale of microbial influences on wider environments, we need to understand how to translate genome and molecular studies into an ecological context relevant to single microorganisms (Fenchel, 2002). New techniques need to be developed even to culture a representative selection of the microorganisms we know are out there (Ash *et al.*, 2002). Also the tools have to be built to integrate the rapidly accumulating genomic data with geographical information before these discoveries could be exploited in a meaningful way (Ash *et al.*, 2002).

Some of the most unlikely nooks and crannies of the planet where microbes have been found growing include : mine drainage with a pH of 1; 1500 meters down in ancient lavas, living off the rock itself; 3.2 km deep in sweltering South African gold mines; and, albeit in suspended animation, preserved for more than 250 million years in ancient brine pockets in subterranean salt (Kerr, 2002).

Microbes seem to thrive wherever there is liquid water at temperatures below the upper lethal limit of 113°C. Almost all the planet's microbes appear to be out of sight beneath their familiar haunts of soil and sea floor. But the growth of most deep sea extremophilic microbes is quite slow.

According to several different types of DNA analyses, more than 95 per cent of the hot-spring microbes are Archaea. Further, gene analysis turned up gene patterns in 95 per cent of these Archaea that were closely related to those of known methanogens, which derive their energy from the reaction of hydrogen with carbon dioxide to produce methane. But only about 100,000 cells per milliliter could be found on the nanomolar concentrations of hydrogen in the springs, as compared with up to 1000 times that abundance of microorganisms in surface waters (Kerr, 2002).

Just how slow and sparse deep life may get is seen in some South African gold mines where at the greatest depth (3.2 kilometers), cells have been found whose DNA marks them as one of the "hyperthermophile" archaeans known from sea-floor hot springs. But on descending from the shallowest depths, biomass decreases rapidly, sometimes to undetectable levels, even though life's currently known upper temperature limit of 113°C is not approached until far below the mines, and also there is plenty of fuel for living.

In general it appears that deep-sediment metabolic activity beyond the near shore is quite low (Kerr, 2002).

The domains Bacteria and Archaea probably contain millions of species whose distribution and abundance in natural environments are now being explored. The prokaryotes are a crucial component of the biosphere–they catalyze processes sustaining all life on Earth and so act as the engines for the biogeochemical cycles. However, only about 4500 species have been characterized, so most of the diversity of prokaryotes is unexplored. One emerging characteristic is that prokaryote diversity in aquatic environments is some orders of magnitude less than in sediments and soils (Torsvik *et al.*, 2002). Estimates of diversity for natural bacterial communities have traditionally depended on the few cultivable species. This has led to an underestimate of their diversity. DNA and RNA analyses imply prokaryotic diversity far greater than hitherto predicted, and are now pointing to the role of bacterial and viral diversity in global ecological cycles. For instance, while most investigations of prokaryotic diversity have been on surface environments, recent research shows that the biota extend deep into Earth's crust, and that the majority of prokaryotic organisms might occur in the oceanic and terrestrial subsurface. The total carbon biomass in subsurface terrestrial microorganisms has been estimated to equal that of all terrestrial and marine plants, and may be the largest constituent of the entire Earth biomass (Whitman *et al.*, 1998).

The genome size (complexity) of prokaryotic organisms can be calculated from the reassociation rate of denatured single-stranded DNA, which depends on the amount of homogeneous DNA present in a sample. It is also used to estimate prokaryotic community genome size (the sum of the sizes of different prokaryotic genomes in a community), which is then used as a measure of the total genomic diversity in a community (see Trevors and van Elsas, 1995). The DNA reassociation method has revealed a high degree of genomic diversity in prokaryotic communities in pristine soil and sediments with high organic content. These samples are estimated to contain ~10^4 different prokaryotic species of equivalent abundances (Torsvik *et al.*, 2002).

Dynamics and Control of Prokaryotic Diversity

Trophic interactions. According to Hutchinson's paradox of the plankton, so many phytoplankton species exist in an apparently homogeneous aquatic environment where all the species present seem to compete for the same mineral nutrient (*i.e.*, a "bottom-up" control of diversity operates). Among prokaryotes, competitors can coexist if some mechanism of selective loss is operating (Fuhrman, 1999) to prevent the most successful competitors from utilizing all the resources. Thus, size-selective predation by protozoa allows bacterial coexistence with phytoplankton of different size classes. Parasitism by host-specific viruses likewise allows coexistence of different bacterial taxa within the bacterial community. Such "top-down" control of diversity at least in theory works even if all bacterial and phytoplankton taxa are limited by the same substrate (*e.g.*, phosphate) (Torsvik *et al.*, 2002).

One consequence of applying a top-down selective-loss mechanism is that large observed differences in total bacterial abundance can be explained by protozoan predation being severely restricted in soils and sediments. Conversely, the proposed similarity of taxon size in aquatic and soil/sediment environments suggests that rates of viral lysis are similar in both.

In a biogeochemical context, a bottom-up perspective implies that high microbial diversity is needed for processing all the different types of substrate molecules (resources) in the system. A top-down model suggests a very different concept for coupling diversity and biogeochemical cycles–the activities of lytic viruses compensate for the high growth rate of some bacterial species (host groups). Should viral lysis be the real mechanism allowing competing bacterial species to coexist, then it is bacterial diversity (in the sense of differences in growth rate between coexisting bacteria) that determines the abundance of viruses present and so determines the reflux of particulate organic matter by viral lysis into dissolved organic matter (Torsvik *et al.*, 2002).

Ecological Diversity in the Bacterial World

For over two decades, taxonomists have been applying whole-genome hybridization as a universal criterion for demarcating species of bacteria–bacterial species were widely recognized as phenotypically distinct groups of strains with 70 per cent or greater annealing of genomic fragments in DNA-DNA hybridization (see Whittam and Ake, 1993; Palys *et al.*, 1997). This criterion has been widely used because the groupings of bacteria based on DNA-DNA hybridization are often the same as those based on phenotypic characters and ecology. However, it is being increasingly felt now that any particular cut-off value (such as 70 per cent) is arbitrary and not guaranteed to yield groups of bacteria that correspond to real ecological units

(Vandamme *et al.*, 1996). We do not know what determines the fraction of genomic segments that anneal in hybridization experiments : is it the fraction of shared genes?

There is another molecular approach that can be a universal criterion for classifying bacterial diversity. It relies on the observation that all living organisms fall into clusters of closely related organisms based on the sequence similarity of shared genes (Mallet, 1995; Ambler, 1996). Bacteria (and other organisms) fall into distinct sequence clusters, where the average sequence divergence between strains of different clusters is much greater than the average divergence between strains of the same cluster. It appears that each sequence similarity cluster observed in the bacterial world might correspond to an ecologically distinct population (Cohan, 1997). Should this conjecture be correct, then a classification system based on sequence clusters would have a theoretical strength that is lacking in the genomic hybridization approach (Palys *et al.*, 1997). Palys *et al.* showed that the DNA sequences of protein-coding genes are more effective than DNA-DNA hybridization for classifying the ecological diversity of bacteria. They have recommended sequence similarity clustering as a primary criterion for demarcating taxa.

The 16S rRNA gene is ubiquitous among cellular organisms, is highly conserved, and can be amplified from any bacterial taxon by universal primers. Consequently, hundreds of bacterial species have been sequenced at this locus, and these sequence data have proved extremely useful in determining the evolutionary tree of all cellular and multicellular life, as well as the trees of bacterial families and genera. The closest relatives of any newly discovered bacterial species may be readily determined by 16S rRNA sequence data.

And yet, the 16S rRNA gene has failed to distingiush many closely related but ecological distinct groups of bacteria. In these cases, there is little or no variation within or between ecological populations.

Very few ecologically distinct populations that are indistinguishable by 16S rRNA data have been surveyed for variation at protein-coding gene loci. Nevertheless, those taxa that have been so surveyed show that ecologically distinct populations that fail to be distinguished by 16S rRNA data fall into separate sequence similarity clusters for protein-coding genes. Examples include *Bacillus atrophaeus*, some strains of *B. subtilis* and *Mycobacterium intracellulare* and *M. avium*.

It appears that the inability of 16S rRNA sequences to distinguish some taxa may be a consequence of the low evolutionary rate of 16S rRNA genes. While 16S rRNA sequence data are useful for distinguishing moderately divergent populations into separate sequence clusters, protein-coding genes provide a better opportunity for distinguishing very closely related ecological populations (Palys *et al.*, 1997).

It has been found that there exist many ecologically distinct populations of bacteria that cannot be distinguished by DNA-DNA hybridization.

A proper understanding of the evolutionary genetics of bacteria furnishes an improved criterion for species classification. The theory of periodic selection predicts that ecologically distinct populations should eventually diverge into distinct DNA sequence clusters at nearly every gene locus. A review of the systematics literature corroborates this prediction–ecologically distinct populations of bacteria nearly always fall into separate clusters based on the DNA sequences of protein-coding gene. Thus, sequence clustering based on protein-coding genes can be a useful criterion for distinguishing ecological populations of bacteria. Moreover, the sequences of protein-coding genes are more effective in distinguishing ecologically distinct populations than either DNA-DNA hybridization or 16S rRNA gene sequences (Palys *et al.*, 1997).

According to Palys *et al.*, molecular criteria for species demarcation should include protein-coding gene sequences, 16S rRNA gene sequences, and DNA hybridization. Specifically, each phenotypically distinguished DNA sequence cluster should be recognized as a separate species or subspecies.

Assessing Microbial Communities in Different Environments

Microorganisms are believed to make up more than one-third of Earth's biomass (Whitman *et al.*, 1998). They perform crucial roles in the cycling of nutrients, interact with animals and plants, and influence Earth's climate. Yet our molecular and functional knowledge of microbes is highly fragmentary in view of the fact that most naturally occurring microbes either cannot be cultivated in the laboratory or have not so far been cultured (Staley and Konopka, 1985).

As the taxonomic composition of environmental communities is a strong pointer to their ecology and function, for characterizing this "unseen majority" of cellular life, the prerequisite is to provide a taxanomic census of microbes in their environments (Schmidt *et al.*, 1991; Pace, 1997). This is done by cloning and sequencing their ribosomal RNA (rRNA) genes (most notably the 16S/18S small subunit rRNA)–an approach that has revealed the overwhelming diversity of microbial life (Hugenholtz *et al.*, 1998). However, this approach suffers from certain limitations due to quantitative errors: The polymerase chain reaction (PCR) step introduces some amplification bias, and generates chimeric and otherwise erroneous molecules that affect phylogenetic analysis (von Wintzingerode *et al.*, 1997).

An alternative approach, *metagenomics* (shotgun sequencing of community DNA), provides a more direct and unbiased access to uncultured organisms (Riesenfeld *et al.*, 2004; Venter *et al.*, 2004; Tringe *et al.*, 2005). This method does not involve PCR amplification. Also, since no specific primers or sequence anchors are required, even very unusual organisms are captured by this technique and its improved versions (Barton *et al.*, 2006).

With a view to utilizing metagenomics data for taxonomic profiling, von Mering *et al.* (2007) analyzed 31 protein-coding marker genes previously shown to provide sufficient information for phylogenetic analysis (they are universal, occur only once per genome, and rarely show horizontal gene transfer (Ciccarelli *et al.*, 2006). They extracted these marker genes from metagenomics sequence data, aligned them to a set of hand-curated reference proteins, and used maximum likelihood to map each sequence to an externally provided phylogeny of completely sequenced organisms. Mapping marker genes from four diverse environmental data sets onto a reference species phylogeny showed that certain communities evolve faster than others. The method also enables determination of preferred habitats for entire microbial clades and proves that such habitat preferences are often remarkably stable over time.

Overall, von Mering *et al.* observed an uneven representation of previously sequenced genomes in naturally occurring communities. Whereas some parts of the tree of life are well covered by published genome sequencing projects, they represent only a small fraction of naturally occurring microbes. Conversely, entire phyla such as the Acidobacteria or the Chloroflexi are poorly represented among the sequenced genomes but are widely abundant in natural communities.

In agreement with Rappe *et al.* (2003), von Mering *et al.* reported Proteobacteria as the most dominant phylum of microbial life in both marine and soil environments. As for other phyla, significant differences within the Proteobacteria were noted: the Rickettsiales were mostly found in the surface-water sample whereas Rhizobiales or Burkholderiales were mostly found in the soil sample.

Although molecular methods usually reveal most phyla to be ubiquitously distributed, Baas-Becking and Beyerinck postulated long ago that microbial taxa have preferred environments: "for microbial taxa, everything is everywhere–but the environment selects" (see Martiny *et al.*, 2006). According to this interesting hypothesis, microorganisms are usually dispersed globally, and they are only later selected by the environments depending on their functional capacities. Existing communities are constantly challenged by intruders from nonspecialist phyla that may occasionally survive simply by chance, acquiring the necessary functionality through horizontal gene transfer (Doolittle, 1999, 2005; Chen *et al.*, 2005). This hypothesis receives considerable support from the work of von Mering *et al.* that there is strong environmental preference along lineages, but with a time-dependent decay.

It appears that most microbial lineages continue to be associated with a certain environment for long time periods, and successful competition in a new environment is usually rare. The latter might require more than just the acquisition of a few essential functions; probably only a limited number of functionalities are self-sufficient enough, and provide enough advantage, to be pervasively transferred (Frigaard *et al.*, 2006). For many other adaptations, very precise regulation and/or subtle changes in the majority of proteins may be needed. As this is quite difficult to achieve, well-adapted specialists may occasionally be challenged in their environment. While this does not rule out the existence of some rare, atypical organisms in each environment, in general most microbial clades seem to have a preferred habitat (von Mering *et al.*, 2007).

Syntrophic Communities of Bacteria

Countless examples are known of symbiotic interactions between distantly related organisms having evolved mutually beneficial relationships for their survival. However, the mechanisms underlying the emergence of symbiotic associations between different species are not well-understood. How different microbes become friendly neighbors?. A clear understanding of the mechanisms governing such friendly neighbourhood can well become a starting point and a set of design principles for engineering "synthetic" multicellular communities having novel structure and function. Kim *et al.* (2008) made a major contribution to this issue of engineering synthetic communities of bacteria by demonstrating how their spatial organization can facilitate the development of syntrophy.

The creation of syntrophic populations of microbes may well be a simple matter of mixing together of strains having capabilities that the other strains lack, which are required for their survival (*e.g.*, the ability to degrade antimicrobial compounds in an antibiotic-rich environment). A reasonable assumption is that if each organism codes for a phenotype that is beneficial to the community, the consortium will become a stable population in which every strain survives. But such an assumption is actually a gross undersimplification–in practice, it is extremely difficult to create stable, mixed populations of *n* bacterial strains, particularly when n ? 3 (Kato *et al.*, 2005). This is because (i) The growth rate of the different species are typically not balanced and the consumption and production of metabolites are not matched. (ii) The fluctuation of physicochemical factors produced by the quorum can bias the growth and survival of some strains (*e.g.*, some species may form biofilms)–it can be bacteriostatic or even bactericidal to other strains (*e.g.*, changes in pH or osmolality). (iii) Certain microbes perceived as competitors may be killed through microbial warfare involving, *e.g.* the production of toxic secondary metabolites. Kim *et al.* (2008) have discussed the importance of "space" and defined how the growth and homoeostasis of a population of unrelated bacteria may be affected (Weibel, 2008).

The role of spatial organization in multicellular populations is well recognized and spatial dynamics are quite important for multicellular interactions (Hansen *et al.*, 2007). Thus, the spatial colocalization of two engineered strains of *Escherichia coli* produces an engineered consortium in which a consensus gene expression response arises (Brenner *et al.*, 2007), and programmed pattern formation occurred between engineered strains of *E. coli* in which one strain sent a chemical signal and the other acted as a receiver.

For their work on the effect of spatial structure on multispecies communities of bacteria, Kim *et al.* developed a microfluidic system to raise populations of three unrelated colonies of certain soil bacteria not known to interact in nature: whereas *Azotobacter vinelandii* fixes dinitrogen into amino acids, *Bacillus licheniformis* hydrolyzes the β-lactam ring of penicillin G and *Paenibacillus curdlanolyticus* hydrolyzes carboxymethyl cellulose to D-glucose. This microfluidic system creates a physical barrier or fence between monoclonal liquid cultures of the 3 organisms but leaves them in fluidic contact as the floor of the compartments is made of porous membrane with nanoscopic holes through which the cells cannot penetrate. Below the membrane a channel filled with liquid connects the colonies together and provides a passage for the movement of small molecules and ions between the cultures. Controlling the distance between the 3 organisms allowed Kim *et al.* to show the emergence of reciprocal syntrophy during growth in nutrient conditions in which the organisms are forced to cooperate to ensure their survival.

The work of Kim *et al.* shows how spatial structure balances the various interactions within a multi-species community of bacteria–stability is not achieved when the distance between the inhabitants is short because mutual consumption of metabolites exceeds the production of secondary metabolites. Increasing the distance between the organisms decreases the consumption rate and the system attains a stable steady state. At even larger distances the system cannot reach a stable steady state which can balance the consumption and production of metabolites (Weibel, 2008).

The conclusion to be drawn from the work of Kim *et al.* is that control over the spatial organization of a multispecies community of microbes can make it possible to override differences in the rate of growth and metabolism of organisms to achieve a stable, reciprocal syntrophic community. The principles highlighted by Kim *et al.* may find application in synthetic biology, the identification of new microbial communities in the biosphere, and the creation or possibly the recapitulation of the emergence of symbiotic relationships (*e.g.*, mitochondria, chloroplasts). Another application could be the isolation, growth, and identification of "unculturable" species of microbes that may require factors for their growth that can be provided by some communities (Riesenfeld *et al.*, 2004).

Role of Microbes in Aquatic Ecosystems

Microbes play a crucial role in aquatic ecosystems. The long-held conventional view of the pelagic food chain (*i.e.* diatoms → crustacean zooplankton → fish) is actually only a small component of a highly complex aquatic food web that is dominated largely by microorganisms (DeLong and Karl, 2005). Phytoplankton cells <3 μm in diameter dominate photosynthetic activities within the open ocean. These picoplankton communities are made of highly diverse minute eukaryotes in addition to a few prokaryotes, such as *Prochlorococcus* and *Synechococcus* (see Reynolds, 2006).

The current concept of "phytoplankton", as adopted by Reynolds (2006) is a 'collective of photosynthetic microorganisms, adapted to live partly or continuously in open water'.

According to Reynolds, the 'microbial loop' organisms (*i.e.* bacteria, flagellates, ciliates etc.) have an extremely important role in waterbodies. Phytoplankton is also highly relevant to our own lives through producing harmful algal blooms, as powerful mediators of atmospheric carbon dioxide and as the base of most aquatic food webs. Indeed, phytoplankton have attracted considerable recent interest in attempts to control global warming by adding iron salts to open ocean so as to stimulate explosive growth of the algae which then mop up anthropogenic carbon dioxide.

Marine Microbes

Molecular analyses of marine microbes have shown how the environmental stresses of living in the deep ocean manifest in the corresponding genomic complements. Derelle *et al.* (2006) sequenced the genome of the green alga *Ostreococcus tauri*, which has an unusually small size (about 1 μm in diameter) as well as a high gene density. This picoeukaryote can pack more than 8000 genes into less than 13 Mb by making the average gene just slightly longer than 1.2 kb and reducing the intergene distance to 0.2 kb. Yet, it still contains entire plantlike metabolic pathways, such as the enzymes for C_4 photosynthesis (an evolutionary adaptation to low CO_2 levels) and for storing glucose as one large starch granule (a pyrenoid) within the single chloroplast. It also has only one copy each of the mitochondrion, a Golgi body, and the nuclear pore.

Bacterial Populations of Urban Aerosols

If we consider the great importance of its potential implications for human health, agricultural productivity, and ecosystem stability, very little is known about the composition or dynamics of the atmosphere's microbial inhabitants. Generally low levels of moisture and nutrients together with fairly high levels of UV radiation render the earth's atmosphere an extreme environment for microbial life and there is a strong need to know more about the atmospheric microbial composition and how it varies by location or meteorological conditions. Plant canopies are significant sources of bacterial aerosols in which upward flux of bacteria is positively impacted by temperature and wind speed (Lindemann and Upper, 1985). Aerosols formed at the surface of water bodies concentrate and carry bacteria through the liquid-air interface. The relationship between environmental conditions and bacterial aerial dispersal suggests that climate change may modify the microbial composition of downwind areas and increase health risk from pathogens or allergenic components of unclassified environmental bacteria. The last decade saw a marked expansion of desertification and a parallel increase in upper atmospheric particulates (Shinn *et al.*, 2003). In parts of Africa, dust storms are blamed for regional outbreaks of meningococcal meningitis caused by *Neisseria meningitides* (WHO, 2003). Recent El Nino weather events have coincided with increased flux of African Dust across the Atlantic which, in turn, has been connected to increased exacerbations of pediatric asthma (Gyan *et al.*, 2005). Because particles from dust storms protect bacterial and fungal passengers from the inactivating effects of UV exposure, global transport of dust can exert profound affects (Brodie *et al.*, 2007).

Natural changes in bacterial composition have implications for atmospheric pathogen monitoring systems engaged in monitoring major U.S. cities for intentional release of biowarfare agents. According to Anda *et al.* (2001), many such pathogens and other related bacteria with undefined pathogenicity have become endemic to the locations that are being monitored; they may interfere with detection networks but very little is known about the frequency or variability of their occurrence. Most aerobiology studies (see Griffin, 2004; Lee *et al.*, 2006) have used culture-based methods for determining microbial composition. Maron *et al.* (2005) applied culture-independent techniques but still very little was revealed about what constitutes the breadth of diversity of "typical" organisms in the atmosphere (as opposed to those which can be grown in laboratory media) and what influences their composition. To address such limitations, Brodie *et al.* (2007) designed a microarray (PhyloChip) for the identification of both bacterial and archaeal organisms. They targeted the variation in the 16S rRNA gene, possessed by all prokaryotes, to know the wide range of microbial diversity that may exist in the atmosphere. This allows bacteria and archaea to be identified and monitored in any type of sample without any need for microbial cultivation (Brodie *et al.*, 2007).

Using a custom high-density DNA microarray, they detected and monitored bacterial populations in two U.S. cities over 17 weeks. These urban aerosols contained at least 1,800 diverse bacterial types, a richness approaching that of some soil bacterial communities. They also revealed the consistent presence of bacterial families with pathogenic members including environmental relatives of select agents of bioterrorism significance. Finally, using multivariate regression techniques, it was demonstrated that temporal and meteorological influences are stronger factors than location in shaping the biological composition of the air we breathe.

The work of Brodie *et al.* involved a thorough molecular analysis of airborne bacterial composition and dynamics. They showed that the atmosphere contains a diverse assemblage of microorganisms probably representing the amalgamation of numerous point sources. The composition of this habitat varies greatly and is probably subject to climatic regulation. A worldwide study of this uncharacterized ecosystem should help determine baselines for bioaerosol transport patterns and the data will enhance our understanding of future anthropogenic impacts, including pollution, bioterrorism, and climate change, in altering the biological composition of the air.

Microbes and Geochemistry

The distribution of microorganisms in the geosphere is remarkably ubiquitous (see Chapelle *et al.*, 2002; Warren and Kauffman, 2003). They have been found in deep-sea hydrothermal vents, highly contaminated, abandoned mines and the Earth's deep subsurface. There are strong linkages between microbes and geochemistry.

Microbial activity is quite significant in aqueous geochemical processes such as mineral precipitation and dissolution, contaminant degradation, sequestration or mobilization, fossilization, and weathering. These linkages can be detected in a wide variety of aqueous systems, including marine, freshwater, groundwater, and subglacial melt.

The geochemical reality of these interactions has recently become quantifiable with high-reslution methods such as x-ray absorption spectroscopy, and molecular biological techniques have revolutionized environmental microbiology by providing genetic perspectives of microbial diversity without requiring organisms to be cultured in the laboratory (Pace, 1997).

Unlike higher eukaryotes, microbial evolution is driven by metabolic diversity, rather than cell complexity and organism structure. Microbes are not restricted by geographic barriers, but their metabolic pathways are constrained by the available redox couples in the biosphere (see Banfield and Nealson, 1997; Finlay, 2002). But microbes can utilise a wide variety of electron acceptors other than molecular oxygen for respiration.

The geochemical influence of microbes manifests in all the important elemental cycles, especially those relevant to life on Earth. Further, microbes are reaction accelerators, speeding up otherwise slow redox reactions. Microbial influence on geochemical processes is widespread and predictable.

Such microbial metabolic or functional activities as the oxidation or reduction of Fe and Mn are spread across the phylogenetic tree. Numerous genetically differentiated microbial strains often have the same geochemical function in differing environments (Finlay, 2002) with potentially differing outcomes. Also, some strains switch metabolic pathways (for example, from sulfate to iron reduction) depending on the prevailing environmental conditions in which they occur. Novel microbial strains have been identified through phylogenetic analyses (mainly of 16S RNA oligonucleotide sequences) from natural systems but the geochemically relevant information gained from such identifications is usually limited. Less than 10 per cent of the microbes that exist in nature have been identified; of that small number, less than 1 per cent have been successfully cultured. This means that despite a swelling phylogenetic databank of environmentally identified microbial strains, we still do not know how they might influence geochemical processes. The central issue, from a geochemical perspective, is how to determine microbial function. Functional gene microarray systems, which look for expression of genes coding for enzymes involved in specific functions, are now under development to tackle this problem.

Another challenge is to pinpoint the role of community microbial activity in shaping geochemical consequences. Geochemical interest in microbial activity is often restricted to a single metabolic function with some relevant geochemical consequence, *e.g.* metal sequestration. But real life situation is that the microbe responsible for that geochemical impact exists within a microbial community or consortium (see Newman and Banfield, 2002). The by-products from one organism's metabolic pathway serve as the nutrients of the next strain. This means that both the metabolic reactants and products associated with a microbial community, and the reaction energetics involved, may differ from one location to the next.

Thus, although commonality of metabolic pathways is conducive to widespread occurrence of certain microbially driven geochemical processes in different environments, variability in microbial consortia and microgeochemical conditions tends to selectively refine that goechemical impact. In doing so, they often create community-specific microbial fingerprints on geochemical processes (Warren and Kauffman, 2003).

Pathogens in the Marine Environment

The discharge of non-disinfected wastewaters into the marine environment is a common worldwide practice, both in developing and highly developed countries. Consequently, the seas are constantly infused with wastewater bacteria, including highly pathogenic ones. In view of the public health significance of this phenomenon, it is surprising how little is actually known concerning the fate of such bacteria once they enter the sea. While numerous studies have addressed the effects of various environmental parameters on colony formation, many of them ignored the fact that bacteria can retain viability and infectivity while losing colony-forming ability (see Belkin and Colwell, 2006).

Biodegradation

Recent advances in genomics, analytical chemistry, and stable isotope probing (SIP) are stimulating progress in biodegradation research. SIP can be used to determine the primary member of a complex community that is responsible for naphthalene catabolism *in situ*. SIP has already been well adopted in biodegradation research; when combined with metagenome sequencing and biodegradation database information it will continue to provide new insights into complex microbial environments (Wackett, 2004).

In physics, the Heisenberg uncertainty principle states that the precise position and momentum of an electron cannot be determined simultaneously. By analogy, any study of the molecular details of an environment can disturb that environment. This environmental uncertainty principle has greatly affected research on biodegradation involving the study of how compounds are broken down in the environment via microbial metabolism. Some have proposed that biodegradation should be studied directly in the soil or water where it occurs. But much has also been learned through laboratory studies. Biodegradation researchers have usually taken some microorganisms isolated from the environment and then studied its genes and enzymes in the laboratory. Such laboratory-based reductionist approaches have thrown much light on the mechanisms of biodegradation. Conversely, modern analytical chemical methods have made it possible to monitor extremely low levels of pollutants in the environment. Although this tells us how much of the pollutant is going away, and how fast, it does not reveal the organisms or the biochemical mechanisms responsible for the disappearance of the pollutant. We do not know how complex mixtures of microbes may interact to accelerate or slow down biodegradation. What dominant microorganism is involved in biodegradation in complex soil or bioreactor environments? New insights are now emerging from advances in biochemistry, genomics, and analytical chemistry (Jeon, 2003).

Stable Isotope Probing

Radajewski *et al.*(2000) fed aerobic soil bacteria with [^{13}C]-CH_4, followed by separation of [^{13}C]-DNA and [^{12}C]-DNA by density-gradient centrifugation. This facilitated identification of the bacterial DNA that was directly oxidizing and incorporating CH_4 into intracellular molecules. This technique is termed stable isotope probing (SIP). It can be applied to several different substrates, *e.g.*, [methyl-^{13}C]-toluene to probe the microbial populations that metabolize toluene anaerobically under denitrifying conditions, or for identifying anaerobic bacteria involved in phenol metabolism in an industrial reactor (Manefield *et al.*, 2002). Van der Hulst (1999) used the technique by feeding [^{13}C]-urea to humans for detecting *Helicobacter pylori*, the causative agent of peptic ulcers in the digestive tract.

Suprisingly, much light is being thrown on understanding environmental biodegradation even through such 'reductionist' approaches as DNA sequencing and stable isotope-incorporation experiments. Many microbial genomes have been sequenced. Genomics is moving beyond the sequencing of microbial species to include the sequencing of the metagenome–the complete complement of clonable DNA from all organisms in a given environment (Rondon *et al.*, 2000). This approach can reveal the complete genomic and metabolic potential of a given environment but does not always inform about what reactions are actually taking place or about the key microbial players responsible for biodegradation (Wackett, 2004).

A number of free, web-accessible databases containing compiled information on DNA sequences, protein function and biodegradation pathways are now available and can be used to analyze the data obtained from sequencing the metagenome. This provides insights into the genetic potential in different environments. SIP can then yield vital information on which genetic components in that environment are most active under real world conditions. These modern approaches can enable us to understand such highly complex environmental processes as biodegradation. This can ultimately speed up biodegradation in natural and engineered environments (Wackett, 2004).

The Evolution of Earth's Atmosphere

The importance of microorganisms arises partly from their production directly or indirectly of nearly all of the oxygen we breathe. Oxygen is produced during photosynthesis by a reaction that can be written as $CO_2 + H_2O \rightarrow CH_2O + O_2$ where "CH_2O" is a shorthand for more complex forms of organic matter. Of course, most photosynthesis on land is carried out by higher plants, not microorganisms; yet terrestrial photosynthesis has little effect on atmospheric O_2 because it is nearly balanced by the reverse processes of respiration and decay. In contrast, marine photosynthesis is a net source of O_2 because a small fraction (~0.1 per cent) of the organic matter synthesized in the oceans is buried in sediments. It is this small leak in the marine organic carbon cycle that accounts for most of our atmospheric O_2 (Kasting and Siefert, 2002).

Although some macroscopic plants (macroalgae *e.g.*, kelp) are found in the oceans, most marine photosynthesis is carried out by microscopic eukaryotic algae such as diatoms and coccolithophorids. Prokaryotic baeteria are also important for another reason. Though they make up only ~1 per cent of marine biomass, cyanobacteria fix nitrogen. Some, such as the filamentous *Anabaena* spp., fix nitrogen only in their heterocysts. Other cyanobacteria fix nitrogen at night and photosynthesize by day. *Trichodesmium* spp. (very abundant in tropical waters), fix nitrogen in the morning and photosynthesize in the afternoon.

In some senses, when it comes to producing oxygen, cyanobacteria constitute the entire story (Kasting and Siefert, 2002). Because they can live anaerobically and aerobically, they have been responsible for the initial rise of atmospheric O_2 around 2.3 billion years ago (Ga). Comparison of ribosomal RNA from cyanobacteria with portions of the DNA inside chloroplasts implies that all eukaryotes, including algae and higher plants, derived their photosynthetic capabilities from cyanobacteria by way of endosymbiosis. *Prochlorococcus* spp., an important component of today's marine ecosystem, could possibly be the living ancestor of the cyanobacterium involved in this event.

Environmental Microbiology-on-a-Chip

Rapid advances in microfabrication, DNA and protein microarray and microfluidic technologies have enabled the development of fully-integrated, miniaturized systems. These 'laboratory-on-a-chip' (LOC) devices perform sample preparation (*i.e.* concentration, separation and purification) together with biochemical reactions and detection steps simply and automatically. LOC technology for environmental microbiology studies is likely to have immediate impacts on microbial-monitoring by achieving detection and identification in just a few minutes at the single-cell level, and on microbial ecology by deepening the understanding of microbial community structure and diversity and correlating these with niche-specific functions within a microspace. In the long run, significant impacts will occur on environmental metagenomics and proteomics.

LOC devices, also known as 'micro total analysis systems (mTAS)' or 'biological microelectronic mechanical systems (BioMEMS)' have become important work platforms for researches in genetics, proteomics, cellomics, clinical and forensic analyses (Andersson and van den Berg, 2003; Figeys, 2002; Verpoorte, 2002) and environmental microbiology. LOC technology offers automated, high-throughput and possibly on-site solutions, from sample preparation, biochemical reactions (*e.g.* immunological reactions, enzymatic reactions and DNA assays) to signal detection (optical or electrochemical). It uses microlithography, micromachining and bonding techniques to create small wells, channels, electrodes and filters on enclosed silicon, glass or polymer substrates (Reyes *et al.*, 2002). These features help to generate microfluidic networks interconnected to fluid reservoirs and reaction compartments where samples and reagents are stored, transported and physical and/or chemical reactions are carried out. Incorporation of micropumps, microvalves and heating devices allows researchers to perform sample and/or reagent dispensing, mixing, incubation, reaction and partitioning for different physical and biochemical reactions before analytes are detected by optical or electrochemical methods. Various microfluidic devices have been developed for sample preparation and product detection either individually or collectively (Pethig and Markx, 1997; Delehanty and Ligler, 2002; Paegel *et al.*, 2003). LOC can greatly advance environmental microbiology studies simply because the sample concentration, reagent volume and reaction and detection time can be manipulated at a level 100-1000 times smaller in size or 10-20 times shorter in time than current practices (Liu and Zhu, 2005).

Rapid Monitoring of Microbial Pathogens in the Environment

Rapid detection of air- or water-borne pathogenic microorganisms present in food is crucial for ensuring environmental quality and public health safety. Culture-free molecular technologies based on nucleic acid and immunological approaches are frequently used for this purpose (Ivnitsky *et al.*, 2003) but require skilled technicians and depend on bench-top equipment, which cannot give on-site detection and monitoring capability for the pathogens. LOC devices are much better systems that provide rapid, sensitive, specific, automated and portable features for on-site use (Liu and Zhu, 2005).

LOC for Immunological Assay

A variety of immunological-based LOC devices for whole-cell or microbial metabolites have been developed for monitoring and diagnostics. Zhu *et al.* (2004) integrated sample trapping and concentration steps with whole-cell immunological assay in a silicon-based LOC consisting of micro-chambers, micro-channels and filter weirs. Through the filter weirs, microbial cells such as *Cryptosporidium parvum* and *Giardia lamblia* (larger in size than the weir gap, 1-2 mm) are trapped, concentrated and labeled during sample injection together with staining solution containing fluorescently labeled antibodies. Consequently, the immunological reaction speeds up with the total analysis time reduced from ~-3 h to 2-5 min, and the total consumption of reagent is reduced by a factor of 20. Delehanty and Ligler (2002) designed an antibody microarray system with continuous fluid flow-through to detect microbial toxins, achieving simultaneous detection of cholera toxin and staphylococcal enterotoxin B at such low levels as 8 and 4 ng/mL, respectively, within 15 min.

LOC for PCR

This method also can rapidly detect microbial pathogens. One simple approach is based on a miniature analytical thermal cycling device on a chip that integrates cell trapping and lysis steps with PCR amplification, real-time detection through fluorescence resonance energy transfer mechanism (*e.g.* TaqMan and molecular beacon probes) and analysis in an array format within a portable suitcase. Multiple detections of targeted pathogenic bacteria and viruses with detection limits in the order of 10^2–10^4 cells/mL can be achieved within 15 min, with a 2-10 times reduction in detection time compared with conventional method (Belgrader *et al.*, 1999). Woolley *et al.* (1996) and Lagally *et al.* (2000) integrated a microfabricated PCR chip with a capillary electrophoresis (CE) chip, to amplify and identify multiple specific gene markers based on fragment size after capillary electrophoresis. Multiple bacterial targets as low as 2-3 cells could be amplified within a 200-nl PCR chamber and subsequently resolved through CE in less than 10 min within a portable system equipped with power supply, electrical control and laser-excited fluorescence detector (Lagally *et al.*, 2004). Yet another good approach may be to combine PCR and DNA-microarray technology inside a microfluidic device.

Impact on Studies of Microbial Diversity

The use of molecular biology technologies has revealed extremely high microbial diversity in many ecosystems (Pace, 1997; Forney *et al.*, 2004). Just one gram of soil can have ~10^{10} cells, representing 4000-7000 different genomes or populations (Torsvik *et al.*, 1990) belonging to different microbial communities which occupy many micro- or macro-habitats differing in physico-chemical and structural characteristics. However, available rRNA-based molecular fingerprinting or clone library methods can only simultaneously monitor or identify the dominant 50-200 populations (Dunbar *et al.*, 1999). The remaining about 95-99 per cent of total populations cannot be identified. The original microbial community structure and diversity existing within individual micro- or macro-habitats is lost in the course of extracting DNA from 1 g of soil, which yields ~ 8.7 mg of total DNA (assuming 9.5×10^{-6} ng of DNA per cell and 10 per cent extraction efficiency). Subsequent experimentation can only reveal the most numerically predominant populations in that particular 1 g of soil but not those in individual micro and/or macro-habitats (Liu and Zhu, 2005).

The above problem can be addressed by employing a fully integrated LOC study of microbial diversity through cloning and sequencing of rRNA genes (rDNA) from mixed microbial assemblages, followed by phylogenetic tree construction (Pace, 1997). It requires a nanoliter DNA processor, submicroliter PCR and a microfluidic 'sequencing factory-on-a-chip'. This last consists of four components for cell transformation and sorting, DNA amplification and Sanger extension, product purification and sequencing through capillary electrophoresis (Paegel *et al.* 2003).

According to Paegel *et al.*, 96 reads of DNA sequences with individual reads > 500 bases in length (>99 per cent accuracy) are obtained within 3 h starting from cell transformation and sorting to sequencing using mCAE. The total reagent volume is only <1 ml for each read. Therefore, up to 500 clones can be screened and sequenced within an rDNA clone library from a sample containing 10^3–10^6 cells in just one day. Hundreds of clone libraries can be made from different micro- and/or macro-habitats from 1 g of soil, to greatly enhance the understanding of soil microbial diversity. Such an indepth resolution can provide much better understanding of the real world of microbial diversity in different microbial ecosystems than the currently used clone library approach.

The 'K'-Selected Oligophilic Bacteria

Much of our knowledge of microbiology has been derived by using pure culture techniques, but a high fraction of the microbial diversity has never been cultivated and is unexplored. There exists a difference of one to two orders of magnitude between direct microscopic count and colony count on any medium. This is known as the 'great plate count anomaly' (Amann *et al.*, 1995). Use of molecular techniques to assess the bacterial diversity independent of culture methods (Ward *et al.*, 1990; Bintrim *et al.*, 1997) has revealed that some 99 per cent of the bacterial diversity remains uncultured and therefore unexplored (Bintrim *et al.*, 1997; Fuhrman and Campbell, 1998).

According to Watve *et al.* (2000), a large proportion of the uncultured diversity from various aquatic and terrestrial habitats are oligophilic (oligotrophic) bacteria that grow only on dilute nutrient media and form very small colonies. Watve *et al.* developed a technique to cultivate and isolate the moderately oligophilic bacteria and isolated 90 cultures, out of which 12 showed high growth yield coefficients and carbon conversion efficiencies at low substrate concentrations and progressively decreased with increasing substrate concentrations. Their growth yields were quite high and approached the theoretical maximum. *Slow* growth rates and *high* yields indicate that they are 'K' selected species. The K-selected bacterial species employ different survival strategies than those employed by fast growing 'r'-selected bacteria. The 'r' selected species are opportunistic fast growers whereas the 'K' selected species, though multiplying slowly, make a more efficient use of the environment and enjoy a larger carrying capacity. The oligophiles fit into the role of the K selected species that enjoy better fitness at saturated or marginally life supporting conditions utilizing the minimum available resources more efficiently.

Oligophiles constitute the majority of marine bacterial communities, and also occur in soil and freshwater. The oligophiles do not seem to be restricted to extreme oligotrophic marine habitats. A large number of moderate oligophiles are found in natural habitats with varying trophicity (Watve *et al.*, 2000).

Temperature Excludes N_2-fixing Heterocystous Cyanobacteria in the Tropical Oceans

Whereas the non-heterocystous cyanobacteria *Trichodesmium* spp. are the dominant N_2-fixing organisms in the tropical oceans (Karl *et al.*, 2002), in freshwater lakes and brackish environments, heterocystous species dominate N_2 fixation. Why are heterocystous cyanobacteria absent in the pelagic of the tropical oceans? *Trichodesmium* can fix N_2, apparently without needing to differentiate heterocysts. Staal *et al.* (2002) showed that differences in the temperature dependence of O_2 flux, respiration and N_2 fixation activity explain how *Trichodesmium* performs better than heterocystous species at higher temperatures. Their results also explain why *Trichodesmium* is not successful in temperate or cold seas.

Primary production in large areas of the oceans is chiefly controlled by the availability of nitrogen, and biological N_2 fixation can overcome this limitation. N_2 fixation in the marine pelagic environment tends to be restricted to (sub)tropical regions. The organisms responsible for most of the N_2 fixation in the tropical oceans are *Trichodesmium* spp. which are filamentous non-heterocystous cyanobacteria that produce massive surface blooms. The numbers of free-living heterocystous cyanobacteria in the marine pelagic environment are low and seem to show very low growth rates (Carpenter and Jansen, 2001).

The virtual absence of heterocystous cyanobacteria in the marine pelagic environment contrasts markedly with their presence in freshwater lakes and brackish environments, where they often form dense blooms. Two questions that need to be addressed are: (1) why are free-living heterocystous cyanobacteria not the dominant N_2-fixing organisms in the tropical oceans? (2) why *Trichodesmium* spp. cannot thrive in marine, brackish or even freshwater environments in temperate and polar regions? According to Staal *et al.*, a glycolipid cell envelope, which acts as an effective diffusion barrier for O_2 in heterocysts, is not of much use in sea water at increased temperatures, and thus heterocystous cyanobacteria are outcompeted by *Trichodesmium* spp. Their results also explain why *Trichodesmium* spp., which possess specialized N_2-fixing cells with high respiration rates but without a glycolipid layer (diazocytes) (see Kana, 1993; Berman-Frank *et al.*, 2001) are not successful in temperate or cold seas.

Indeed, measurements of nitrogenase activity by the acetylene reduction assay in the heterocystous *Nodularia spumigena*, *Anabaena* sp. and *Trichodesmium* sp. strain IMS 101 have showed that nitrogenase activity in the dark (N_d) and in the light (N_{tot}) increases with temperature. In the light, photosynthetic electron transport adds to the respiratory production of ATP; therefore, at saturating irradiances nitrogenase is not limited by ATP and so achieves its maximum attainable activity (N_{tot}). When N_{tot} does not increase, notably under anaerobic conditions, it can be considered to be the potential nitrogenase activity.

Due to a limited influx of O_2 in N_2-fixing cells, respiratory ATP production limits N_2 fixation in the dark. Because the flux of O_2 into heterocysts and diazocytes is crucial for regulating N_2 fixation, it is important to consider the factors on which it depends.

The potential nitrogenase activity (N_{tot}) was found to increase relatively more with temperature than did N_d, resulting in an increase in the ratio N_{tot}/N_d with temperature. The ratio N_{tot}/N_d (^{3} 1) may be used as an indicator of the relative ATP limitation of nitrogenase activity in the dark. A higher N_{tot}/N_d points to a higher ATP limitation of nitrogenase in the dark.

Staal *et al.* examined several different heterocystous species isolated from various marine, brackish and freshwater environments. The average N_{tot}/N_d at 20°C for all of these organisms was 3.2 ± 1.07, suggesting that there is an optimal N_{tot}/N_d in heterocystous cyanobacteria in the range of 2.5-4. They calculated that the flux of O_2 is increased by about 25 per cent when the salinity is decreased from 35 (sea water) to 0 (fresh water), underscoring the need for an efficient diffusion barrier for the N_2-fixing cells in fresh or brackish waters. The same is true for marine environments, which are frequently oversaturated with O_2. This explains the presence of heterocystous cyanobacteria in these environments (Staal *et al.*, 2003).

Microbial Biogeography

Most macroscopic animals and plants have a limited distribution range on the surface of Earth. Over geological time, taxonomic groups have evolved within continents or even within certain lakes, mountain ranges, or oceanic islands. Individuals of these groups remained there as a result of natural barriers to migration. Microbes are different. About a century ago, the Dutch microbiologist Beijerinck concluded that any bacterial species can occur wherever its environmental requirements are met: The distribution of microbes can be understood solely in terms of habitat properties. Most taxonomists have assumed that this also applies to unicellular eukaryotes (protozoa, microalgae), but the topic has recently drawn renewed interest (Finlay, 2002). For sexual, outbreeding eukaryotes, a theoretically based species concept exists, at least in principle, and so "cosmopolitan distribution" has a precise meaning–something that is less obvious in the case of bacteria. Numbers explain cosmopolitan distribution of microbes–microbial population sizes are enormous. The probability of dispersal is high and the probability of local extinction extremely low. In the absence of effective migration barriers and local extinctions, every habitat will contain a pool of bacterial species that do not thrive locally, but may grow if the environment turns more favourable. About 10^{18} viable bacteria annually are transported through the atmosphere between continents (Griffin *et al.*, 2002). Bacteria can be isolated from places where they would not be expected, such as thermophilic bacteria from cold seawater (Isaksen *et al.*, 1994). Indeed, species of aquatic and soil microbes are found wherever their needs are met. But Whitaker *et al.* (2003) and Papke *et al.* (2003) have challenged this view on the basis of their studies of thermophilic prokaryotes from geothermal springs (Fenchel, 2003).

Whitaker *et al.* studied the archaebacterium *Sulfolobus solfataricus* which thrives in water and mud surfaces of hot springs at 50°-57°C at a pH of about 4. It grows by oxidizing elemental sulfur to sulfate or by metabolizing organic matter. Strains were isolated from several geothermal springs in remote geographic locations (Russia, USA and Iceland). The strains seem to belong to the same species because they have almost identical 16S ribosomal RNA (rRNA) sequences. But when the nine protein-coding loci were sequenced, it was found that genetic distances between populations increased proportionally with geographic distance. Although the genetic distances are slight, results indicate that there is not a continuous rapid exchange of cells between these remote habitat patches.

Papke *et al.* studied two groups of thermophilic cyanobacteria: the unicellular *Synechococcus* and the filamentous *Oscillatoria*. These are oxygenic phototrophs. In hot springs they form microbial mats on surfaces at temperatures up to about 70°C. Material was collected in hot springs from Yellowstone National Park, Oregon, Japan, Italy, and New Zealand. Sequencing of rRNA genes revealed several lineages, each consisting of a cluster of genotypes. These clusters may possibly be regarded as "species". The results revealed some geographical patterns. Notably, representatives for two of the clusters were found only at North American sites, but in other cases, strains from New Zealand and Japan were clustered together. For both studies, genotypes did not correlate with their occurrence in particular microhabitats (water chemistry, temperature, pH). Phenotypically the strains appeared identical within the genotype clusters, showing that genetic variation does not necessarily indicate adaptive fine-tuning and that the recorded variation probably reflects neutral or near-neutral mutations (Fenchel, 2003).

The two reports point to some limited exchange of cells between the study sites. The small total area constituted by geothermal springs on Earth's surface, their mutual remoteness, and the fact that the specialized obligate thermophiles suffer under more normal ambient temperatures are unusual properties of microbial biota and may account for the limited transfer of cells from one site to the other.

In panmictic (randomly mating) populations of outbreeding organisms, genetic variation between individuals is constrained although as a rule there are no two genetically identical individuals. When species populations are structured into metapopulations or have wide geographical ranges, spatial genotypic variation arises through local selection pressure and genetic drift, regardless of limited gene flow between subpopulations. But genetic differentiation and speciation certainly require some genetic isolation. Bacteria have clonal evolution (occasional horizontal gene transfer notwithstanding), and there is no such constraint on genetic diversification. The progeny of a given cell continuously accumulate neutral mutations, and different isolates of nominal species often show great mutual genetic distances, whereas natural selection maintains particular phenotypes on adaptive peaks. Such neutral genetic variation is once in a while purged from populations, *e.g.*, when a carrier of a favourable mutation and its descendants outcompete other lineages, resulting in a clustering of genotypes–and such clusters may be considered bacterial "species" (Cohan, 2002). The number of possible genotypes within such "species" is infinite. Locally, some particular clone may dominate at a given time, and gene sequences of bacteria sampled from different sites can lead to the impression of geographical structuring (Fenchel, 2003).

Geomicrobiology

In the mid-1980s, Derek Lovley and Kenneth Nealson created the new discipline of geomicrobiology that deals with microbes that live off metals.

According to Finlay (2002), the abundance of individuals in microbial species is so high that dispersal is rarely restricted by geographical barriers. This large dispersal warrants an alternative view of the scale and dynamics of biodiversity at the microbial level, wherein global species number is relatively low and local species richness is always sufficient to drive ecosystem functions.

Whereas many animal species–especially the larger ones–are known to have restricted geographical distributions and isolation led to their speciation, resulting in distinctive island faunas, this does not seem to hold good for microbes. Enrichment culture techniques revealed that diverse types of bacteria could be cultured from almost any type of natural material, and species recorded from a particular habitat type located in geographically distant places are usually quite similar to each other.

There is also no evidence that flagellated protozoan morphospecies have biogeographies–communities from adjacent sites are no more similar to each other than they are to those from more distant places. The same planktonic foraminiferan morphospecies are common to both Arctic and Antarctic waters, and some of these are also genetically identical (Darling *et al.*, 2000). Most freshwater ciliated protozoan morphospecies identified from a volcanic crater lake in Australia in the late 1990s had already been recorded from Northern Europe by the mid-1930s (see Finlay, 2002).

It is now emerging that protozoa and several other microbial eukaryotes do not have biogepgraphies, probably because they are so abundant that continuous large-scale dispersal sustains their global distribution.

Should ubiquitous dispersal be typical of most microbial eukaryotes, we would expect relatively low global species richness. This has turned out to be true for the free-living ciliates and independent estimates place the species richness of all free-living protozoa somewhere in the range of 10,000 to 20,000 (Raven and Williams, 1999).

Finlay and Clarke (1999) found 80 per cent of the global species total in the flagellate genus *Paraphysomonas* in <0.1 cm^2 of sediment from a freshwater pond in England. High local/global species ratios will also produce relatively flat species-area curves, distinguishing them from animal groups with locally restricted species, such as the insects.

Random dispersal is primarily driven by high absolute abundance. And since organism size and abundance are inversely related, there should be some size range where ubiquitous dispersal becomes less likely and where species are more likely to be geographically restricted. The organism size range where this change occurs is generally thought to be about 1 to 10 mm (see Finlay, 2002).

The Concept of Morphospecies

This is a useful concept for the protozoa because of the strong connection between body form (particularly the structure of the feeding apparatus) and the way the protozoon functions in the natural environment (see Levin, 2001). But morphospecies may conceal variations of different type, including clonal or sexual lifestyles. In sexual populations, there is periodic sex between individuals in closed gene pools known as sibling species (see Wilson, 1999). Different sibling species can be extremely divergent genetically, or some others can be genetically identical to each other, meaning that in some ciliates at least, genetic isolation and genetic divergence are not correlated (Finlay, 2002).

According to Finlay, free-living microbial eukaryotes having body sizes less than about 2 mm appear to be sufficiently abundant to have worldwide distribution. As prokaryotes are far smaller and much more abundant, they are even less likely than microbial eukaryotes to be restricted by geographical barriers (Fenchel, 1994). Secondly the species richness that makes up the microbial "seedbank" of any recognizable ecosystem is quite large meaning that microbially mediated ecosystem functions are never compromised by lack of microbial diversity, so long as the particular ecosystem remains within the physicochemical limits of microbial life. Finally, the basic properties of biodiversity at the microbial level (astronomical abundance of individuals within species populations, potentially very short generation times, high dispersal rates, low speciation rates, and relatively flat species-area curves) differ greatly from those of macroscopic organisms, and it may not be wise to add such alien organisms into any of the contemporary neutral theories of biodiversity and biogeography (Hubbell, 2001; Finlay, 2002).

Shewanella for Bioremediation

It is the microbes found in soils and waters that carry out the task of removing pollutants from the environment. The extraordinary respiratory versatility of one such metal ion-reducing bacterium, *Shewanella oneidensis*, is potentially promising for the development of new bioremediation strategies for removing toxic chemicals from groundwater. The genome of this bacterium has been fully sequenced (Heidelberg *et al.*, 2002).

Microbes involved in cleansing the environment of pollutants may either be managed or genetically changed to do their jobs better and more quickly. Many pollutants are difficult to remove, immobilize, or transform into non-toxic products. Simpler, more obvious remediation approaches have failed to clean up many of the polychlorinated biphenyls (PCBs), polynuclear aromatic hydrocarbons (PAHs), chlorinated solvents, heavy metals, and radionuclides that contaminate the environment. This warrants more fundamental and multidisciplinary approaches (Tiedje, 2002). Bioremediation is the most versatile and lowest-cost solution for some pollutants, and genomics allows one to assess what resources a microbe possesses or lacks, and thus how it might be managed or improved. Genomic analysis offers new insights in two crucial areas: (i) understanding the catalytic capabilities of the microbe and how they might be enhanced, and (ii) how to ensure that the microbe successfully carries out its work over the time and area intended (Tiedje, 2002).

S. oneidensis strain MRI is a bacterium that can use over ten different respiratory electron acceptors (including Fe(III), Mn(IV), S^o, sulfite, nitrate, O_2, Cr(VI), and U(VI). Of these electron acceptors, nitrate, chromium, and uranium are pollutants for which regulatory standards exist. *S. oneidensis* MR-1 converts nitrate to ammonium and the soluble forms of U(VI) and Cr(VI) to the reduced, insoluble oxides V(IV) and Cr(III). So this and other metal-reducing bacteria, such as the species of *Geobacter* and *Desulfotobacterium*, can halt or inhibit the groundwater transport of uranium, chromium, and technetium (Tc^{99}, another mobile radionuclide at weapons production sites) and stop their leaching into rivers and potable groundwater supplies (Lloyd and Lovley, 2001). Understanding of both the biochemical and ecological features of these organisms, as well as site conditions, is essential to make them do their job effectively and produce immobilized products that remain stable for long periods of time (Tiedje, 2002).

Strain MR-1 has a multicomponent, branched electron transport system that includes 39 *c*-type cytochromes, 32 of which were hitherto unknown. Diversity in the use of electron acceptors is one of the most important ecological attributes of microbes, because these reductions are central to the balancing of the Earth's geochemical cycles. This strain is particularly adept at using solid-phase electron acceptors (in nature, usually iron or manganese oxides).

The sequencing of the *S. oneidensis* genome and the ensuing proteomics and functional studies have the potential to answer several questions about electron-acceptor versatility and to enhance our understanding of how the organism has evolved to exploit solid-phase chemistry for its energetics; also what possible impact this could have exerted on Earth's geochemistry.

Genome-based analysis of other microbes is also underway so as to explore their potential uses in environmental biotechnology. Other metal-transforming strains whose genomes have been sequenced include two anaerobes, *Geobacter sulfurreducens* and *G. metallireducens* and an aerobe, *Ralstonia metallidurans*. *Deinococcus radiodurans* (not a metal reducer) can withstand high doses of radiation by correctly re-assembling its genome after radiation-induced breakage (Makarova *et al.*, 2001). This process is now being studied through genomic analysis with a view to carrying bioremediation traits into highly radioactive sites (Tiedje, 2002).

Some microbes use halogenated organic compounds as metal acceptors and in the process degrade halogenated pollutants. Genome sequences are already available for two such microbes: *Dehalococcoides* and *Desulfotobacterium*. These sequences contain a large number of dehalogenases and dehalogenase-like genes, of unknown functions. Exploiting this information in the engineering of new strains is promising for bioremediation research.

Finally, the genomes of several aerobic pollutant-degrading organisms are being (or have been) sequenced: *Burkholderia* strain LB400 is the most effective PCB degrader known, and *Rhodococcus* strain RHA1 is a versatile Gram-positive PCB degrader. As microbes cannot grow on PCBs, genomic information will open up approaches that may overcome the metabolic blocks to growth. The genomes of *Pseudomonas putida* and *P. fluorescens*, that are highly versatile in using aromatic hydrocarbons as their carbon source and thereby mineralize these pollutants, have been sequenced. The objectives of improving expression, recruiting missing or improved enzymatic activities, enhancing bioavailability, and improving cell transport and survival may be achieved by the availability of increasing amounts of genomic information (Tiedje, 2002).

Bacterial Respiration of Minerals

By attaching to and transforming minerals, microbes facilitate the weathering of rocks near the surface and perhaps even at depth (see MSA, 1997).

One of the most fascinating examples of microbial interactions with rocks is the use of minerals for respiration. Lower *et al.* (2001) used a modified atomic force microscope (AFM) to observe bacteria while they respire minerals.

Bacteria have been for billions of years respiring minerals. In respiration, energy is harvested by transferring electrons from an electron donor to an electron acceptor. Typically, this transfer occurs down a respiratory chain embedded in the cell membrane: Specific molecules pass on electrons from one end to the other, thereby generating a potential across the membrane that can *e.g.*, store chemical energy in the form of adenosine triphosphate (Newman, 2001).

For respiration to succeed, a terminal electron acceptor, such as oxygen, must exist to receive the electrons. Most terminal electron acceptors used by bacteria for respiration, such as oxygen, nitrate, and sulfate, are soluble. They can thus make their way to the cell to receive electrons from the membrane-bound molecules of the respiratory chain. But this route is not open to microbes that use solids like hematite (α-Fe_2O_3) and goethite (α-FeOOH) as electron acceptors because these minerals are effectively insoluble under environmentally relevant conditions (Morel and Hering, 1993). Several mechanisms have been proposed: (1) Bacteria may solubilize the minerals by producing chelating molecules. (2) They may use soluble shuttles, such as organic compounds with quinone moieties, to transfer electrons from the cell to the mineral. These shuttles may be either exogenous substances, or may be produced by the organisms themselves. (3) They directly transfer electrons from the cell surface to the mineral. A variety of biomolecules (including cytochromes, quinones, and dehydrogenasees) have been identified as part of this electron transfer pathway (see Newman, 2001). Of these, several are located on the outer membrane of the cell and presumably come in contact with the mineral directly.

Lower *et al.* (2001) reported the first quantitative measurements of the nanoscale interactions between *Shewanella oneidensis,* a well-studied mineral-respiring microorganism and two different minerals.

They measured the approach and retraction forces between an individual cell of *S. oneidensis* and goethite (α-FeOOH) or diaspore (α-AIOOH). Although these minerals have the same crystal structure, goethite is used by *S. oneidensis* as a terminal electron acceptor whereas diaspore is not. This is because the Fe(III) in goethite can receive an electron but Al(III) in diaspore cannot.

The affinity of *S. oneidensis* for goethite happens to be strongest under anoxygenic conditions for which electron transfer from the bacterium to the mineral is expected. Similar affinities are not observed for diaspore. According to Lower *et al.*, a 150-kD protein in the outer membrane of the cell specifically interacts with the goethite surface to facilitate electron transfer. This protein, along with others in the outer membrane of *S. oneidensis,* has been identified as a putative electron carrier to iron minerals. This exciting result opens up the possibility of using nanomechanical measurements to test biochemical and mineralogical hypotheses about what controls mineral respiration. Attempts are underway to combine nanoscale force measurements with molecular genetics and mineralogy, to find out which components of the electron transfer pathway in the cell are most important for direct electron transfer to minerals.

This is being done by knocking out genes thought to encode outer-membrane proteins involved in electron transfer, and then comparing the interactive forces between the mutant and a mineral to those between the wild type and the same mineral (see Newman, 2001). Detection of substantial differences in measurements would yield evidence for that particular protein's role in direct electron transfer to the mineral surface.

Once the components of the electron transfer system have been identified the next challenge will be to determine how the relevant proteins work. Are they similar to other electron transfer proteins that participate in different respiratory metabolisms? Which residues in the proteins are critical to electron transfer? Are the proteins used by one species more efficient than those used by another, and can this be correlated with their environmental niche? What structural properties make minerals good electron acceptors? Lower *et al.*'s work provides scientists with an exciting new technique with which to approach these questions.

Thermophilic Sulphate-reducing Bacteria and Mild Steel Corrosion

Some microbes and their metabolic activities cause what is known as microbiologically influenced corrosion (MIC) of metals and alloys. According to Beech and Sunner (2004), diverse bacteria, *e.g.* sulphate-, iron- and CO_2-reducing and sulphur-iron- and manganese-oxidizing bacteria, are associated with the MIC of metals and alloys. Among them, sulphate-reducing bacteria (SRB) represent a major group involved in anaerobic corrosion. These SRB can coexist in naturally occurring biofilms with a wide bacterial community, including fermentative bacteria and often form synergistic communities (consortia) that influence electrochemical processes through cooperative metabolism. The sulphate reductive activity of SRB may account for >75 per cent of the corrosion in productive oil wells, and for >50 per cent of the failures of buried pipelines and cables (Walch, 1992).

The strictly anaerobic SRB carry out terminal mineralization of organic matter in anoxic environments. These prokaryotes have been divided into four groups based on rRNA sequence analysis: Gram-negative mesophilic SRB, Gram-positive spore-forming SRB, thermophilic bacterial SRB, and thermophilic archaeal SRB (Castro *et al.*, 2000). Many SRB in oil-field environments are mesophilic microorganisms and grow optimally between 20°C and 40°C.

Miranda *et al.* (2005) studied the role of *Desulfotomaculum capillatus* isolated from an oil field separator in corrosion. The temperature of condenser tubes in cooling towers usually ranges from 50°C to 70°C. The incomplete description of thermophilic SRB species in industrial environments has greatly hindered the understanding of their diversity and their role in cycling of matter or nutrient materials. Although some studies have been done on thermophilic SRB and their role on corrosion process in heat exchangers (Villanueva *et al.*, 2006), the corrosion behaviour of thermophilic SRB in the Indian process industry has not been studied (see Anandkumar *et al.*, 2009). These authors have worked on SRB present in cooling towers/oil field environments to understand the mechanism of corrosion at high temperatures. They used sensitive and specific molecular techniques for the identification of SRB with their functional gene sequences and analysis of corrosion behaviour of *D. geothermicum* in mild steel corrosion.

The role of this species in mild steel corrosion was evaluated by electrochemical and surface analysis techniques. In the presence of *D. geothermicum*, the corrosion rate was higher as compared to control. The bacteria enhanced cathodic reaction and suppressed anodic reaction. X-ray diffraction studies revealed that the presence of FeS enhanced the cathodic reaction by the formation of a protective film on the metal surface. Pitting, detected by confocal microscopy, may be due to cathodic depolarization. The study suggests the importance of *D. geothermicum* in the corrosion of cooling towers of the petroleum refinery (see Anandkumar *et al.*, 2009).

Sanchez *et al.* (2001) studied the effect of some thermophilic SRB on pitting behaviour. Anandakumar *et al.* observed that the thermophilic strain of *D. geothermicum* can survive at 50ºC by forming spores. The vegetative forms probably develop and multiply when they encounter optimum temperature. Sulphate is converted to sulphide at 50ºC; at lower temperature, sulphide production was not detected. On the basis of critical experiments done on the effect of sulphate, and on lactate metabolism of SRB, Anandkumar *et al.* proposed a mechanism for the pitting corrosion by SRB on the metal surface, which is illustrated in Figure 11.1, with reference to the role of adenosine phospho sulphate reductase, dissimilatory sulphite reductase in sulphate reduction for energy metabolism with electron transfers, carbon source (lactate) metabolism for biomass production, role of hydrogenase in the production of H_2S and corrosive nature of H_2S on metal surface.

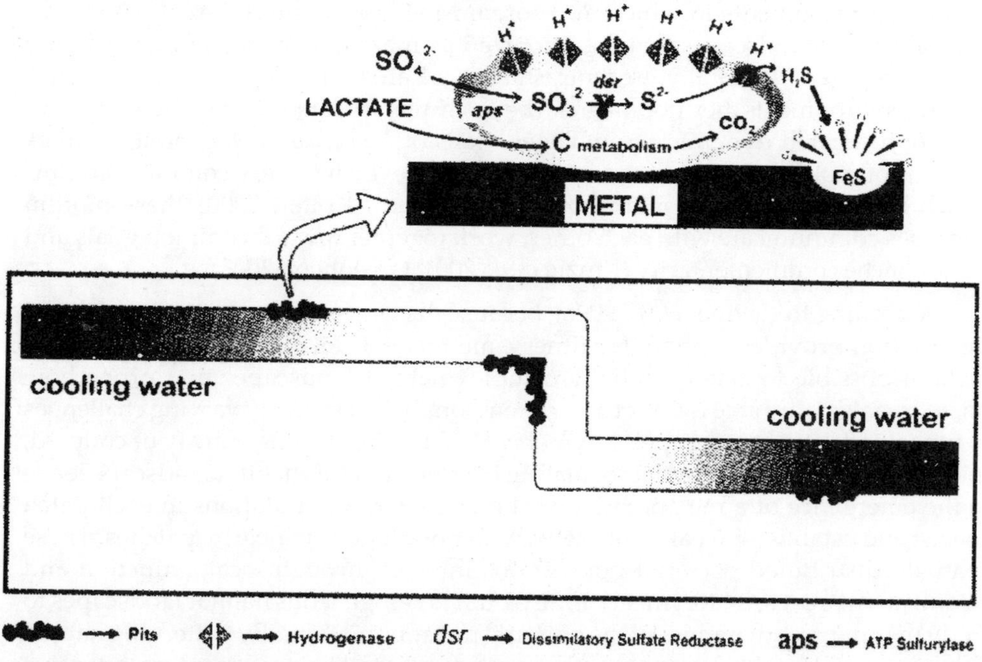

Figure 11.1: Pitting Corrosion Mechanism of SRB on Metal of Cooling Towers (After Anandkumar *et al.*, 2009).

Microbes have been boring into rocks since times immemorial, perhaps to avoid competitors or predators, to escape from environmental extremes or find a site that provides safety from turbulent waters. Another possibility is that they are mining minerals for essential nutrients. Some microbes that even tunnel towards a particular mineral, point to the occurrence of nutrient mining.

Walton (2008) described microscopic tubes in submarine glassy basalts from Hawaii that exhibit the highly complex features of microbial borings. The microbial boreholes converge on olivine microcrystals but strictly avoid plagioclase.

Olivine contains nickel, copper and chromium which are essential micronutrients for many microbes–these metals form the reactive centres in metalloenzymes and cofactors that catalyse important steps in vital metabolic pathways. As these are quite sensitive to oxygen and sulphides, their bioavailability probably changed when Earth's surface environment became more oxygenated and periodically more or less sulphidic. It appears that the microbes have been mining olivine for rare metals (Buick, 2008).

Bacteria in Nanofabricated Landscapes

Natural habitats are typically patchy and aggregate at several scales, creating a discrete habitat landscape (Levin, 1992). Such landscapes provide communities with a distribution of characteristic temporal and spatial scales which can be used to partition the landscapes (Tilman and Kareiva, 1997). A metapopulation (population of populations) generally develops over such habitat landscape, characterized by local population extinctions and colonizations.

The environments in which microorganisms live are no different. However, natural habitat landscapes that are colonized by macroorganisms cannot be easily approached experimentally. In contrast, bacteria are better experimental systems, and the environments they populate can be easily manipulated. Bacteria self-organize into sophisticated dynamic assemblages. Cells of *Escherichia coli* exhibit complex patterns of motility. Individual bacteria associate even into very complex biofilms which resemble a human metropolis (Watnick and Kolter, 2000)–these biofilm microbes communicate with each other, work together toward common goals and exploit niche complementarity (Kinzig *et al.*, 2001; Greenberg, 2003).

According to Keymer *et al.* (2006), because the landscape ecology of the habitat distribution provides a proxy for fitness, methods of micro- and nanofabrication make it possible to create spatially complex habitat landscapes that probe how microorganisms are able to adapt to both temporally and spatially varying challenges to fitness (adaptive landscape). Keymer *et al.* (2006) built a linear array of coupled, microscale patches of habitat. Inoculating bacteria into this habitat landscape leads to the emergence of a metapopulation. Local bacterial populations in each patch coexist and establish a weak connection with populations in nearby patches. These spatially distributed bacterial populations interact through local extinction and colonization processes. Keymer *et al.* also built heterogeneous habitat landscapes to study the adaptive dynamics of the bacterial metapopulations. By patterning habitat differences across the landscape, they could physically implement an adaptive landscape. In landscapes with higher niche diversity, rapid adaptation to large-scale, low-quality (high-stress) areas was observed. These results clearly reveal the potential lying at the interface between nanoscale biophysics and landscape evolutionary ecology (Keymer *et al.*, 2006).

Establishing effective collaborations between landscape ecologists and experimental (nanoscale) physicists may conceivably prove helpful in designing more complex (temporal and spatial) gradients and possibly realize the cherished wish of many biologists to achieve directed evolution of microbial populations.

Polythioesters

Polythioesters (PTEs) are a recently discovered group of chemicals formed by certain bacteria when fed with particular nutrients. Their potential importance stems from their high compatibility with human body tissues–they are less likely to trigger adverse reactions than plastics commonly used for surgical implants. They also seem to be less prone to infection by pathogenic microbes.

PTEs were discovered when it was noticed (see Steinbuchel, 2001) that the bacterium *Ralstonia eutropha*, when fed with nutrients containing the acid thiodipropionate (TDP), form a polymer containing sulfur in its backbone. The sulfur gives the polymer antibacterial properties while the polymer itself is highly compatible with organic tissues.

Ability to produce these unusual polymers containing sulfur means that we may soon have surgical implants that are able to prevent bacteria from growing and causing infection.

This would be highly desirable since bacteria grow easily on plastics used at present. Infections, especially by some antibiotic-resistant superbugs found in hospitals, are often difficult to treat. Furthermore, the human body reacts to current implants, such as catheters and heart valves, as alien objects and, though not rejecting them like some organ implants, triggers an inflammatory response which can make the patient feel ill and reduce the life of the implant. Problems like these can increase costs of operations such as hip replacements tenfold where hospitalization is prolonged.

Fishing for Microbes

Chlorinated ethenes and ethanes are some of the commonest industrial contaminants of soils and groundwaters all over the world (EPA, 2002). Most are either suspected or known to be carcinogens. Sun *et al.* (2002) discovered a bacterium that derives energy by degrading 1,1,1-trichloroethane (TCA). This anaerobic microbe uses TCA as a respiratory electron acceptor to oxidize molecular hydrogen for energy production, reducing TCA in the process.

This bacterium is the latest in an ever-growing list of bacteria that eke out a living by reducing chlorinated two-carbon solvents. This bacterium is the first microbe found to degrade TCA (Gossett, 2002).

In the last decade, many chloroethene–and chloroethane-respiring anaerobic bacteria have been isolated (for example, see Maymo-Gatell *et al.*, 1997); also new aerobes that derive energy through oxidation of chlorinated aliphatic solvents using oxygen as an electron acceptor (Gossett, 2002).

Characterization of bacterial isolates is being aided by molecular biology which enhances our understanding of degradation genes and their expression, degradation enzymes and their regulation, and the evolution and adaptation of contaminant-degrading microbes. Many contaminant- degrading bacteria have something else in common with Sun *et al.'s* microbe–they produce energy for growth only through dehalorespiration. An interesting but puzzling question is, how did these bacteria survive before around 1950, when chlorinated solvents largely replaced supposedly "harmful" hydrocarbon solvents as cleaners and degreasers?

Isolation of new bacteria leads to better understanding of how sub-surface bioremediation works. Studying the degradation enzymes of these microbes and their regulation yields new insights into bioremediation.

The failure of all but a few microbes to be grown in culture may reflect unsuitable conditions, or the complex but obligate interdependence of coinhabiting species in soil. The search for new contaminant-degrading bacteria sometimes *begins* with a molecular fishing expedition: Genetic material is extracted from contaminated sites and analyzed to see which categories (guilds) of organisms are found. With luck, a scientist may be able to isolate new microbes based on the hints provided by molecular biology. But even if a new microbe cannot be isolated, molecular techniques allow the study of bioremediation in mixed microbial cultures. It has even become possible to look for and study enzymes without ever isolating the organisms that produce them (Gossett 2002). In the near future, it should be possible to use array-based probes containing thousands of markers to detect contaminant-degrading bacteria, the ones that help them or compete with them, and to determine whether they express degradation enzymes. This kind of information will enable us to make informed decisions about remedial options, leading to less-expensive site-characterization and more-reliable bioremediation.

The death-defying bacterium *Deinococcus radiodurans* is virtually indestructible. It shows extremely high resistance' tolerance to high energy radiation. It can survive exposure to as heavy a dose as 1.5 million rads of gamma rays, which is about 3000 times the dose that kills humans!

Some extremophilic microorganisms can die by dehydration in highly saline media. *Halobium* lives happily. *D. radiodurans* dehydrates but does not die even when so dehydrated! *D. radiodurans* was described in 1956 by Dr. Arthur Anderson of the Oregon Agricultural Experimental Station at Corvallis, Oregon, U.S.A. Its tenacity is indeed terrible. It is resistant to most mutagenic chemicals which inactivate the genes of other organisms; it eats toxic wastes, survives dehydration, can withstand vacuum, and has been suggested as an organisms that might help colonise Mars! Military researchers are trying to determine whether it can be used to clean up nuclear wastes.

High doses of high-energy short-wavelength radiation are harmful to all life forms. High energy ultraviolet rays can scald the skin in no time. X-rays are more damaging and the ultrahigh energy gamma rays are the most lethal. Radioactive sources also emit alpha and beta rays that ionize materials and destroy them.

Radiation-induced damage can be in the form of breaking chemical bonds so that the molecule is destroyed and so loses its function or activity. It can also generate free radicals (reactive atoms) that, in turn, damage molecules and render them ineffective. Reactive oxygen species (ROS) *e.g.*, cause havoc in the form of oxidative damage. If such damage occurs to the proteins, fats, and other components of the cell, they could be removed and replaced by newly synthesized ones. Such turnover control exists in the genes in most cells. But if the genes or the DNA itself gets damaged, the consequences can be deadly. Most cells have evolved to have built-in preventive mechanisms and repair tools. DNA damage can be repaired by certain enzymes. One of these scans the length of the strand and alerts on where damage is done. Another set clips off the damaged part and stitches in the right parts. Prevention is always better than treatment and can be achieved by means of antioxidant enzymes and molecules which sense the presence of ROS, grabbing and neutralizing them. Yet another strategy is to have redundancy of information *i.e.*, if one gene is damaged, another identical one held in store, somewhat like the spare tyre in the car, can operate. *D. radiodurans* appears to use all these mechanisms, and possibly some others as well. It has one of the highest activities of the protectant enzyme catalase, reported for any bacteria; this can remove a particular form of ROS. Secondly, it maintains as many as eight copies of its genome per cell, so that if its genetic information contained in one is damaged, there are back-up copies, which can function as the damage is getting repaired.

DNA repair occurs fairly efficiently in the bacterium¾*D. radiodurans* accumulates particularly high amounts of manganese in its cells, and low levels of iron. The preferential manganese enrichment over iron is particularly advantageous. *D. radiodurans* is not unique in doing so; some bacteria that colonise our bodies (*Enterococcus, Lactobacillus*) also do so. Manganese is essential for the cell to overcome oxidative stress. It forms part of the enzyme superoxide dismutase, which detoxifies one type of ROS (catalase, the other enzyme the *D. radiodurans* is rich in, gets rid of another type). Manganese is also safer than iron in another respect: while iron can itself generate ROS, manganese is not known to do so.

Microbial Ecology for Management of Environmental Biotechnology Processes

There exists an intimate connection between microbial ecology and environmental biotechnology. While microbial ecology forms the scientific foundation for the processes used to achieve the practical goals of environmental biotechnology, processes in environmental biotechnology provide interesting ecosystems which microbial ecologists study to advance their concepts and methods. Indeed, environmental biotechnologists usually apply the concepts and tools of microbial ecology to manage their processes. Whereas the concepts and tools of microbial ecology form the basis for managing processes in environmental biotechnology, these processes in turn provide interesting ecosystems to nurture the concepts and tools of microbial ecology. Exciting advancements in molecular tools for revealing the structure and function of microbial communities enhance the power of microbial ecology. A push from current researches in modern materials and a pull from a societal need for greater sustainability enable environmental biotechnology to create novel processes. Five principles illuminate the manner how the two fields work together: (i) aim for big benefits; (ii) develop and apply stronger tools to understand microbial communities; (iii) follow the electrons; (iv) retain slow-growing biomass; and (v) integrate thoroughly.

Among the above five, 'follow the electrons' is particularly important because microbially catalyzed reactions are usually oxidations and reductions. Thus, following the electrons as they move through the microbial ecosystem effectively translates knowledge about the structure and function of the microbial community into practice.

Services Provided by Environmental Biotechnology

Microbial communities are nothing but self-organizing, self-sustaining assemblages of different taxa of microorganisms. When properly managed, they provide several different services reliably, continuously, and economically (Rittmann, 2006). Some of the services are listed below.

- ✰ Detoxification of contaminated water, wastewater, sludge, sediment, or soil;
- ✰ Capture of energy, water and other renewable resources;
- ✰ Sensing contaminants or pathogens in the environment; and
- ✰ Protecting the public from risks of exposure to pathogens.

In fact, we can define environmental biotechnology as managing microbial communities to provide services to society. Processes that depend on microbial communities have been known for over a century (Rittmann and McCarty, 1999). Two better known processes are activated-sludge treatment of wastewater and anaerobic digestion of sludge; both these have been widely used long before their microbiological bases were understood. However, great improvements became possible after the microbiological and ecological principles were recognized. The best example is biological nutrient removal (BNR)–cycling the microbial community through a series of aerobic, anoxic, and anaerobic stages enables complete removal of nitrogen (N) and phosphorus (P) from wastewater. The different stages can be used to select for three distinct groups of bacteria that can oxidize NH_3-N to NO_3-N, reduce NO_3-N to N_2 gas, and store extra PO_4-P, respectively (Rittmann, 2006).

Certain modern materials are well suited to the development of new and better processes. In 1970s, the availability of lightweight, high-strength plastics made biological towers possible: biological towers represented the first biofilm processes with high surface areas and small footprints. The decades of the 1980s and 1990s saw lightweight biofilm carriers in the form of gravel-sized pellets, making even more compact 'high-rate processes' possible. Today, micro-filtration membranes are used for converting activated sludge systems to membrane bioreactors (MBRs), which improve effluent quality and are much more reliable and compact (Daigger *et al.*, 2005). The more recent use of the membrane biofilm reactor (MBfR) has allowed utilizing the multiple advantages of hydrogen gas as an electron donor to reduce nitrate, perchlorate, and several other oxidized contaminants in drinking water, ground water, and wastewater (Nerenberg and Rittmann, 2004; Rittmann, 2006a).

Microorganisms often live in close contact with the materials as biofilms which tend to self-assemble to provide the community optimal access to substrates besides protecting them from toxicity, predation, desiccation, and wash out. By establishing gradients of substrates, biofilms also create specialized niches where microorganisms with different, apparently incompatible metabolic functions can co-exist in the same biofilm. One good example is the hydrogen-based MBfR for reducing several oxidized contaminants at the same time. Another very exciting example is the microbial fuel cell (MFC). This is used to generate electricity from organics: the bacteria live attached to an anode and transfer electrons to the anode instead of directly to a soluble electron acceptor (Kim *et al.*, 2002; Liu *et al.*, 2004).

In an MFC, the organic fuel, *e.g.*, glucose, is oxidized by bacteria living as a biofilm on the anode. The bacteria transfer the electrons to the anode, which collects the electrons. The electrons move through the electrical circuit and reduce O_2 to H_2O at the cathode. The corresponding protons (H^+) move to the cathode by another route, *e.g.* by using a polymer electrolyte membrane. While the net reaction is $C_6H_{12}O_6 + 6O_2$ ®$6CO_2 + 6H_2O$, the oxidation and reduction reactions occur at different electrodes, creating electrical energy. One notable merit of an MFC over a conventional fuel cell is that organic fuels, such as $C_6H_{12}O_6$, can be used directly, allowing direct use of renewable organic materials such as wastes, as fuels to generate electrical energy (Rittmann, 2006).

The Scientific Foundation of Microbial Ecology

Environmental biotechnology rests on the foundation of microbial ecology which aims to understand microbial communities and how they interact with their environment. Microbial ecology focuses on four core issues:

- ✰ The structure of the community (*i.e.*, the microorganisms present in the community).
- ✰ The phenotypic potential of the community (the capabilities of the microorganisms for carrying out reactions).
- ✰ The function of the community (the reactions that the community members actually perform).
- ✰ The inter-relationships among the individuals of the community and between it and their environment.

Some recent techniques such as selective, rapid amplification of DNA by means of the polymerase chain reaction (PCR) and hybridization with DNA oligonucleotides have enabled ecologists to directly interrogate the genetic information of individual microorganisms and entire communities (Lane *et al.*, 1985). Whereas the small sub-unit rRNA (SSU rRNA) yields information on the phylogenetic identity of the microorganisms, amplifying and detecting specific genes in the chromosome catalogues the phenotypic potential (Marsh, 1999; Muyzer, 1999), and the mRNA shows which genes are being expressed and, hence, which functions are being performed (Freeman *et al.*, 1999). Similarly, microscopic visualization of RNA targets with fluorescence in situ hybridization (FISH) facilitates exploration of spatial organization (Amann *et al.*, 2001). Selected examples of some common genomics tools employed in molecular microbial ecology include hybridization after RNA extraction (target RNA); FISH (fluorescence in situ hybridization) (target rRNA); DGGE (denaturing gradient gel electrophoresis) (for specific gases); T-RFLP (terminal restriction fragment length polymorphism) (for specific genes); cloning and sequencing (for specific genes); and RT-PCR (reverse-transcription polymerase chain reaction) (for mRNA).

Biomass is a large storehouse for electrons, particularly for high-energy reactions such as the aerobic oxidation of glucose; but it is a small storehouse for low-energy reactions, such as methanogenesis.

Biochemical oxygen demand (BOD), the well known water pollutant is simply a misplaced electron donor. When BOD enters receiving water, bacteria oxidize it, consuming dissolved oxygen in stoichiometric proportion. If too much BOD is oxidized, dissolved oxygen depletes, leading to serious water-quality problems, including odor, color, fish mortality, and mobilization of metals and nutrients (Rittmann, 2006).

Oxidized contaminants perform an opposite role to BOD because they are electron acceptors. For removing oxidized contaminants, electrons have to be transferred to them in proportion to the change in oxidation state from the oxidized pollutant to a reduced form which is harmless (Rittmann *et al.*, 2004). Oxidized contaminants can be detoxified by supplying a bioavailable electron donor, and hydrogen is the most universal donor for microbial reductions.

Since electrons carry the energy value of organic matter, energy value of BOD can be recovered by transferring them from complex organic matter to CH_4. Each 64 g of BOD contains eight electron equivalents, *i.e.*, the requirement to generate one mole of CH_4. Combusting CH_4 to generate electricity captures ~33 per cent of the energy content of the original BOD.

Moving the electrons to bioethanol or biohydrogen is particularly important: the bioethanol can be used as a gasoline fuel additive to reduce emissions that produce smog, and the biohydrogen may be used to power a conventional fuel cell (Logan, 2004)–64 g BOD can generate four moles of hydrogen because each hydrogen molecule has two electron equivalents: $H_2 \rightarrow 2H^+ + 2e^-$.

A traditional fuel cell captures up to about 50 per cent of the energy content of the BOD, which is routed first to hydrogen and then to the anode.

An attractive option today for energy capture from BOD is the microbial fuel cell (MFC). This can potentially generate electricity directly from the oxidation of organic compounds provided the bacteria transfer the electrons to an anode and not directly to a traditional electron acceptor. It offers the greatest potential for renewable fuels (Rittmann, 2006).

The Microbial Biodegradation and Biotransformation Gene Pool

According to Galvao *et al.* (2005), only 0.1-1 per cent of microorganisms can be cultivated by using current techniques and therefore very little is known about the enormous range of microbial lifestyles. Venter *et al.* (2004) stated that even though the current rate of discovery of new genes in environmental DNA might exceed one million per year, few, if any, new biochemical reactions have been found since the late 1970s (Dong *et al.*, 2004). Habitats polluted with toxic chemicals and industrial wastes are veritable biotechnological gold mines because they include special niches for microorganisms that have evolved the necessary enzymes to use these exotic, frequently xenobiotic compounds as their carbon and energy source (Galvao *et al.*, 2005). Microbes perform enzymatic reactions that are uncommon in other domains of life but which, when exploited, can add much value to selected substrates. Indeed, microbial biodegradative pathways for recalcitrant chemicals are both highly appealing for remediation of environmental pollution and also rich sources of novel catalytic activities useful in green chemistry and white biotechnologies (see Timmis and Pieper, 1999; Schmid *et al.*, 2001). It would be no exaggeration to say that the processes of biodegradation and biotransformation reflect the unlimited power of microbial biochemistry.

The microbial metagenome being the largest reservoir of genes that determine enzymatic reactions, novel and robust techniques are being developed to identify the genes that underlie many valuable chemical biotransformations carried out by microbes, particularly in pathways for biodegradation of recalcitrant and xenobiotic molecules. It is on the basis of the knowledge of catabolic pathways built on research during the past few decades that we can attempt to explore and identify some little-explored pathways that exist in nature.

The two most complete databases of microbial biocatalytic reactions on pollutant compounds (Ellis *et al.*, 2003; Pazos *et al.*, 2005) list about 140 pathways, 930 reactions, 880 compounds and intermediates, 590 enzymes, 340 microorganism entries, 250 biotransformation rules and 50 organic functional groups. Because these data are based mostly on work done on pure cultures by using the relatively small number of fast-growing microorganisms that are easily cultured, uncultured microbes are still largely an unexplored reservoir of enzymes for pollutant biodegradation. Today, molecular techniques enable researchers to directly study contaminated sites, so that information about environmentally relevant reactions can be gathered irrespective of whether the hosts can or cannot be cultured. Some methods for identifying the enzymatic diversity of specific microbial consortia in polluted sites are sequence-dependent and sequence-independent approximations, the latter involving genetic traps for revealing specific reactions from amongst otherwise highly complex biodegradation gene pools.

Many modern techniques to identify genes for given catalysts in microbial communities are based on sequence similarity to known enzymes; in these, some information on the amino acid sequence of the pursued protein or on the corresponding DNA is needed before looking for new variants. Applicable techniques in this context include fluorescence *in situ* hybridization (FISH) of radioactive or fluorescent oligonucleotide probes, straight PCR of environmental DNA (or Real Time RT-PCR of mRNA) for amplifying target sequences, including monitoring their levels of presence or expression (see Liang, 2002) and *in situ* RT-PCR to the same end (Galvao *et al.*, 2005). The use of DNA-dependent procedures (*e.g.* those based on PCR or oligonucleotides) facilitates simultaneous environmental monitoring for the presence and activity of genes for biodegradative enzymes (Galvao *et al.*, 2005).

Fromin *et al.* (2002) reviewed the DNA fingerprinting methods based on gel denaturing gradients or restriction polymorphisms denaturing gradient gel electrophoresis (DGGE), temperature gradient gel electrophoresis (TGGE), amplified ribosomal intergenic spacer analysis (ARISA), and restriction fragment length polymorphism (RFLP) that have been extensively used to map the phylogenetic distribution of complex communities based on rRNA gene sequences, but their application for mapping catabolic gene sequence diversity is still in infancy. Fingerprinting of catabolic genes can group sequences in a manner which reveals classes of substrate specificity and/or enzyme processivity as an alternative to sequencing of full genes. By means of appropriate statistical analysis, fingerprinting of catabolic genes (amplified from environmental DNA), similar to that of rRNA genes could conceivably become a robust tool for the systematic study of gene/transcript sequence and abundance in heterogeneous conditions (see Junca and Pieper, 2003).

Another class of approaches finds environmental DNA segments that are candidates to encode biocatylysts in a sequence-independent manner. Examples are cloning of genes expressed only in conditions of induction (differential display), direct detection of new reactions (optimally, associated with a color change in the substrate, density enrichment of the genomes of those members of the community which metabolize a certain substrate labeled with ^{13}C [stable isotope probing (SIP)] (see Wackett, 2004) and genetic traps to translate the desired enzymatic reaction into a selectable or scorable property of the surrogate host of environmental DNA (Galvao *et al.*, 2005).

Gene Traps and Metagenomic Libraries

Metagenomic libraries are built from total environmental DNA and facilitate archiving of the full genetic complement of a specified, usually complex, microbial community. These libraries represent the most complete environmental sequence reservoir available for biotechnological applications, coding for synthesis of pharmacological compounds or enzymes with catalytic activities of industrial benefit (Daniel, 2004). They enable culture-independent screening of novel activities, often without any sequence knowledge. Yet, extraction of high-quality DNA from harsh environmental scenarios continues to be quite difficult because of biases in cell recovery and owing to inefficient enzymatic manipulations because of contamination with humic acids. Despite these problems, novel methods for recovery of intact nucleic acids from natural sources are being developed; the construction of an mRNA library from environmental samples was reported by Poretsky *et al.* (2005).

DNA from environmental samples is normally cloned into plasmids, bacterial artificial chromosomes (BACs), or cosmids, and is propagated in a host such as *E. coli* which can then be screened for the desired enzymatic activity. Some organisms being used as library hosts are *Streptomyces* and *Pseudomonas* (Wang *et al.*, 2000; Martinez *et al.*, 2004). Successful screening of metagenomic libraries depends on the following important parameters: (i) the abundance of a gene in the library; (ii) the relationship between insert size and gene or operon length; (iii) an expression system that might include a vector with its own inducible strong promoter, and occasionally code for the environmental sequence as a transcriptional or translational fusion; (iv) actors in *trans* such as inducers, chaperones, cofactors, post-translational modifications, secretion and transcription factors; and (v) optimization of plating assays and/or use of trait-deficient strains for screening (Galvao *et al.*, 2005).

Brisk research activity using the several new techniques for understanding environmental microbiology is uncovering the remarkable variety of sequences relevant in biodegradation and biotransformation but also conditions in which these processes occur in their natural habitats. Strategies aimed at characterizing microbial communities in the environment, whether by large-scale sequencing of metagenomes or by exploring biodegradative potential by mRNA-type techniques, may go hand in hand with pure culture systems for critical functional studies of biodegradative functions, to provide information about microbial enzymatic specificity and processivity (Galvao *et al.*, 2005).

Chapter 12
Microbiology from the Bottom Up

Introduction

W. Szybalski first coined the term synthetic biology in 1974 to describe the modification of organisms by adding and subtracting genes. Three decade-long experience has shown that the risks involved in genetic modification of organisms are negligible whereas the rewards can be enormous. Today, the field of synthetic biology is expanding. In some laboratories, bacteria are being created to produce pharmaceutical intermediates. Craig Venter, innovater of contemporary genomics, has even proposed reorganization of the natural parts of natural genomes. Some of these restructured microbes are so scrambled that they may be considered as new species.

The remit of synthetic biology has also widened greatly. Eric Kool of Stanford University, California, defined synthetic biology as the construction by chemists of unnatural molecules that can operate within natural living systems. Drew Endy of the Massachusetts Institute of Technology considers it to mean the process of creating, mostly by modifying existing biomolecules, units that can function as interchangeable parts in larger assemblies. To Stephen Wolfram and others "artificial life" is a computer programme whose output behaviour is analogous to the behaviour of living systems (see Benner, 2008).

The synthetic approach to biology involves intentional creation of new forms of life from the bottom up rather than the top down. It generates new possibilities to test nature. Building something requires a deep understanding of its parts. Only rarely are data collected neutrally during analyses by researchers–some data may be discarded in the belief that they are wrong if they do not meet their expectations. Synthesis helps manage this trouble.

Creation of synthetic genetic systems can enhance understanding of how natural genetic systems work; by attempting to create synthetic metabolisms, we could learn more about how natural metabolisms work; by creating synthetic regulatory circuits, we learn about how natural regulatory circuits work (Regis, 2008; Benner, 2008).

Tinkering with Biological Systems

For understanding any system one needs to disturb it to see how the disturbance affects the system. A proper understanding of biological systems has been facilitated by suitable tools that may be employed to perturb them (Stockwell, 2004). Molecular genetics enables researchers to eliminate specific proteins by 'knocking out' genes; to increase the concentrations of specific proteins either by increasing the number of copies of the corresponding genes or by means of a more active promoter on such genes; or to alter the function of a protein by introducing specific mutations in the corresponding gene (Hartwell, 1991).

These methods have been applied successfully in model organisms such as *Saccharomyces cerevisiae* and *Drosophila melanogaster* but not in mammals; the latter cannot be studied by using genetic-screening approaches because of their slow reproduction, large physical size and large genomes. For mammals, small organic molecules are used instead of mutations. This so-called chemical genetics approach reveals the molecular mechanisms underlying biological processes (Stockwell, 2000; Schreiber, 2003). Small molecules alter protein functions by binding to them and inhibiting or activating their normal functions, thereby perturbing living systems and revealing the molecular 'wiring diagrams' of these systems.

Employing small molecules complements gene-based methods of perturbing protein function, and is sometimes superior to such methods. If a protein has several functions in a cell, a deletion mutation can eliminate all these functions. But certain small molecules perturb only one of several functions of a protein, so providing valuable information about protein function that is not obtainable through gene-based perturbation. Another merit of small molecules is the ease in exerting temporal control of protein function because they can be added to induce some effect and then washed away to return a cell to its wild-type state. Although most small molecules are not drugs, the occasional development of a small molecule into a drug motivates researchers to use small-molecules as tools to study biology (Stockwell, 2004).

To realize the full potential of chemical genetics, collections of small molecules suited to modulating the functions of many different proteins need to be created. Usually, each protein class requires a different type of small-molecule modulator; also it is necessary to analyze both protein binding of, and phenotypic responses to, small molecules. For this, we need to design synthetic chemicals, perform protein-binding and phenotypic assays, and ensure quality control so as to make a comprehensive map of biological activity space using small molecules. Some other complementary systematic approaches are the use of RNA interference (RNAi), in which synthetic RNA fragments interfere with the expression of specific genes, or antibodies (Hannon, 2002; Moore and Clayton, 2003).

For some proteins, no ligand is known and screening of chemical libraries has to be undertaken for possible identification of compounds that bind to the protein with reasonable affinity. Two kinds of complementary approaches are applied: experimental (usually high-throughput) screening, and structure-based virtual screening (Stockwell, 2004). In some experimental screening, the protein is expressed, purified; and used in a high-throughput screen to detect small molecules that bind to it. In structure-based virtual screening X-ray crystallography or nuclear magnetic resonance (NMR) spectroscopy is used to obtain an atomic resolution structure of the protein. This protein structure is used in a computer-based experiment to detect small molecules predicted to bind to the protein. By means of certain programs, millions of compounds may be examined *in silico* for their ability to interact with the target protein, and the relative fit of each candidate is scored (Obsterberg *et al.*, 2002; Oprea and Matter, 2004). In a high-throughput screen, several different chemicals or test reagents can be tested in the same biological test for their effects on a protein or cellular process. The term 'screen' indicates that many different chemicals are tested but only a few of them may show activity. The term 'high-throughput' suggests that many chemicals are evaluated through this process in a short period of time.

In the pharmaceutical industry, virtual screening is widely used as a 'filter' to select compounds from very large virtual libraries for experimental screening. This filtering process can use various types of information (*e.g.*, the crystal structures of the protein itself), so as to enrich the library that is experimentally screened with 'active' structures. Computational filters may also be used to eliminate compounds showing inappropriate properties from the screening library (Bajorath, 2004).

Effective chemical libraries for chemical genetics should have compounds which can affect specific proteins and phenotypes while not affecting other closely related proteins and phenotypes. The design of highly effective libraries is facilitated by assessments of the specificity and diversity of existing libraries, and of each new chemical library that is designed and synthesized. Optimal libraries for a given purpose may be rationally assembled from members of other libraries (Stockwell, 2004).

Protein-binding Assays

The ability of small molecules to bind to specific proteins can be measured by employing a variety of different methods (see Salemme, 2003). Some of these protein-binding assays use labelled compounds whereas others are label-free. A label means a fluorescent or radioactive group added to a test compound. There are several methods available to measure the binding of a small molecule to a protein or to some related proteins but few, if any, methods can systematically measure the binding of small molecules to hundreds or thousands of proteins. Such high-throughput protein-binding measurements are indispensable for detecting the range of activities shown by small molecules. A few examples of methods involving the use of labelled compounds are fluorescence polarization, fluorescence correlation spectroscopy, small molecule microassay, protein microassay, and radioligand binding assay. Label-free methods are exemplified by fluorescence perturbation, nuclear magnetic resonance, mass spectrometry and differential scanning calorimetry (see MacBeath *et al.*, 1999; Timasheff *et al.*, 1991; Hicks, 2001; Burke *et al.*, 2003; Stockwell, 2004).

Metabolic Engineering–First Autonomous Synthetic Life Form

All living species arise from the same five nucleotides that form the building blocks for DNA and RNA, and 20 amino acids that serve as building blocks for proteins. (At least two additional natural amino acids are made by a few odd species). These ingredients limit the chemical reactions that can occur inside cells and so constrain what life can do. This constraint was relieved in 2001, when Lei Wang, Peter G. Schultz and others at the Scripps Research Institute in La Jolla, Calif., successfully added to *Escherichia coli* all the genetic components the cells need to decode the three-nucleotide DNA sequence TAG into unnatural amino acids of various kinds.

That made the modified bacterium the first synthetic life form with a chemistry unlike anything found in nature. To make proteins with unnatural amino acids, however, the new amino acid had to be incorporated into the bacterium's growth medium. Since then, Schultz's team has gone one step further by engineering an *E. coli* that not only incorporates a 21st amino acid into its makeup but also manufactures the compound by itself (see Service, 2003). Amino acids by themselves have relatively few uses. The chief objective to modify cells so that they not only synthesize artificial amino acids but also string them together with natural amino acids is to make proteins that no other form of life can make. Schultz's group has actually done just that with *E. coli* and more recently with yeast.

These exciting results provide a novel means of exploring evolution. One may ask how can this organism evolve, now that it is able not only to use a nonnatural building block but also for creating that building block.

For enabling *E. coli* to expand its standard amino acid repertoire, Schultz had to reengineer the basic machinery that cells use to make proteins. That machinery starts with codon triplets of nucleotide bases that make up DNA. Each codon directs a particular amino acid to be added to a protein chain. To make its first synthetic bug, the research team utilized the little-used *amber* stop codon and reengineered the cellular enzymes to add a new amino acid, *O*-methyl-L-tyrosine, whenever it saw that codon.

In their latest work, they followed the same strategy with a new amino acid, *p*-aminophenylalanine. To enable the microbes to make the new amino acid on their own, they also spliced a trio of enzyme-producing genes from a strain of *Streptomyces* into the *E. coli* called chorismate into *p*-aminophenylalanine, which protein-building enzymes then picked up and inserted into growing proteins (see Service, 2003).

To find out what difference the altered chemistry makes, Schultz is randomly inserting amber stop codon mutations into the *E. coli* genome. Next, they intend to put these into *E. coli* and the 20-amino-acid variety under selective pressure by changing their food supply and other factors, to know whether the 21 amino acid bugs fare better or worse than natural ones.

Six kinds of unnatural amino acids have been produced in yeast. The scientists intend to adapt the systems to work in human kidney cells and in roundworms. Attempts are underway in designing a system that can make two different unnatural amino acids and put them in the same protein-a formidable task because in order to do that, the cell has to decode a four-nucleotide DNA sequence, which no cell has ever done.

Such an advance could generate tremendous biomedical potential : white blood cells could make novel proteins to destroy pathogens or cancerous cells more quickly. The technology is already producing new research tools, such as proteins that include fluorescent amino acids or that change behaviour upon exposure to light. It allows researchers to attach polymers to therapeutic proteins, so as to enhance their action as drugs (see Gibbs, 2004).

Molecular Prodigality

In living cells, macromolecules, especially proteins, occur in many forms but we do not know how many of these molecular variants are accidental products of RNA splicing and posttranslational modification, and how many are genetically designed and of evolutionary significance (Bray, 2003). *Escherichia coli* uses a receptor (Tar) to detect temperature, and the same protein–the chemotaxis receptor–to detect the amino acid aspartate. Even the sense of the temperature response depends upon the concentration of aspartate in the environment. When the medium lacks this amino acid, Tar acts as a detector of warmth and *E. coli* swims toward a distant source of heat. But if the ambient medium contains dissolved aspartate, then Tar becomes a cold detector and the bacterium swims away from heat. At the molecular level this change in behaviour involves methylation of the cytoplasmic domains of the Tar protein (Bray, 2003). The receptor becomes highly methylated in the course of its adaptation to aspartate; this seems to change the conformational response of the receptor to a rise in temperature.

Besides Tar, *E. coli* can form three other types of methyl-accepting receptors. Between them, they mediate attractant and repellent responses to perhaps 50 distinct chemicals, as well as to pH and temperature. The receptors are multifunctional, cluster together on the cell surface, and influence each other's responses (Ames *et al.*, 2002). This means that the aspartate temperature system could be applied to many other combinatorial pairs, such as serine and aspartate, or ribose and pH. Thousands of transmembrane chemotaxis receptors of *E. coli* aggregate together in the plasma membrane by binding to downstream signaling proteins.

The clusters of *E. coli* chemotaxis receptors are highly complex–each receptor can be methylated at up to eight positions and can adopt at least two possible conformations. Within a cluster, thousands of receptors, of four different types, can interact with each other in sets of three. So, the number of distinct receptor trimers–each having a different response to multiple environmental influences–is extremely large.

Many other proteins *e.g.* of the mammalian myofibril are found in multiple forms as a consequence of alternative RNA splicing. Indeed, troponin T, which affects the calcium sensitivity of muscle, exists in over 80 different forms; these forms can change their relative abundance depending on the type of muscle, growth, and exercise. Another interesting case is that of the histone proteins in chromatin, whose amino-terminal tails become altered at multiple sites by acetylation, phosphorylation, or methylation. The enzymes responsible for these modifications are specific for particular amino acid positions and are themselves influenced by a confluence of signaling pathways and by the local concentrations of other modified histones. In fact, these expensive modifications have prompted the suggestion that they constitute a "histone code" that extends the information contained in the DNA sequence itself (Jenuwein and Allis, 2001).

Many structural genes in humans may be alternatively spliced, and it appears that one-half of all proteins are modified by phosphorylation. The brain appears to be a particularly rich source of molecular diversity. Humans have just three genes that code for neuronal cell surface proteins called neurexins but through alternative promoters and alternative splicing, the three genes yield thousands of different isoforms expressed in different combinations on the surfaces of different neuronal cell types (see Bray, 2003). When considered along with various other sources of molecular variation, it is likely that each cell in an organism–even the estimated 10^{11} nerve cells of the human nervous system–is chemically unique (see Bray, 2003).

While most variations in protein structure arise from chance encounters between enzymes and their substrates, other variations may be caused by such environmental influences as temperature, stress, and the instantaneous concentrations of ions and small molecules. There also must be a genetic component underlying these structural variations. Each gene, either by itself or together with some others can elaborate a complex pattern of protein products in the same way that computer simulations generate highly complex patterns by the repeated application of a few simple rules (see Wolfram, 2002).

In the immune system, variations in protein structure are manifestly functional. Mammals make millions of lymphocytes, each of which makes a distinct type of randomly generated antibody. Clonal selection proliferates those few lymphocytes that recognize a specific antigen; the remaining lymphocytes die. Antibody diversity is generated by the combinatorial joining of gene segments, random variations during gene segment joining, combinatorial joining of antibody light and heavy chains, and somatic hypermutation in the antibody gene. This basket of mechanisms allow a human being to create some 10^{12} different antibody molecules. The great success of this strategy raises the question of whether analogues of this strategy might also exist in nonimmune processes. Conceivably, there may exist mechanisms of protein selection in cells that enhance some protein isoforms at the expense of others (Bray, 2003).

Exploitable Microbiology–Synthetic Biology for Microbial Biofuels

Among the several serious problems confronting the world, three pressing examples are (1) developing a cheap antimalarial drug that could save millions of lives in the developing world; (2) climate change; and (3) the need for abundant renewable fuels. The three problems are currently attracting headlines–re-engineering of certain microbes can potentially enable conversion of agricultural products into gasoline, diesel, and aviation (jet) fuel. J. Keasling (Univ. California, Berkeley) had, a few years ago, engineered *Escherichia coli* and baker's yeast to produce a costly plant-derived antimalarial compound called artemisinin. More recently Keasling and others working at the Joint BioEnergy Institute (JBEI) are creating classes of compounds, such as alkanes, that are key components of gasoline and other transportation fuels. These alkanes are, like artemisinin, also hydrocarbons.

Synthetic biologists start from scratch for harnessing synthetic biology to enable *E. coli*, yeast, and other easily grown microorganisms to produce mixtures of compounds that can be used to make various things, including gasoline, jet fuel, plastics, and cosmetics. Certain algae, for instance, produce oils that are converted into biodiesel. The combination of rapidly improving bioengineering technology and the huge demand for cheap, low-net carbon fuels has aroused tremendous excitement.

Some start-up companies are collaborating with major energy companies such as Chevron and BP, and are producing fuels which, however, are more expensive than conventional petroleum. If technology improves rapidly, the prices may become competitive.

For the near-term future, most of the new biofuel producers are planning on displacing ethanol rather than gasoline and diesel. According to the Renewable Fuels Association, in 2007, the world produced over 50 billion liters of ethanol fuel. Most of it is made by fermenting food crops–corn kernels in the United States and sugarcane in Brazil. This method, however, is escalating food costs. The production of corn-based ethanol requires about as much energy from fossil fuels to drive the tractors, produce fertilizer, and run the ethanol plants as what comes out in the alcohol at the end. The end result could at best be a marginal reduction of CO_2. Also, ethanol cannot be shipped through existing oil pipelines because it mixes easily with water that accumulates in the pipelines (Service, 2008).

In the face of so many negative points against ethanol, synthetic biologists wish to engineer microbes to produce fuels that are essentially identical to existing fossil fuels. For this, they need organisms that can grow on the sugars from agricultural waste and other "cellulosic" materials, as this will reduce the need to use scarce agricultural land to grow fuels.

Keasling *et al.* succeeded in reengineering *E. coli* to produce gasoline-type molecules. They had earlier produced artemisinin by focusing on metabolic pathways in E. coli and yeast that normally produce small amounts of isoprenoids (precursors for building many pharmaceuticals such as artemisinin). By making many genetic changes–adding genes for additional enzymes, promoters, and so on, the organisms were coaxed to convert these intermediate compounds to artemisinin.

Similar hardwork and more genetic changes may be needed to make microbial fuel cost-effective. The objective is to engineer the microbes to increase the flux of the starting material (sugar) through the microbes' fuel-producing pathway. An *E. coli* has been engineered to express a protein scaffold for a trio of enzymes that work in the isoprenoid pathway and bind them in a way that more efficiently transforms starting compounds to the final product, leading to a greatly increased flux through the isoprenoid pathway.

Others are interested in reengineering *E. coli* and other organisms to make the so-called renewable petroleum. Rather than co-opting the microbe's isoprenoid pathway, the focus is on the route that converts sugars to fatty acids, which may then be converted to biodiesel (Service, 2008).

James Liao (Univ. of Calif., Los Angeles) has taken a third approach. *E. coli* has been engineered to produce isobutanol, a longer chain alcohol than ethanol. The longer chain energizes isobutanol significantly and enables it to be separated from water more easily. The molecule can be easily converted to fuels that can be blended with gasoline as well as transformed into other commodity chemicals. The focus of Liao *et al.* was on the metabolic route that organisms use to convert the common starting materials (alpha-keto acids) to amino acids. They chose the route because it is already adapted to cope with large fluxes of hydrocarbons. Liao's group is also making progress on inducing photosynthetic bacteria to produce isobutanol. The idea here is that bacteria manufacture fuel simply by absorbing sunlight and CO_2.

One biofuel start-up company (Solazyme in San Francisco) is working with natural and engineered algal strains to produce 'renewable' biodiesel. Instead of growing the algae outside in the sunlight, they are grown in dark stainless steel fermenters, in which the algae convert sugars to oils. Turning off the photosynthetic apparatus enables the algae to produce oils more efficiently and makes it less costly to recover. The company is already producing large volumes of biodiesel whose separate versions have been certified as jet fuel and biodiesel.

The long-term advantages of the next-generation biofuels may be plenty. Synthetic biology researchers can test hundreds of potential improvements in a short time. Unlike such renewable-energy technologies as wind and hydropower, the economics of the technology improve as it is scaled up, allowing companies to take advantage of large-scale production efficiencies.

For the near term, however, microbial biofuels will continue to depend on food-based agriculture to generate the sugars needed to feed the organisms. And even if cellulosic wastes can be economically converted to sugars, some wastes will need to be left behind on crop fields and forest land to return needed nutrients to the soils.

Microbial Surface-display for Environmental Remediation and Biofuels Production

Detailed and deep understanding of microbial genomes and proteomes coupled with advances in recombinant technology have enhanced the ability to manipulate microorganisms for biotechnological applications considerably. The ability to express heterologous proteins on the cell surface has become crucial for many important medical and environmental applications. Surface display was first used in the development of vaccines by using a filamentous phage-coat protein and the *Escherichia coli* membrane protein LamB to express antigens on the cell surface (Smith, 1985; Charbit *et al.*, 1986). Surface proteins naturally anchor 'passenger' proteins on cell membranes. It was soon realized that display of functional proteins on the cell surface works as a useful technique for the improvement of conventional biocatalysts. This led to the development of different host cells, carrier proteins and applications. The chief merits of surface-expressed, compared with intracellularly expressed proteins, are the elimination of the target compound crossing the membrane barrier, thus removing the rate-limiting step and increasing the overall kinetics; stabilization of enzymes and proteins through attachment to cells to increase yield and simplification of the protein purification step by centrifugation or settling of cells (Wu *et al.*, 2008). Several novel display methods have been designed for avoiding the generation of genetically modified microorganisms. Cells with surface display are used as biocatalysts, biosorbents and biostimulants. Microbial cell-surface display has turned out to be extremely important for many applications, ranging from combinatorial library screening and protein engineering to bioremediation and biofuels production. Current interest in surface display research is focused on expressing large multimeric proteins greater than 60 kDa, displaying multiple proteins and alleviating the spread of genetically modified organisms (GMOs) in live-vaccine and field applications (Wu *et al.*, 2008).

Microbial cell-surface display attracts several applications such as industrial catalysts, sorbents, sensors, vaccine-delivery vehicles and screening platforms. Remediation of pollutants and biofuels production effectively exploit surface-displayed biocatalysts and biosorbents. Clean-up of polluted sites attracts much attention all over the world. Most pollutants can be categorized into inorganic compounds and recalcitrant organic compounds. Microbes with surface-expressed proteins can be used to hasten the adsorption of heavy-metal containments and the degradation of organic compounds. Also, viable alternative fuel sources are badly needed to reduce continued tensions over fossil-fuel supplies. Biocatalysts with surface-displayed enzymes that can convert plant biomass substrates into ethanol and butanol are potentially promising. Alcohol-based fuels also lessen petroleum by-product contamination of the environment (Wu *et al.*, 2008).

Becker *et al.* (2004), Desvaux *et al.* (2006), and Jose (2006) have described different types of microbial cell surface-display systems for Gram-positive bacteria, Gram-negative bacteria and yeast, as well as current applications (see Paterson and Mitchell, 2007). This technology also has potential application for bioremediation and biofuels production.

Recent advances in recombinant molecular engineering have facilitated and catalyzed the exploitation of surface display of catalytically active enzymes or stable peptide sequences on various host organisms. Sources of new display anchor-protein systems and passenger proteins are being reported (see Wachet, 2006). Most of the proteins used for surface display are involved in pathogenesis. Unfortunately, the mechanism and precise functioning of many of the surface-proteins are obscure. This calls for critical studies to understand the cellular processes that involve surface proteins in bacteria, archaea and fungi. Unravelling the molecular mechanisms underlying pathogenesis and other functions will lead to the identification of more useful anchor proteins (Wu *et al.*, 2008).

There is an urgent need to find suitable methods for xenobiotic remediation and renewable energy generation. Microbial surface display shows tremendous promise in improving biosorption and biocatalytic capabilities.

Microbial Fuel Cells

Microbial fuel cells (MFCs) are potentially promising as a good method of water treatment and as power sources for environmental sensors. The power produced by these systems is rather limited by high internal (ohmic) resistance. However, improvements in the system architecture can result in power generation that depends on the capabilities of the microorganisms. The bacterial communities that develop in these systems are highly diverse, ranging from primarily δ-proteobacteria that predominate in sediment MFCs to communities composed of other diverse proteobacteria, Firmicutes and uncharacterized clones in other types of MFCs. Understanding the physiology of bacteria capable of exocellular electron transfer, collectively defined as a community of 'exoelectrogens' may facilitate harnessing of this emerging technology.

Although as early as 1911, Potter had reported that bacteria generate electricity, the idea has attracted substantial interest only recently. Using bacteria to generate electricity is appealing in view of the growing need for new sources of energy. In an MFC, bacteria are separated from a terminal electron acceptor at the cathode so that they can only respire by transfer of electrons to the anode. The flow of electrons to the cathode occurs as a result of the electrochemical potential between the respiratory enzyme and the electron acceptor at the cathode. Electron transfer from the anode to the cathode is matched by an equal number of protons moving between these electrodes to maintain electroneutrality (Logan and Regan, 2006; Heilmann and Logan, 2006).

Kim *et al.* (1999, 2004) showed that electricity could be produced by a natural consortium of bacteria in the absence of exogenous mediators. By simply placing an anode into anoxic sediment and a cathode in overlying water, enough current can be produced to power subsurface devices. That electricity may even be generated from any biodegradable material, ranging from acetate, glucose, cysteine, bovine serum albumin and ethanol to complex mixtures of organic matter including domestic (human), animal, food-processing and meat-packing wastewaters (Cheng *et al.*, 2006; Min *et al.*, 2005; Liu *et al.*, 2005; Oh and Logan, 2005).

By making small modifications to the MFC structure and operation, hydrogen can be produced instead of electricity in quite high yields. Removing oxygen from the cathode and applying a small additional voltage to the circuit evolves hydrogen gas from the cathode. Logan and Grot (2005) termed this a bio-electrochemically assisted microbial reactor (BEAMR). It is analogous to the bacterial electrolysis of organic matter because the protons and electrons come from the organic matter, not water. The highly endothermic water electrolysis requires, in practice, the application of 1.8 V. But bacterial electrolysis of organic matter is exothermic, provides energy for the bacteria, and produces an anode potential of approximately–0.3 V (versus a normal hydrogen electrode). At neutral pH, this is not enough to generate hydrogen gas from the protons and electrons produced in this system. Adding ~0.25 V makes it possible to produce hydrogen gas at the cathode. Yields of hydrogen per mole of acetate have reached 2.9 moles per mole of acetate for an energy input equivalent to 0.5 moles of hydrogen per mole of acetate for an energy input equivalent to 0.5 moles of hydrogen, representing a net energy gain by a factor of 5.8 in terms of the electricity needed for biohydrogen production by the BEAMR process, versus a net loss of energy needed for water electrolysis (Logan and Regan, 2006). The net energy input for biohydrogen can therefore, be sustained by the organic matter degraded by the bacteria.

The surprising fact is that the most significant block to achieving high power densities in MFCs is the system architecture, not the composition of the bacterial community. Improvements in system architecture and operation increase power densities greatly (Min *et al.*, 2005a). Platinum is usually used as a cathode catalyst. Power densities as high as 4.31 W/m^2 can be achieved, although this requires a non-renewable chemical reaction at the cathode involving ferricyanide. Power production can be boosted through system architecture modifications which reduce the internal resistance of the MFC, *e.g.*, by reducing electrode spacing, providing flow through a porous anode and increasing solution conductivity (Cheng *et al.*, 2006).

Some important electrochemically active bacteria in MFCs are iron-reducing bacteria such as *Shewanella* and *Geobacter* species, as well as several other species found in the biofilm community (Phung *et al.*, 2004).

Some microbes that produce hydrogen significantly include facultative anaerobic bacteria such as the Gram-negative *Alcaligenes faecalis* and Gram-positive *Enterococcus gallinarum*. *Aeromonas hydrophila* can reduce iron and produce current by using glucose in an MFC (Pham *et al.*, 2003). Some proteobacteria dominate in sediment MFCs. In these systems, the anode is immersed in anoxic sediment and strict anaerobic conditions are maintained. By contrast, oxygen can diffuse into the anode chamber in other kinds of MFCs. In a marine sediment MFC, 71 per cent of the sequences obtained in a 16S rDNA clone library from an anode electrode were δ-Proteobacteria, and 70 per cent of these belonged to the family Geobacteraceae (Bond *et al.*, 2002). A comparison of marine, salt marsh and freshwater sediments in five laboratory and field tests showed the predominance of the δ-Proteobacteria in these samples (54-76 per cent of gene sequences recovered from the anode), with the next most predominant sequences from γ-Proteobacteria (3 cases, 9-10 per cent), Cytophagales (33 per cent) and Firmicutes (11.6 per cent) (Holmes *et al.*, 2004).

Wireless and Wired Communities

A wide diversity of bacteria can be evolved in MFC reactors by employing different operating conditions. This shows the versatility of bacteria that can either transfer electrons to the electrode or can live in the reactor symbiotically with electricity-producing bacteria. It appears that microbes that are not in direct contact with the anode can be integral members of the community–bacteria away from the anode can transfer electrons to the surface.

The ability of bacteria to transfer electrons to a distant surface appears to be mediated by the presence of bacterial mediators or electron shuttles (wireless communities), or highly conductive nanowires produced by the bacteria (wired communities). *Pseudomonas aeruginosa* produces phenazines that transfer electrons between cells and surfaces; the addition of these compounds in MFCs can increase power (Rabaey *et al.*, 2005).

Both *Shewanella* and *Geobacter* species produce highly conductive nanowires. This means that cells not directly in contact with a surface can drive electron transfer to that surface (Gorby *et al.*, 2006). The nanowires produced by *Geobacter sulfurreducens* are long thin strands. Those made by *S. oneidensis* MR-1 are cable like bundles of nanowires containing many individual conductive filaments. Inoculating an MFC with a pure culture of *S. oneidensis* makes the bacteria colonize the electrode and rapidly produce several nanowires that touch both the surface and other cells.

Electrochemically active bacteria are quite abundant in various types of samples used to inoculate MFCs, *e.g.* wastewaters, sludges, and river and marine sediments. Liu and Logan (2004) inoculated an MFC with domestic wastewater and observed a repeatable cycle of power production in only 3-4 days when the reactor was emptied and refilled with fresh medium daily.

The ability of a wide range of bacteria to transfer electrons exocellularly to iron and other metal oxides to carbon electrodes and possibly to other bacteria necessitates a new term to classify this functional capability of bacteria. Logan and Regan (2006) suggested 'exoelectrogens'–this term points to the ability of bacteria to generate and transfer electrons outside of the cell. Lovley (2006), on the other hand, proposed that these bacteria should be called 'electricigens'–a term that focuses on electricity generation.

Potential Applications

The MFCs currently in existence are exciting systems for studying microbial communities and improving the understanding of how bacteria transfer electrons to solid substrates. Harnessing that power economically, however, remains a greater challenge. The first applications of MFCs are likely to be as power sources for monitoring devices in the environment and for water treatment (Logan and Regan, 2006a). Large systems have shown sufficient power production by sediment MFCs for long-term, unattended power sources for data-collecting devices (Lowy *et al.*, 2006). Electricity-consuming wastewater-treatment bioreactors are currently employed but their operation is expensive. Their replacement with the cheaper electricity-producing MFC-based systems is attracting much interest.

Renewable electricity production using MFC-based technologies is still a few years in the future. However, with advances, MFCs could become practical as a suitable method for producing a mobile fuel from renewable biomass sources. In fact, with some suitable improvements, MFCs could be exploited for hydrogen production from the same materials envisioned for ethanol production (sugar and cellulosic wastes) and from other biodegradable materials. The electricity produced by MFCs may either be used for the electrolysis of water or, better still, hydrogen might be produced directly from biomass sources by using the BEAMR process (see Zuo *et al.*, 2005).

References

Abrescia, N.G.A. *et al. Nature* 432: 68-74 (2004).

Achtman, M., Morelli, G., Schwuchow, S. *J. Bacteriol.* 135: 1053 (1978).

Acinas, S. *et al. Nature* 430: 551 (2005).

Ackermann, M., Chao, L. *Curr. Biol.* 14, R73-R74 (2004).

Adachi, J., Hasegawa, M. *Mol. Phylogenet. Evol.* 6: 72 (1996).

Adamo, S.A. *Behav. Cogn. Neurosci. Rev.* 5: 128-140 (2006).

Agogue, H. *et al. Nature* 456: 788-791 (2008).

Akashi, H., Gojobori, T. *PNAS* (USA) 99: 3695-3700 (2002).

Alby, K., Schaefer, D., Bennett, R.J. *Nature* 460: 890-893 (2009).

Aldridge, P., Hughes, K.T. *Curr. Opin. Microbiol.* 5: 160-165 (2002).

Aldsworth, T.G., Sharman, R.L., Dodd, C.E.R. *CMLS, Cell. Mol. Life Sci.* 56: 378-383 (1999).

Alekshun, M.N., Levy, S.B. *Cell* 128: 1037 (2007).

Almazán, F. *et al. PNAS* (USA) 97: 5516-5521 (2000).

Amann, R. *et al. Curr. Opin. Biotechnol.* 12: 231-236 (2001).

Amann, R.I., Ludwig, W., Schleifer, K. *Microbiol. Rev.* 59: 143-169 (1995).

Ambler, R.P. *J. Mol. Evol.* 42: 617-630 (1996).

Ames, P. *et al. PNAS* (USA) 99: 7060 (2002).

Amsellem, Z., Cohen, B.A., Gressel, J. *Nature Biotechnol.* 20, 1035-1039 (2002).

An, Y.H., Dickinson, R.B., Doyle, R.J. In: *Handbook of Bacterial Adhesion: Principles, Methods, and Applications*, pp. 1-27. Humana Press Inc., Totowa, N.J. (2000).

Anandkumar, B. *et al. Curr. Sci.* 97: 342-348 (2009).

Anda, P. *et al. Emerg. Infect. Dis.* 7: 575-582 (2001).

Anderson, R.G. *Annu. Rev. Biochem.* 67: 199 (1998).

Anderson, R.M. *et al. Nature* 423: 181-185 (2003).

Anderson, S.G.E., Kurland, C.G. *Trends Microbiol.* 6: 263-268 (1998).

Andersson, H., van den Berg, A. *Sens. Actuators B Chem.* B92: 315-325 (2003).

Andersson, S.G.E. *et al. Nature* 396: 133-140 (1998).

Archer, D.B., Turner, G. In: *The Mycota. A Comprehensive Treatise on Fungi as Experimental Systems for Basic and Applied Research*, vol. 13, pp. 75-96. Brown, A.J.P. (Ed.). Springer, Berlin (2006).

Archibald, J.M. *et al. PNAS* (USA) 100: 7678-7683 (2003).

ARI (Agharkar Research Institute). *Annual Report 2007-2008*. ARI, Pune (2008).

Arnold, D.L. *et al. Trends Genet.* 23: 293-300 (2007).

Arp, A.J., Childress, J.J. *Science* 219: 295 (1983).

Arrigo, K.R. *Nature* 437: 349-355 (2005).

Asad, L.M. *et al. Mutat. Res.* 383: 137-142 (1997).

Asad, N.R. *et al. Genet. Mol. Biol.* 27: 291-303 (2004).

Ash, C., Hanson, B., Norman, C. *Science* 296: 1055 (2002).

Ausubel, F.M. *Nature Immunol.* 6: 973-979 (2005).

Avery, O.T. *et al. J. Exp. Med.* 79: 137-158 (1944).

Azam, F. *Science* 280: 694 (1998).

Azam, F., Worden, A.Z. *Science* 303: 1622-1624 (2004).

Babic, A. *et al. Science* 319: 1533-1536 (2008).

Babu, S.T. *et al. Plant Cell Physiol.* 44: 1320-1329 (2003).

Bae, J.-W., Park, Y.-H. *Trends Biotechnol.* 24: 318-323 (2006).

Bailey, B; Collins, R., Anderson, J. *Weed Sci.* 48, 776-785 (2000).

Bajorath, J. *Nature Rev. Drug Discov.* 1: 882-894 (2002).

Baker, S. *et al. Mol. Microbiol.* 66: 1207-1218 (2007).

Baker-Austin, C., Dopson, M. *Trends Microbiol.* 15: 165-171 (2007).

Bakkenist, C.J., Kastan, M.B. *Cell* 118: 9-17 (2004).

Baldauf, S.L. *Science* 300: 1703 (2003).

Bamford, D.H., Burnett, R.M., Suart, D.I. *Theor. Popul. Biol.* 61: 461-470 (2002).

Bamford, D.H., Grimes, J.M., Stuart, D.I. *Curr. Opin. Struct. Biol.* 15: 655-663 (2005).

Banfield, J.F., Nealson K.H. (Eds.). *Geomicrobiology: Interactions Between Microbes and Minerals*. Mineral. Soc. America, Washington, D.C., 35:5-34 (1997).

Bapteste, E., Boucher, Y. *Trends Microbiol.* 16: 200-207 (2008).

Barbour, A.G., Carter, C.J., Sohaskey, C.D. *Infection Immunity* 68: 7114-7121 (2000).

Barry, E.R., Bell, S.D. *Microbiol. Mol. Biol. Rev.* 70: 876-887 (2006).

Bartee, E., McCormack, A., Früh, K. *PLoS Pathogens* 2: e107 (2006).

Bartkova, J. *et al. Nature* 434: 864-870 (2005).

Barton, H.A. *et al. J. Microbiol. Methods* 66: 21 (2006).

Bartosch, B. *et al. J. Biol. Chem.* 278: 41624-41630 (2003).

Basu, S. *et al. Nature* 434: 1130-1134 (2005).

Becker, A. *et al. Nature* 427: 117-120 (2004).

Becker, S. *et al. Curr. Opin. Biotechnol.* 15: 323-329 (2004).

Beech, I.B., Sunner, J. *Curr. Opin. Biotechnol.*, 15: 181-186 (2004).

Belcher, C.E. *et al. PNAS* (USA) 97: 13847-13852 (2000).

Belgrader, P. *et al. Science* 284: 449-450 (1999).

Belkin, S.S., Colwell, R.R. *Oceans and Health: Pathogens in the Marine Environment*. Springer (2006).

Belyaev, I.Y. *et al. Mutat. Res.* 358: 223-230 (1996).

Beman, J.M., Arrigo, K.R., Matson, P.A. *Nature* 434: 211-214 (2005).

Bemporad *et al. EMBO J.* 27: 10.1038/emboj.2008.82 (2008).

Benner, S. *Nature* 452: 692-694 (2008).

Bennett, R.J., Johnson, A.D. *EMBO J.* 22: 2505-2515 (2003).

Bennett, R.J., Johnson, A.D. *Annu. Rev. Microbiol.* 59: 233-255 (2005).

Bentley, S.D., Parkhill, J. *Annu. Rev. Genet.* 38: 771-791 (2004).

Berg, D.E. In: *Gene Transfer in the Environment*, pp. 99-137. Levy, S.B., Miller, R.V. (eds.). McGraw–Hill, New York (1989).

Berg, I.A. *et al. Science* 318: 1782- (2007).

Berman-Frank, I. *et al. Science* 294: 1534-1537 (2001).

Bernander, R. *Mol. Microbiol.* 66: 557-562 (2007).

Berry, R.M., Armitage, J.P. *Science* 320: 1599-1600 (2008).

Beudeker, R.F., Gottschal, J.C., Kuenen, J.G. *Antonie van Leeuwenhoek* 48: 39-51 (1982).

Beveridge, T.J. *Microbe* 1(6): 279-284 (2006).

Bhalla, U.S., Iyengar, R. *Science* 283, 381 (1999).

Bianchi, M.E. *J. Leukoc. Biol.* 81: 1 (2007).

Biddle, J.F. *et al. PNAS* (USA) 103: 3846 (2006).

Bigot, S. *et al. Mol. Microbiol.* 64: 1434-1441 (2007).

Bintrim, S.B. *et al. PNAS* (USA) 277-282 (1997).

Bird, A.P. *Trends Genet.* 11: 94-100 (1995).

Birge, E.A. *Bacterial and Bacteriophage Genetics.* 3rd ed. Springer–Verlag, New York (1994).

Blair, K.M. *et al. Science* 320: 1636 (2008).

Blakemore, R.P. *Annu. Rev. Microbial.* 36: 217-238 (1982).

Blanton, H.L. *et al. PLoS Genet.* 1: e40 (2005).

Blattner, F.R. *et al. Science* 277: 1453-1462 (1997).

Blomfield, I.C. *Adv. Microb. Physiol.* 45: 1-49 (2001).

BNGPC (Brazilian National Genome Project Consortium). *PNAS* (USA) 100: 11660-11665 (2003).

Bohannon, J. *Science* 315: 1486-1487 (2007).

Bolwell, P.G. *et al. J. Exp. Bot.* 53: 1367-1376 (2002).

Bond, D.R. *et al. Science* 295: 483-485 (2002).

Bond, P.L., Banfield, J.F. *Microb. Ecol.* 41: 149-161 (2001).

Bonhoeffer, S. *et al. Science* 306: 1547 (2004).

Boone, D.R. *et al. Bergey's Manual of Systematic Bacteriology*, 2nd edn. Springer (2001).

Boreham, D.R., Mitchel, R.E.J. *Radiat. Res.* 135: 365-371 (1993).

Boucher, D.H. (Ed.). *The Biology of Mutualism.* Croom Helm, London pp. 29-39 (1985).

Boucher, Y. *et al. Annu. Rev. Genet.* 37: 283-328 (2003).

Bour, S. *et al. J. Virol.* 70: 820-829 (1996).

Branda, S.S. *et al. Trends Microbiol.* 13: 20-26 (2005).

Brantl, S., Behnke, D., Alonso, J.C. *Nucl. Acids Res.* 18: 4783–4790 (1990).

Braslavsky, I. *et al. PNAS* (USA) 100: 3960 (2003).

Bray, D. *Nature* 376: 307-312 (1995).

Bray, D. *PNAS* (USA) 99: 7-9 (2002).

Bray, D. *Science* 299: 1189-1190 (2003).

Bray, D., Bourret, R.B., Simon, M.I. *Molec. Biol. Cell* 4: 469-482 (1993).

Breitbart, M. *et al. PNAS* (USA) 99: 14250-14255 (2002).

Brenner, D., Staley, J., Krieg, N. *Bergey's Manual of Systematic Bacteriology.* Springer, New York (2000).

Brenner, K. *et al. PNAS* (USA) 104: 17300-17304 (2007).

Bresnahan, W.A., Shenk, T. *Science* 288: 2373 (2000).

Brinkmann, V. *et al. Science* 303: 1532 (2004).

Brochier, C. *et al. BMC Evol. Biol.* 5: 36 (2005).

Brodie, E.L. *et al. PNAS* (USA) 104: 299-304 (2007).

Brooks, D.R. *SSED* 1: 1-29 (2001).

Broothaerts, W. *et al. Nature* 433: 629-633 (2005).

Brown, E.J. *et al. Nature* 369: 756-758 (1994).

Brown, J.R., Doolittle, W.F. *Microbiol. Mol. Biol. Rev.* 61: 456 (1997).

Brown, S.M. *et al. J. Virol.* 71: 9442-9449 (1997).

Bryant, P.E. *Int. J. Radiat. Biol.* 27: 95-102 (1975).

Bryant, P.E. *Prog. Phys. Theor. Chem.* 6: 305-313 (1979).

Buckley, R.H. *Lancet* 360: 1185-1186 (2002).

Buick, R. *Nature* 455: 569 (2008).

Bult, C.J. *et al. Science* 273: 1058-1073 (1996).

Burch, C.L., Chao, L. *Genetics* 151: 921-927 (1999).

Burnett, R.M. *J. Mol. Biol.* 185: 125-143 (1985).

Burnett, R.M. *PNAS* 103: 3-4 (2006).

Burton, B.M. *et al. Cell* 131: 1301-1312 (2007).

Buttner, D., Bonas, U. *Trends Microbiol.* 10: 186-192 (2003).

Button, D.K. *Appl. Environ. Microbiol.* 57: 2033 (1991).

Button, D.K. *et al. Appl. Environ. Microbiol.* 70: 5511 (2004).

Camacho, A.G. *et al. Biol. Chem.* 383: 1701 (2002).

Canback, B., Andersson, S.G., Kurland, C.G. *PNAS* (USA) 99: 6097 (2002).

Canfield, D.F. *et al. Mar. Geol.* 113: 27 (1993).

Cann, I.K.O. *PNAS* (USA)105: 18653-18654 (2008).

Capone, D.G. *et al. Science* 276: 1221-1229 (1997).

Carpenter, E.J., Capone, D.G. *Nitrogen in the Marine Environment.* Academic Press, New York (1983).

Carpenter, E.J., Jansen, S. *Phycologia* 40: 105-110 (2001).

Carson, C.T. *et al. EMBO J.* 22: 6610-6620 (2003).

Casjens, S.R. *et al. J. Bacteriol.* 186: 1818-1832 (2004).

Caspar, D.L., Klug, A. *Cold Spring Harbor Symp. Quant. Biol.* 27: 1-24 (1962).

Castanie-Cornet, M.P. *et al. J. Bacteriol.* 181: 3525-3535 (1999).

Castro, H.F., Williams, N.H., Ogram, A. *FEMS Microbiol. Ecol.* 31: 1-9 (2000).

Caumette, P. In: *Microbial Mats*, pp. 283-304. Amer. Soc. Microbiol., Washington, D.C. (1989).

Cavalier-Smith, T. *Biosystems* 14: 461-481 (1981).

Cavalier-Smith, T. *Nature* 326: 332 (1987).

Chalfie, M. *et al. Science* 263: 802-805 (1994).

Chankova, S.G., Bryant, P.E. *Radiat. Biol. Radioecol.* 42: 600-603 (2002).

Chankova, S.G. *et al. Radiat. Environ. Biophys.* 46: 409-416 (2007).

Chapelle, F.H. *et al. Nature* 415: 312 (2002).

Characklis, W.G., Marshall, K.C. (eds.). *Biofilms*. Wiley Intersci., N. York (1990).

Charbit, A. *et al. EMBO J.* 5: 3029-3037 (1986).

Charlebois, R.L., Beiko, R.G., Ragan, M.A. *Nature* 421: 217 (2003).

Chen, D.Q., Purcell, A.H. *Curr. Microbiol.* 34: 220-225 (1997).

Chen, I., Christie, P.J., Dubnau, D. *Science* 310: 1456 (2005).

Cheng, S. *et al. Environ. Sci. Technol.* 40: 2426-2432 (2006).

Cho, J., Tiedje, J.M. *Appl. Environ. Microbiol.* 66: 5448 (2000).

Choi, B.K. *et al. PNAS* (USA) 100: 5022 (2003).

Christensen, S.K. *et al. PNAS* (USA) 98: 14328 (2001).

Christensen, S.K., Gerdes, K. *Mol. Microbiol.* 48: 1389 (2003).

Christie, J.M., Briggs, W.R. In: *Handbook of Photosensory Receptor*, pp. 277-304. Briggs, W.R., Spudich, J.L. (eds.). Wiley-VCH, Weinheim, (2005).

Chuah, M.K. *et al. Blood* 101: 1734-1743 (2003).

Ciccarelli, F.D. *et al. Science* 311: 1283-1287 (2006).

Clardy, J., Walsh, C.T. *Nature* 432: 829-837 (2004).

Clark, L.L., Ingall, E.D., Benner, R. *Nature* 393: 426 (1998).

Clark, M.E. *et al. Hydrometallurgy* 83: 3-9 (2006).

Clarke, A.J., Warren, G.J. *Annu. Rev. Genet.* 13: 69–125 (1979).

Claverie, J.M. *et al. Virus Res.* 117: 133-144 (2006).

Clewell, D.B. (Ed.) *Bacterial Conjugation*. Plenum, New York (1993).

Coenye, T. *et al. FEMS Microbiol.* Rev. 29: 147-167 (2005).

Cohan, F.M. *Annu. Rev. Microbiol.* 56: 457-487 (2002).

Cohan, F.M. In: *Evolutionary Genetics from Molecules to Morphology*, Singh, R., Krimbas, C. (ed.). Cambridge University Press, Cambridge (1997).

Cohan, F.M., King, E.C., Zawedzki, P. *Evolution* 48: 81-95 (1994).

Cohen, J. *Science* 298: 351-353 (2002).

Collier, J.D., Muller, S.J. In: Farre, G., Oksala, T. (eds.) *Emergence, Complexity, Hierarchy and Organization*, pp. 1-30. *Acta Polytech. Scand.* (1998).

Collins, L.J., Penny, D. *Mol. Biol. Evol.* 22: 1053 (2005).

Coleman, J.R. *et al. Science* 320: 1784-1787 (2008).

Colwell, R.R., Grimes, D.J. (eds.). *Nonculturable Microorganisms in the Environment.* pp.1-6. American Society for Microbiology, Washington, D.C. (2000).

Connon, S.A., Giovannoni, S.J. *Appl. Environ. Microbiol.* 68: 3878-3885 (2002).

Cooper, S., Helmstetter, C.E. *J. Molec. Biol.* 31: 519-540 (1968).

Corredor, J.E. *et al. Biogeochemistry* 46: 163-178 (1999).

Cortes, S. *et al. Biophysical Chemistry* 120: 168-177 (2006).

Cossart, P. *et al.* (eds.). *Cellular Microbiology.* Am. Soc. Microbiol., Washington, DC (2000).

Costerton, B. *et al. Antimicrob. Agents Chemo.* 38: 2803-2809 (1994).

Costerton, J.W. *Annu. Rev. Microbiol.* 41: 435-464 (1987).

Costerton, J.W., Stewart, P.S., Greenberg, E.P. *Science* 284: 1318-1322 (1999).

Covacci, A. *et al. Science* 284: 1328-1333 (1999).

Craig, N.L. *et al.* (eds.) *Mobile DNA II.* ASM Press, Washington, D.C. (2002).

Creager, A.N.H. *The Life of a Virus: Tobacco Mosaic Virus as an Experimental Model.* Univ. Chicago Press (2001).

Crossman, L. *et al. Nature Rev. Microbiol.* 2: 616-617 (2004).

Crosson, S., Rajagopal, S., Moffat, K. *Biochemistry* 42: 2 (2003).

Cubonova, L. *et al. J. Bacteriol.* 187: 5482-5485 (2005).

Cullen, B.R. *Nature Genet.* 38(Suppl.): S25-S30 (2006).

Cullen, B.R. *Nature Immunol.* 7: 563-567 (2006a).

Cullen, D. *Nature Biotechnol.* 25: 189-190 (2007).

Cummings, C.A., Relman, D.A., *Science* 296, 1976-1979 (2002).

Curtis, T.P. *et al. PNAS* (USA) 99: 10494 (2002).

Curtis, T.P., Sloan, W.T. *Science* 309: 1331-1333 (2005).

D'Costa, V.M. *et al. Science* 311: 374 (2006).

D'Hondt, S. *et al. Science* 306: 2216 (2004).

Daga, R.R. *et al. Nature Cell Biology* 8: 1108-1113 (2006).

Dahl, J. *et al. J. Virol.* 79: 13007-13017 (2005).

Daigger, G.T. *et al. Environ. Sci. Technol.* 39: 399A-406A (2005).

Dale, C. *et al. PNAS* (USA) 98: 1883-1888 (2001).

Daniel, R. *Curr. Opin. Biotechnol.* 15: 199-204 (2004).

Daniel, R. *et al. J. Virol.* 79: 2058-2065 (2005).

Danovaro, R. *et al. Trends Ecol. Evol.* 16: 505 (2001).

Dantas, G. *et al. Science* 320: 100-103 (2008).

Darbinyan, A. *et al. J. Virol.* 78: 8593-8600 (2004).

Darling, K.F. *et al. Nature* 405: 43 (2000).

Darwin, C. *On the Origin of Species by Means of Natural Selection*. John Murray, London (1859).

Daubin, V., Moran, N.A., Ochman, H. *Science* 301: 829-832 (2003).

David, H., Akesson, M., Nielsen, J. *Eur. J. Biochem.* 270: 4243-4253 (2003).

Davidson, J.F. *et al. PNAS* (USA) 93: 5116 (1996).

Davies, D. *Nature Rev. Drug Discov.* 2: 114-122 (2003).

Davies, J. *Science* 264: 375 (1994).

Dawson, S.C., Pace, N.R. *PNAS* (USA) 99: 8324-8329 (2002).

De Hoff, P., Hirsch, A.M. *Plant-Microbe Interact.* 16: 371-376 (2003).

De Kievit, T.R., Iglweski, B.H. *Infect. Immun.* 68: 4839-4849 (2000).

Dean, F.B. *et al. PNAS* (USA), 99: 5261 (2002).

Dechesne, A., Or, D., Smets, B.F. *FEMS Microbiol. Ecol.* 64: 1-8 (2008).

Dedysh, S.N. *et al. Science* 282: 281-284 (1998).

Delaney, M.L. *Global Biogeochem. Cycles* 12: 563 (1998).

Delehanty, J.B., Ligler, F.S. *Anal. Chem.* 74: 5681-5687 (2002).

Dell'Anno, A., Danovaro, R. *Science* 309: 2179 (2005).

DeLong, E.F. *Trends Biotechnol.* 15: 203-207 (1997).

DeLong, E.F. *Natural History* 112(4): 40-46 (May 2003).

DeLong, E.F. *Science* 306: 2198-2200 (2004).

DeLong, E.F. *et al. Science* 311: 496 (2006).

DeLong, E.F., Karl, D.M. *Nature* 437: 336-342 (2005).

Deng, Z. *et al. J. Virol.* 79: 4640-4650 (2005).

Derelle *et al. PNAS* (USA) 103: 11647 (2006).

Desjardins, P.R., Olsen, G.B. *Cyanophage Strains, Stability on Phages and Hosts, and Effects of Environmental Factors on Phage-Host Interactions [Viral Control of Nuisance Cyanobacteria (Blue-green algae) II]*. California Water Resource Center, University of California, Davis (1983).

Desvaux, M. *et al. FEMS Microbiol. Lett.* 256: 1-15 (2006).

Deutsch, C. *et al. Glob. Biogeochem. Cycles* 15: 483-506 (2001).

Diaz-Torres, M.L. *et al. Antimicrob. Agents Chemother.* 47: 1430-1432 (2003).

Dibdin, G.H. *et al. J. Antimicrob. Chemo.* 38: 757-769 (1996).

Dimova, E.G., Bryant, P.E., Chankova, S.G. *Genet. Mol. Biol.* (Brazil) 31: 396-408 (2008).

Dingler, C., Oelze, J. *Arch. Microbiol.* 147: 291-294 (1987).

DiRita, V.J. *Science* 289: 1488-1489 (2000).

Dmitriev, B., Toukach, F., Ehlers, S. *Trends Microbiol.* 13: 569-574 (2005).

Döbereiner, J. *Pl. Soil* 15: 211-216 (1961).

Dobrindt, U., Hacker, J. *Curr. Opin. Microbiol.* 4: 550-557 (2001).

Dobrindt, U. *et al. Nature Rev. Microbiol.* 2: 414-424 (2004).

Dodd, C.E.R. *et al. Trends Food Sci. Technol.* 8: 238-241 (1997).

Dong, C. *et al. Nature* 427: 561-565 (2004).

Donlan, R.M., Costerton, J.W. *Clin. Microbiol. Rev.* 15: 167-193 (2001).

Doolittle, W.F. *Science* 284: 2124-2128 (1999).

Doolittle, W.F. *Trends Cell Biol.* 9: M5 (1999).

Doolittle, R.F. *Nature* 416: 697-700 (2002).

Doolittle, R.F. *Curr. Opin. Struct. Biol.* 15: 248 (2005).

Douglas, A.E., Raven, J.A. *Philos. Trans. R. Soc. London* B 358: 5-18 (2003).

Doulatov, S. *et al. Nature* 431: 476-481 (2004).

Doyle, M. *et al. Science* 315: 251 (2007).

Duckworth, D.H. In: *Phage Ecology*, pp. 1-43. Goyal, S.M., Gerba, C.P., Bitton, G. (eds.). Wiley, New York (1987).

Dufresne, A., Garczarek, L., Partensky, F. *Genome Biol.* 6: R14 (2005).

Dumas, F. *et al. PNAS* (USA) 98: 485-490 (2001).

Dunbar, J. *et al. Appl. Environ. Microbiol.* 65: 1662-1669 (1999).

Dunfield, P.F. *et al. Nature* 450: 879-882 (2007).

Dunlap, J.C., Loros, J.J. *Curr. Opin. Microbiol.* 9: 579 (2006).

Dunny, G.M., Leonard, B.A. *Annu. Rev. Microbiol.* 51: 527-564 (1997).

Dunny, G.M., Winans, S.C. (eds.). *Cell-cell Signalling in Bacteria.* ASM Press, Washington (1999).

Dupre, J. In: *Species: New Interdisciplinary Essays*, pp. 3-20. Wilson, R.A. (Ed.). Camb. Univ. Press (1999).

Duran, N., Menck, C.F. *Crit. Rev. Microbiol.* 27, 201-222 (2001).

Ecker, R.E., Schaechter, M. *Biochim. Biophys. Acta* 76: 275-279 (1963).

Eigen, M., Oswatitsch, R.W. In: *Molecular Evolution: Computer Analysis of Protein and Nucleic Acid Sequences.* Doolittle, R.F. (Ed.). Academic Press, New York (1990).

Eigen, M., *et al. Biochemistry* 30: 11005-11018 (1991).

Eigen, M. *Scientific American* 42-49 (July,1993).

Eilers, H. *et al. Appl. Environ. Microbiol.* 66: 3044-3051 (2001).

Elena, S.F., Lenski, R.E. *Nature* 390: 395-398 (1997).

Ellis, L.B. *et al. Nucleic Acids Res.* 31: 262-265 (2003).

Ellis, R.J. *Trends Biochem. Sci.* 26: 597 (2001).

Eltsov, M., Dubichet, J. *J. Bacteriol.* 1878047-8054 (2005).

Embley, T.M., Martin, W. *Nature* 440: 623-630 (2006).

EPA. U.S. Environmental Protection Agency's National Priorities List, 2 October 2002, www.epa.gov/superfund/sites/npl/.

Evans, M.C.W., Buchanan, B.B., Arnon, D.I. *PNAS* (USA) 55: 928 (1966).

Falcon, L.I. *et al. Appl. Environ. Microbiol.* 68: 5760-5764 (2002).

Falkowski, P.G. *et al. Science* 305: 354-360 (2004).

Feinendegen, L.E. *et al. Mutat. Res.* 358: 199-205 (1996).

Feinendegen, L.E. *et al. C.R. Acad. Sci.* III 322: 245-251 (1999).

Feinendegen, L.E. *Br. J. Radiobiol.* 78: 3-7 (2005).

Feinendegen, L.E., Neumann, R.D. *Hum. Exp. Toxicol.* 25: 11-17 (2006).

Feling, R.H. *et al. Angew. Chem. Int. Ed. Engl.* 42: 355-357 (2003).

Felsenstein, J. *Inferring Phylogenies.* Sinauer (2004).

Fenchel, T. *Philos. Trans. R. Soc. London Ser. B* 343: 51 (1994).

Fenchel, T., Finlay, B.J. *Ecology and Evolution in Anoxic Worlds.* Oxford Univ. Press, Oxford (1995).

Fenchel, T. *Science* 301: 925-926 (2003).

Fenster, C.B., Galloway, L.F., Chao, L. *Trends Ecol. Evol.* 12: 282-286 (1997).

Fiegna, F. *et al. Nature* 441: 310-314 (2006).

Field, C. *et al. Science* 281: 237 (1998).

Field, D., Wills, C. *PNAS* (USA) 95: 1647-1652 (1998).

Fields, S., Johnston, M. *Science* 296, 671-672 (2002).

Figeys, D. *Proteomics* 2: 373-382 (2002).

Filee, J., Siguier, P., Chandler, M. *Trends Genet.* 23: 10-15 (2006).

Finlay, B.J. *Science* 296: 1061-1063 (2002).

Finlay, B.J., Clarke, K.J. *Nature* 400: 828 (1999).

Fisher, C.R., Girguis, P.R. *Science* 315: 198-199 (2007).

Fleischmann, R.D. *et al. Science* 269: 496-512 (1995).

Föllmi, K.B. *Earth Sci. Rev.* 40: 55 (1996).

Follows, M.J. *Science* 315: 1843-1846 (2007).

Forche, A. *et al. PLoS Biol.* 6: e110 (2008).

Forney, L.J. *et al. Curr. Opin. Microbiol.* 7: 210-220 (2004).

Fortunato, E.A. *et al. PNAS* (USA) 97: 863-858 (2000).

Foster, K.R. *J. Evol. Biol.* 17: 1058-1072 (2004).

Foster, K.R. *Science* 308: 1269-1270 (2005).

Foster, K.R., Ratnicks, F.L. *Trends Ecol. Evol.* 20: 20: 363-364 (2005).

Francis, C.A. *et al. PNAS* (USA) 102: 14683-14688 (2005).

Frank, S.A. *The Foundations of Social Evolution.* Princeton Univ. Press, Princeton (1998).

Franke, A.E., Clewell, D.B. *J. Bacteriol.* 145: 494–502 (1981).

Fraser, C.M. *et al. Science* 270: 397-403 (1995).

Fraser, C.M. *et al. Science* 315: 476 (2007).

Fred, E.B., Baldwin, I.L., McCoy, E. *Root Nodule Bacteria and Leguminous Plants.* Univ. of Wisconsin Press, Madison, (1932).

Freeman, W.M. *et al. Biotechniques* 26: 112-125 (1999).

Freiberg, S., Zhu, X.X. *Int. J. Pharm.* 282: 1-18 (2004).

Frick, D.N., Richardson, C.C. *Annu. Rev. Biochem.* 70: 39-80 (2001).

Friedmann, T. *Nature Med.* 2: 144-147 (1996).

Friedrich, T., Scheide, D. *FEBS Lett.* 479: 1-5 (2000).

Friend, T. *The Third Domain: The Untold Story of Archaea and the Future of Biotechnology.* Joseph Henry Press, Washington, D.C. (2007).

Frigaard, N. *et al. Nature* 439: 847-850 (2006).

Frischer, M.E., Stewart, G.J., Paul, J.H. *FEMS Microbiol. Ecol.* 15: 127–136 (1994).

Fromin, N. *et al. Environ. Microbiol.* 4: 634-643 (2002).

Frost, L.S. *et al. Nature Rev. Microbiol.* 3: 722-732 (2005).

Fuhrman, J.A., McCallum, K., Davis, A.A. *Nature* 356: 148-149 (1992).

Fuhrman, J.A., Campbell, L. *Nature* 393: 410-411 (1998).

Fuhrman, J.A. *Nature* 399: 541-548 (1999).

Fuqua, W.C., Winans, S.C., Greenberg, E.P. *Annu. Rev. Microbiol.* 50: 727-751 (1996).

Fuqua, W.C., Winans, S.C., Greenberg, E.P. *J. Bacteriol.* 176: 269-275 (1994).

Fux, C.A. *et al. Trends Microbiol.* 13: 34-40 (2005).

Gabaldon, T., Huynen, M.A. *Science* 301: 609 (2003).

Gagiano, M., Bauer, F.F., Pretorius, I.S. *FEMS Yeast Res.* 2: 433 (2002).

Gajendiran, N., Jeevanram, R.K. *Indian J. Exp. Biol.* 40: 95-100 (2002).

Galm, U. *et al. Chem. Biol.* 11: 173-183 (2004).

Galvao, T.C., Mohn, W.W., de Lorenzo, V. *Trends Biotechnol.* 23: 497-506 (2005).

Gans, J., Wolinsky, M., Dunbar, J. *Science* 309: 1387-1390 (2005).

Garcia-Pichel, F. *Origins Life Evol. Biosphere* 28: 321-347 (1998).

Garczarek, L., Poupon, A., Partensky, F. *FEMS Microbiol. Lett.* 222: 59-68 (2003).

Gavrilets, S. *Fitness Landscapes and the Origin of Species*. Princeton University Press, Princeton (2004).

Gelvin, S.B. CAMBIA technology landscape paper (<http://www.bios.net/Agrobacterium>) (2003).

Gelvin, S.B. *Nature* 433: 583-584 (2005).

Gemmill, T.R., Trimble, R.B. *Biochim. Biophys. Acta* 1426: 227 (1999).

Gera, C., Srivastava, S. *Curr. Sci.* 90: 666-676 (2006).

Gerdes, K. *J. Bacteriol.* 182: 561 (2000).

Gerhart, E. *et al. Trends Biochem. Sci.* 23: 451-454 (1998).

Gerhart, J., Kirschner, M. *Cells, Embryos, and Evolution*. Blackwell Science, Malden (1997).

Gil, R. *et al. PNAS* (USA) 99: 4454-4458 (2002).

Gill, J., Abedon, S.T. APSnet (http://www.apsnet.org/online/feature/phages/) (2003).

Giovannoni, S.J. *et al. Science* 309: 1242-1245 (2005).

Girguis, P.R., Childress, J.J. *J. Exp. Biol.* 209: 3516 (2006).

Girigoswami, B.K., Ghosh, R. *Radiat. Environ. Biophys.* 44: 131-137 (2005).

Gleason, H.A. *Ecology* 3: 158-162 (1922).

GOC (Gene Ontology Consortium). *Nucleic Acids Res.* 32: D258 (2004).

Gogarten, J.P., Doolittle, W.F., Lawrence, J.G. *Mol. Biol. Evol.* 19: 2226 (2002).

Gold, T. *PNAS* (USA) 89: 6045 (1992).

Goldenfeld, N., Woese, C. *Nature* 445: 369 (2007).

Golyshina, O.V., Timmis, K.N. *Environ. Microbiol.* 7: 1277-1288 (2005).

Goodman, A.L. *et al. Dev. Cell* 7: 745-754 (2004).

Goodridge, L.D. *Trends Biotechnol.* 22: 384-385 (2004).

Gorby, Y.A. *et al. PNAS* (USA) 103: 11358-11363 (2006).

Gorgoulis, V.G. *et al. Nature* 434: 907-913 (2005).

Gossett, J.M. *Science* 298, 974-975 (2002).

Gottesman, S. *Annu. Rev. Microbiol.* 58: 273-301 (2004).

Gottesman, S. *Trends Genet.* 21: 399-404 (2005).

Göttlinger, H.G. *Nature* 451: 306-408 (2008).

Gottwein, J.M., Bukh, J. *PNAS* (USA) 104: 13215-13216 (2007).

Grady, R., Hayes, F. *Mol. Microbiol.* 47: 1419 (2003).

Graham, J.B., Istock, C.A. *Mol. Gen. Genet.* 166: 187–290 (1978).

Graham, P.H. *Principles and Applications of Soil Microbiology*, pp. 322-345. Sylvia, D.M. *et al. (eds.).* (1988).

Grant, S.R. *et al. Annu.* Rev. *Microbiol.* 13: 425-449 (2006).

Gray, M.W. *Curr. Opin. Genet. Develop.* 9: 678-687 (1999).

Gray, M.W., Burger, G., Lang, B.F. *Science* 283: 1476 (1999).

Greenberg, E.P. *ASM News* 63: 371-377 (1998).

Greenberg, P. *Nature* 424: 134 (2003).

Greene, K. *Science* 296: 1000 (2002).

Greenleaf, W.J., Block, S.M. *Science* 313: 801 (2006).

Gregg, W.W. *et al. Deep-Sea Res.* II 50: 3143 (2003).

Gregory, M.A. *et al. Angew. Chem. Int. Ed.* 44: 4757-4760 (2005).

Gressel, J. *Trends Biotechnol.* 19: 149-154 (2001).

Gressel, J., *Molecular Biology of Weed Control.* Taylor and Francis, London (2002).

Griffin, D.W. *et al. Am. Sci.* 90: 228 (2002).

Griffin, D.W. *Aerobiologia* 20: 135-140 (2004).

Griffith, F. *Jour. Hygiene* 27: 113-156 (1928).

Grill, S.W. *et al. Science* 301: 518-521 (2003).

Gupta, R.S. *et al. PNAS* (USA) 91: 2895-2899 (1994).

Gupta, R.S. *Microb. Molec. Biol. Rev.* 62: 1435-1491 (1998).

Gupta, R.S. *FEMS Microbiol. Rev.* 24: 367-402 (2000).

Guruprasad, K.P. *et al. Mutagenesis* 17: 108 (2002).

Guttman, D.S. *Trends Ecol. Evol.* 12: 16-21 (1997).

Gyan, K. *et al. Int. J. Biometeorol.* 49: 371-276 (2005).

Gyohda, A. *et al. Adv. Biophys.* 38: 183-213 (2004).

Hacker, J. Kaper, J.B. *Annu. Rev. Microbiol.* 54, 641-679 (2000).

Hacker, J., Hentschel, U., Dobrindt, U. *Science* 301: 790 (2003).

Haefner, B. *Drug Discov. Today* 8: 536-544 (2003).

Hakansta, C. *Biotech. Develop. Monitor* No. 34: 17-19 (1998).

Haldar, S., Chattopadhyay, A. *J. Phys. Chem.* B111: 14436-14439 (2007).

Haldar, S., Chattopadhyay, A. *J. Biosci.* 34: 169-172 (2009).

Hall, R.M., Collis, C.M. *Molec. Microbiol.* 15: 593-600 (1995).

Halliwell, B., Gutteridge, J.M. *Free Radicals in Biology and Medicine.* 2 ed. Clarendon Press, Oxford (1989).

Hall-Stoodley, J.L., Costerton, W., Stoodley, P. *Nature Rev. Microbiol.* 2: 95-108 (2004).

Hamilton, S.R. *et al. Science* 301: 1244 (2003).

Hamilton, S.R. *et al. Science* 313: 1441-1443 (2006).

Hanage, W.P. *et al. J. Theor. Biol.* 239: 210-219 (2006).

Handelsman, J. *et al. Chem. Biol.* 5: R245-249 (1998).

Handelsman, J. *Microbiol. Mol. Biol. Rev.* 68: 669-685 (2004).

Handelsman, J. *Nature Biotechnol.* 23: 38-39 (2005).

Hannon, G.J. *Nature* 418: 244-251 (2002).

Hansell, D.A., Carlson, C.A. *Glob. Biogeochem. Cycles* 12: 443 (1998).

Hansen, S.K. *et al. Nature* 445: 533-536 (2007).

Hanski, I.A., Gilpin, M.E. (eds.). *Metapopulation Biology: Ecology, Genetics, and Evolution.* Elsevier, Burlington, MA (1997).

Hanson, B.J., Watson, L.E., Barnum, S.R. *J. Mol. Evol.* 58: 390-399 (2004).

Hanson, R.S., Hanson, T.E. *Microbial. Rev.* 60: 439-471 (1996).

Harold, F.M. *Microbe* 1(6): 295 (2006).

Harrington, L.C., Rogerson, A.C. *J. Bacteriol.* 172: 7263 (1990).

Harris, T.D. *et al. Science* 320: 106-107 (2008).

Harrison, S.C. *et al. Nature* 276: 368-373 (1978).

Hartman, H., Fedorov, A. *PNAS* (USA) 99: 1420 (2002).

Hartwell, L.H. *Genetics* 4: 975-980 (1991).

Hastings, J.W., Greenberg, E.P. *J. Bacteriol.* 181: 2667-2668 (1999).

Hatzenpichler, R. *et al. PNAS* (USA) 105: 2134-2139 (2008).

Hayes, F. *Science* 301: 1496-1499 (2003).

Hayes, F. In: *E. coli Plasmid Vectors: Methods and Applications*, pp. 1-17. Preston, A., Casali, N. (eds.). Humana Totowa, NJ (2003).

Hayes, W. *The Genetics of Bacteria and Their Viruses.* 2 ed. Blackwell, London (1968).

He, Y., Duan, W., Tan, S.L. *Drug Discovery Today* 12: 209-217 (2007).

Heidelberg, J.F. *et al. Nature Biotech.* 20: 1118-1123 (2002).

Heilmann, J., Logan, B.E. *Water Environ. Res.* 78: 531-537 (2006).

Heintzelman, M.B., Schwartzman, J.D. *J. Mol. Biol.* 271: 139-146 (1997).

Heintzelman, M.B. *Curr. Biol.* 13: R57-R58 (2003).

Henderson, B. *et al.* (eds.). *Cellular Microbiology: Bacteria-Host Interactions in Health and Disease.* Wiley, West Sussex (1999).

Hendrix, R.W. *et al. PNAS* (USA) 96: 2192-2197 (1999).

Hendrix, R.W. *Theor. Popul. Biol.* 61: 471-480 (2002).

Hendrix, R.W. *PNAS* (USA) 101: 7495-7496 (2004).

Hengge-Aronis, R. In: *Bacterial Stress Responses*, pp. 161-178. Storz, G., Hengge-Aronis, R. (eds.). ASM Press (2000).

Hengge-Aronis, R. *Microbiol. Mol. Biol. Rev.* 66: 373-395 (2002).

Henne, A. *et al. Appl. Environ. Microbiol.* 65: 3901-2907 (1999).

Hennes, K.P., Simon, M. *Appl. Environ. Microbial.* 61: 333-340 (1995).

Hentze, M.W. *Science* 293: 1058-1059 (2001).

Henz, S.R. *et al. Bioinformatics* 21: 2329-2335 (2005).

Herndl, G.J. *et al. Appl. Environ. Microbiol.* 71: 2303-2309 (2005).

Herter, S. *et al. J. Biol. Chem.* 277: 20277 (2002).

Hess, W.R. *Curr. Opin. Biotechnol.* 15: 191-198 (2004).

Hetzer, M., Gruss, O.J., Mattaj, I.W. *Nature Cell Biology* 4: E177-184 (2002).

Hewinson, R.G. *et al. Vet. Microbiol.* 112: 127-139 (2006).

Hicks, R.P. *Curr. Med. Chem.* 8: 627-650 (2001).

Higgins, D.A. *et al. Nature* 450: 883-886 (2007).

Hillova, J., Drasil, V. *Int. J. Radiat. Biol. Relat. Stud. Phys. Chem. Med.* 12: 201-208 (1967).

Hjort, K., Bernander, R. *Mol. Microbiol.* 40: 225-234 (2001).

Heitman, J. Love the one you're with. *Nature* 460: 807-808 (2009).

Hoek, J. *et al. Geobiology* 1: 119-127 (2003).

Hofseth, L.J. *Carcinogenesis* 25: 1787-1793 (2004).

Holmes, A.J. *et al. Environ. Microbiol.* 5: 383-394 (2003).

Holmes, D.E. *et al. Microb. Ecol.* 48: 178-190 (2004).

Horner-Devine, M.C. *et al. Nature* 432: 750-753 (2004).

Horner-Devine, M.C. Carney, K.M., Bohannan, B.J.M. *Proc. R. Soc. Biol. Sci.* B271: 113-122 (2004a).

Horsburgh, M.J. *et al. Mol. Microbiol.* 44: 1269 (2002).

Horsley, R.J., Laszlo, A. *Int. J. Radiat. Biol. Relat. Stud. Phys. Chem. Med.* 20: 593-596 (1971).

Howard, A., Cowie, F.G. *Radiat. Res.* 75: 607-616 (1978).

Hu, C.-D., Chinenov, Y., Kerppola, T.K. *Mol. Cell* 9: 789-798 (2002).

Huang, G. *et al. Curr. Biol.* 19: 330-334 (2009).

Hubbell, S.P. *The Unified Neutral Theory of Biodiversity and Biogeography.* Princeton Univ. Press, Princeton, NJ (2001).

Huber, R.G. *et al.* In: *Microbial Growth on C_1 Compounds,* pp. 45-51. van Verseveld, H.W., Duine, J.A. (ed.). Martinus Nijhoff, Dordrecht (1986).

Huffaker, A. *et al. PNAS* (USA) 103: 10098-10103 (2006).

Hugenholtz, P. *Genome Biol.* 3: 3 (2002).

Hugenholtz, P., Goebel, B.M., Pace, N.R. *J. Bacteriol.* 180: 4765-4774 (1998).

Hughes, K.A., Sutherland, I.W., Jones, M.V. *Microbiology* 144: 3039-3047 (1998).

Humes, E. *Monkey Girls: Evolution, Education, Religion, and the Battle for America's Soul.* Ecco, New York (2007).

Hunter, R.C., Beveridge, T.J. *J. Bacteriol.* 187: 7619-7630 (2005).

Hutchinson, C.A. *et al. Science* 286: 2165-2169 (1999).

Hutchinson, G.E. *Am. Nat.* 95: 137 (1961).

Iarmarcovai, G. *et al. Mutagenesis* 20: 425-432 (2005).

Iarmarcovai, G. *et al. Mutat. Res.* 615: 18-27 (2007).

Ignacimuthu, S. *Current Science* 78: 1284-1285 (2000).

Imlay, J.A., Linn, S. *J. Bacteriol.* 166: 519-527 (1986).

Inagaki, F. *et al. Appl. Environ. Microbiol.* 69: 7224 (2003).

Inagaki, F. *et al. PNAS* (USA) 103: 2815 (2006).

Ingalls, A.E. *et al. PNAS* (USA) 103: 6442-6447 (2006).

Ingham, C.J. *et al. PNAS* (USA) 104: 18217-18222 (2007).

Isaacs, A., Lindermann, J. *Proc. R. Soc. Lond.* B147: 258-273 (1957).

Isaksen, M.F., Bak, F., Jorgensen, B.B. *FEMS Microbiol. Ecol.* 14: 1 (1994).

Islam, T. *et al. PNAS* (USA) 105: 300-304 (2008).

Israel, D.A. *et al. PNAS* (USA) 98, 14625 (2001)

Ivnitski, D. *et al. Biotechniques* 35: 862-869 (2003).

Iyer, L.M. *et al. Virus Res.* 117: 156-184 (2006).

Jacob, F., Brenner, S., Cuzin, F. *Cold Spring Harbor Symp. Quant. Biol.* 28: 329-348 (1963).

Jaffe, A., Ogura, T., Hiraga, S. *J. Bacteriol.* 163: 841 (1985).

Jain, R., Rivera, M.C., Lake, J.A. *PNAS* (USA) 96: 3801-3806 (1999).

Jankovics, F., Brunner, D. *Developmental Cell* 11: 375-385 (2006).

Jazwinski, S.M. *Annu. Rev. Microbiol.* 56: 769-792 (2002).

Jensen, E.C. *et al Appl. Environ. Microbial.* 64: 575-580 (1998).

Jenuwein, T., Allis, C.D. *Science* 293: 1074 (2001).

Jeon, C.O. *et al. PNAS* (USA) 100: 13591-13596 (2003).

Jeong, H. *et al. Nature* 407, 651 (2000).

Jett, J.H. *et al. J. Biomol. Struct. Dyn.* 7: 301 (1989).

Johnsen, J. *Arch. Microbiol.* 115: 271 (1977).

Johnson, D.B. *FEMS Microbiol. Ecol.* 27: 307-317 (1998).

Johnson, D.B., Hallberg, K.B. *Res. Microbiol.* 154: 466-473 (2003).

Joiner, M.C. *Int. J. Radiat. Biol.* 65: 79-84 (1994).

Joiner, M.C., Lambin, P., Marples, B. *CR Acad. Sci.* III 322: 167-175 (1999).

Joksic, G. *et al. Cell Mol. Life Sci.* 57: 842-850 (2000).

Jones, K. *J. Ecol.* 62: 553-565 (1974).

Jones, S.A. *et al. Redox Report* 4: 291-299 (1999).

Jopling, C.L. *et al. Science* 309: 1577-1581 (2005).

Jordan, D.C. *Int J. Syst. Bacteriol.* 32, 136-139 (1982).

Jørgensen, B.B. *Philos. Trans. R. Soc. London Ser. B* 298: 543-561 (1982).

Jørgensen, B.B., Des Marais, D.J. *FEMS Microbiol. Ecol.* 38: 179-186 (1986).

Jorgensen, B.B., D'Hondt, S. *Science* 314: 932-934 (2006).

Jorgensen, B.B., D'Hondt, S.L., Miller, D.J. In: *Proc. Ocean Drilling Prog.*, 201: *Scientific Results*. pp. 1-45. In: Jorgensen, B.B. *et al.* (eds.) ODP, College Station, TX (2006).

Jose, J. *Appl. Microbiol. Biotechnol.* 69: 607-614 (2006).

Joset, F., Guespin–Michel, J. *Prokaryotic Genetics: Genome Organization, Transfer and Plasticity*. Blackwell Scientific, Oxford (1993).

Jovtchev, G., Stergios, M. *Compt. Rend. Acad. Bulg. Sci.* 56: 75-80 (2003).

Junca, H., Pieper, D.H. *J. Microbiol. Methods* 55: 697-708 (2003).

Jurica, M.S., Moore, M.J. *Cell* 12: 5 (2003).

Kaeberlein, T., Lewis, K., Epstein, S.R. *Science* 296: 1127-1129 (2002).

Kaiser, D., Losick, R. *Sci. Am.* 276: 68-73 (1997).

Kaluzhnaya, M. *et al. Syst. Appl. Microbiol.* 24: 166-176 (2001).

Kana, T.M. *Limnol. Oceanogr.* 38: 18-24 (1993).

Kaper, J.B,., Nataro, J.P., Mobley, H.L.T. *Nature Rev. Microbiol.* 2: 123 (2004).

Kappe, S. *et al. J. Cell. Biol.* 147: 937-943 (1999).

Karaolis, D.K.R. *et al. Nature* 399: 375-379 (1999).

Karavolos, M.H. *et al. Microbiology* 149: 2749 (2003).

Karl, D. *et al. Biogeochemistry* 57/58: 47-98 (2002).

Karner, M.B., DeLong, E.F., Karl, D.M. *Nature* 409: 507-510 (2001).

Karsenti, E. *Interdisciplin. Sci. Rev.* 32: 163-175 (2007).

Kashefi, K. *et al. Int. J. Syst. Bact.* 52: 719 (2002).

Kashefi, K., Lovley, D.R. *Science* 301: 934 (2003).

Kashi, Y., King, D.G. *Trends Genet* 22: 253-259 (2006).

Kasting, J.F., Siefert, J.L. *Science* 296: 1066-1067 (2002).

Katlinka, M.D. *et al. Nature* 414: 450-453 (2001).

Kato, S. *et al. Appl. Environ. Microbiol.* 71: 7099-7106 (2005).

Kauffman, S., Levin, S. *J. Theor. Biol.* 128: 11-45 (1987).

Kearns, D. *et al. Mol. Microbiol.* 55. 739 (2005).

Kearns, D. *et al. Science* 320: 1636-1638 (2008)

Keim, P. *et al. J. Clin. Microbiol.* 39, 4566 (2001).

Keller, L., Surette, M.G. *Nature Rev. Microbiol.* 4: 249-258 (2006).

Kemp, M. *Nature* 396: 123 (1998).

Kennedy, A.D. *et al. PNAS* (USA) 105: 1327 (2008).

Kennedy, M.J. *et al. Microbiology* 140: 2513-2529 (1994).

Kennis, J.T.M., Crosson, S. *Science* 317: 1041-1042 (2007).

Kerr, R.A. *Science* 296: 1056-1058 (2002).

Keymer, J.E. *et al. PNAS* (USA) 103: 17290-17295 (2006).

Khayat, R. *et al. PNAS* (USA) 102: 18944-18949 (2005).

Kilroy, C., Bergey, L. *Water and Atmosphere (NIWA)* 7(3): 13-16 (1999).

Kim, B.H. *et al. Appl. Microbiol. Biotechnol.* 63: 672-681 (2004).

Kim, H.J. *et al. Enzyme Microb. Technol.* 30: 145-152 (2002).

Kim, H.J. *et al. PNAS* (USA) 105: 18188-18193 (2008).

Kim, H.J., Kim, W., Cunningham, C.W. *Mol. Biol. Evol.* 16: 423-427 (1999).

Kimura, M. *Theor. Popul. Biol.* 37: 150-158 (1990).

Kimura, M. *The Neutral Theory of Molecular Evolution*. Camb. Univ. Press, Cambridge (1983).

Kinzig, A., Pacala, S., Tilman, D. *The Functional Consequences of Biodiversity*. Princeton Univ. Press, Princeton (2001).

Kistler, H.C. In: *Microbial Control of Weeds*, pp. 151-170. TeBeest, D.O.(Ed.) Chapman and Hall, New York (1991).

Kitano, H. *Science* 295, 1662 (2002).

Kitano, H. *Nature Rev. Genet.* 5: 826-837 (2004).

Klappenbach, J.A., Dunbar, J.M., Schmidt, T.M. *Appl. Environ. Microbiol.* 66: 1328 (2000).

Klasson, L., Andersson, S.G.E. *Trends Microbiol.* 12: 37 (2004).

Kleidon, A., Mooney, H.A. *Glob. Change Biol.* 6: 507 (2000).

Knight, P. *ASM News* 57 (11): 567–570 (1991).

Knoll, A.H. *Science* 256: 622-627 (1992).

Ko, M. *et al. Mutat. Res.* 597: 11-17 (2006).

Kobayashi, I. *Nucleic Acids Res.* 29: 3742 (2001).

Koch, P.L. *et al.* (eds.) *Stable Isotopes in Ecology and Environmental Science*. Blackwell, Oxford (1994).

Kokjohn, T.A. In: *Gene Transfer in the Environment*, pp. 73-97. Levy, S.B., Miller, R.V. (eds.). McGraw–Hill, New York (1989).

Kolenbrander, P.E. *et al. Trends Microbiol.* 13: 11-15 (2005).

Kollmann, M. *et al. Nature* doi:10.1038/nature04228 (2005).

Kolter, R., Greenberg, E.P. *Nature* 441: 300-302 (2006).

Kondrashov, A.S. *Nature* 336: 435-440 (1988).

Könneke, M. *et al. Nature* 437: 543-546 (2005).

Konstantinidis, K.T., Tiedje, J.M. *PNAS* (USA) 102: 2567-2572 (2005).

Koonin, E.V. *et al. Mol. Microbiol.* 25: 619-637 (1997).

Koonin, E.V. *Annu. Rev. Genomics Hum. Genet.* 1: 99-116 (2000).

Koonin, E.V. Virology: *Curr. Biol.* 15: R167-R169 (2005).

Koonin, E.V. *et al. Annu. Rev. Microbiol.* 55: 709-742 (2001).

Koonin, E.V., Aravind, L. *Cell Death Differ.* 9: 394 (2002).

Koslow, T. *The Silent Deep: The Discovery, Ecology and Conservation of the Deep Sea.* Univ. of Chicago Press (2007).

Kravchenko, V.V. *et al. J. Biol. Chem.* 281: 2418-2432 (2006).

Kreft, J.U. *Microbiology* 150: 2751-2760 (2004).

Kruis, W. *et al. Gut* 53: 1617 (2004).

Krulwich, T.A. *et al. J. Bacteriol.* 162: 768-772 (1985).

Kuenen, J.G. *Plant and Soil* 43: 49-76 (1975).

Kuenen, J.G. In: *Microbial Mats. Pp. 349-365.* Cohen, Y., Rosenbeg, E. (eds.). Amer. Soc. Microbiol., Washington, D.C. (1989).

Kumar, H.D. *Introductory Phycology*. 2ed. Affil. East West Press, New Delhi (1999).

Kumar, H.D., Kumar, S. *Modern Concepts of Microbiology*. Vikas Publishing House, N. Delhi (1998, 2000).

Kunkel, T.A., Bebenek, K. *Ann. Rev. Biochem.* 69: 497-529 (2000).

Kupzig, S. *et al. Traffic* 4: 694-709 (2003).

Kurland, C.G., Canback, B., Berg, O.G. *PNAS* (USA) 100: 9658-9662 (2003).

Kurland, C.G., Collins, L.J., Penny, D. *Science* 312: 1011-1014 (2006).

Lagally, E.T. *et al. Sens. Actuators B Chem.* 63: 138-146 (2000).

Lagally, E.T. *et al. Anal. Chem.* 76: 3162-3170 (2004).

Lai, M.M.C. *PNAS* (USA) 97: 5025-5027 (2000).

Lake, J.A. *Nature* 331: 184-186 (1988).

Lake, J.A. *Nature* 446: 983 (2007).

Lake, J.A., Rivera, M.C. *Mol. Biol. Evol.* 21: 681-690 (2004).

Lam, G.Y., Brumell, J.H. *Nature* 455: 1186-1187 (2008).

Lane, D.J. *et al. PNAS* (USA) 82: 6955-5959 (1985).

Lang, B.F. *et al. Curr. Biol.* 12: 1773-1778 (2002).

Larkin, J.M., Strohl, W.R. *Annu. Rev. Microbiol.* 37: 341-367 (1983).

Lau, A. *et al. Nature Cell Biol.* 7: 493-500 (2005).

Lawrence, J.G., Ochman, H. *PNAS* (USA) 95: 9413-9417 (1998).

Lax, A.J. *Nature Rev. Microbiol.* 3: 343 (2005).

Lazazzera, B.A., Grossman, A.D. *Trends Microbiol.* 7: 288-294 (1998).

Lederberg, J., Tatum, E.L. *Cold Spring Harb. Symp. Quant. Biol.* 11: 113 (1946).

Lederberg, J., Tatum, E.L. *Cold Spr. Harb. Symp. Quant. Biol.* 11: 113–114 (1946a).

Lederberg, J., Tatum, E.L. *Nature* 158: 558 (1946b).

Lee, G.H., Stotzky, G. *Korean J. Microbiol.* 28: 210–218 (1990).

Lee, P.C., Bochner, B.R., Ames, B.N. *PNAS* (USA) 80: 7496-7500 (1983).

Lee, T. *et al. Indoor Air* 16: 37-47 (2006).

Lee, T.I. *et al. Science* 298, 799 (2002).

Lehmann, L., Bargum, K., Reuter, M. *J. Evol. Biol.* 19: 1507-1516 (2006).

Leibold, M.A. *et al. Ecol. Lett.* 7: 601-613 (2004).

Lenski, R.E. *Evolution* 42: 433-440 (1988).

Lenski, R.E. *Int. Microbiol.* 1: 265-270 (1998).

Leon, L.L. *et al. J. Antimicrob. Chemother.* 48, 449-450 (2001).

Levesque, M. *et al. Curr. Biol.* 13, 129-133 (2003).

Levin, S. (Ed.). In: *Encyclopedia of Biodiversity.* Vol. 4: 901-915 Academic Press, San Diego (2001).

Levin, S.A. *Ecology* 73: 1943 (1992).

Levine, N.D. *et al. J. Protozool.* 27: 37-58 (1980).

Levit, M.N., Stock, J.B. *J. Biol. Chem.* 277: 36760-36765 (2002).

Lewinski, M.K., Bushman, F.D. *Adv. Genet.* 55: 147-181 (2005).

Lewis, K. *Antimicrob. Agents. Chemother.* 45: 999-1007 (2001).

Lewis, K., Ausubel, F.M. *Nature Biotechnol.* 24: 1504-1507 (2006).

Ley, R.E. *et al. Nature* 444: 1022-1023 (2006).

Ley, R.E. *et al. Science* 320: 1647-1651 (2008).

Li, N.J. *et al. Mol. Microbiol.* 59: 1097-1113 (2006).

Li, W.-H. *Molecular Evolution*, pp. 185-190. Sinauer, Sunderland, MA (1997).

Li, X., Qin, L. *Trends Biotech.* 539-543 (2005).

Liang, P. *Biotechniques* 33: 338-346 (2002).

Liba, C.M. *et al. J. Appl. Microbiol.* 101: 1076-1086 (2006).

Lilley, C.E. *et al. PNAS* (USA) 102: 5844-5849 (2005).

Lilley, C.E., Schwartz, R.A., Weitzman, M.D. *Trends Microbiol.* 15: 119-126 (2007).

Lin, L.-H. *Geochim. Cosmochim. Acta* 69: 893 (2005).

Lin, S.J. *et al. Nature* 418: 344-348 (2002).

Lin, Y.P., Gregory, V., Bennett, M., Hay, A. *Virus Res.* 103: 47-52 (2004).

Lindahl, T. *Nature* 362: 709-715 (1993).

Lindas, A.-C. *et al. PNAS* (USA) 105: 18942-18946 (2008).

Lindemann, J., Upper, C.D. *Appl. Environ. Microbiol.* 50: 1229-1232 (1985).

Lindenbach, B.D. *et al. Science* 309: 623-626 (2005).

Lindstrom, S.E., Cox, N.J., Klimov., A. *Virology* 328: 101-119 (2004).

Lininger *et al. Nature* 456: 712-714 (2008).

Litchman, E. *et al. Biogeosciences* 3: 585 (2006).

Liu, H. *et al. Environ. Sci. Technol.* 38: 2281-2285 (2004).

Liu, H. *et al. Environ. Sci. Technol.* 39: 658-662 (2005).

Liu, H., Logan, B.E. *Environ. Sci. Technol.* 38: 4040-4046 (2004).

Liu, L. *et al. J. Biol. Chem.* 276: 45484-45490 (2001).

Liu, W.-T., Zhu, L. *Trends Biotechnol.* 23: 174-179 (2005).

Lloyd, J.R., Lovely, D.R. *Curr. Opin. Biotechnol.* 12: 248-253 (2001).

Lloyd, R.G., Buckman, C. *Genetics* 139: 1123 (1995).

Lo, I. *et al. Nature* 446: 537-541 (2007).

Logan, B.E. *Environ. Sci. Technol.* 38: 160A-167A (2004).

Logan, B.E., Grot, S. A bioelectrochemically assisted microbial reactor (BEAMR) that generates hydrogen gas. Patent application 60/588,022 (2005).

Logan, B.E., Regan, J.M. *Trends Biotechnol.* 24: 512-518 (2006).

Logan, B.E., Regan, J.M. *Environ. Sci. Technol.* 40: 5172-5180 (2006a).

Lopez-Garcia, P. *et al. Nature* 409: 603-607 (2001).

Lorenz, M.G., Wackernagel, W. *Microbiol. Rev.* 58: 563–602 (1994).

Losos, J.B. *Nature* 408: 847-850 (2000).

Lovley, D.R. (ed.). *Environmental Microbe-Metal Interactions.* ASM Press, Washington, D.C. (2000).

Lovley, D.R. *Nature Rev. Microbiol.* 4: 497-508 (2006).

Low, K.B. *Methods Enzymol.* 204: 43 (1991).

Lower, S.K., Beveridge, T.J., Hochella, M.F. Jr., *Science* 292: 1360 (2001).

Lowy, D.A. *et al. Biosens. Bioelectron.* 21: 2058-2063 (2006).

Lozupone, C.A., Knight, R.K. *PNAS* (USA) 104: 11436-11440 (2007).

Luckow, V.A. In: *Recombinant DNA Technology and Applications*, pp. 97-151. Prokop, A., Bajpai, R.K., Ho, C. (eds.). McGraw–Hill, New York (1991).

Lundgren, M., Bernander, R. *Curr. Opin. Microbiol.* 8: 662-668 (2005).

Luppens, S.B. *et al. Appl. Environ. Microbiol.* 68: 4194-4200 (2002).

Lynch, J.M. In: *Bacterial Genetics in Natural Environments*. pp. 172–181. Fry, J.C., Day, M.J. (eds.). Chapman and Hall, London (1990).

Lynch, M. *PNAS* (USA) 99: 6118 (2002).

MacBeath, G., Koehler, A.N., Schreiber, S.L. *J. Am. Chem. Soc.* 121: 7967-7968 (1999).

Machida, M. *et al. Nature* 438: 1105-1115 (2005).

Macnab, R.M. *Annu. Rev. Microbiol.* 57: 77 (2003).

Madshus, I.H. *Biochem. J.* 250: 1-8 (1988).

Maeda, S. *et al. Nature* 315: 592–594 (1985).

Maine, G.T., Sinha, P., Tye, B K. *Genetics* 106: 365-385 (1984).

Majdalani, N. *et al. PNAS* (USA) 95: 12462-12467 (1998).

Majdalani, N. *et al. Mol. Microbiol.* 46: 813-826 (2002).

Makarova, K.S. *et al. Microbiol. Mol. Biol. Rev.* 65: 44-79 (2001).

Malakoff, D. *Science* 295: 2352-2353 (2002).

Mallet, J. *Trends Eccl. Evol.* 10· 294-299 (1995).

Manefield, M. *et al. Appl. Environ. Microbiol.* 68, 5367-5373 (2002).

Marches, O. *et al. Mol. Microbiol.* 50: 1553 (2003).

Margolin, W. *Nature Rev. Mol. Cell Biol.* 6: 862-871 (2005).

Margulis, L. *Symbiotic Planet*. Basic Books, Long Branch, New Jersey (1998).

Mariani, R. *et al. J. Virol.* 74: 3859-3870 (2000).

Marians, K.J. *Annu. Rev. Biochem.* 61: 673-719 (1992).

Marinsek, N. *et al. EMBO Rep.* 7: 539-545 (2006).

Markert, S. *et al. Science* 315: 247 (2007).

Maron, P.A. *et al. Atmos. Environ.* 39: 3687-3695 (2005).

Marsh, T. *Curr. Opin. Microbiol.* 2: 323-327 (1999).

Marshal, E. *Science* 294: 1640 (2001).

Marshall, K.C. In: *Microbial Mats*. Pp. 239-245. In: Cohen, Y., Rosenberg, E. (eds.). Amer. Soc. Microbiol., Washington, D.C. (1989).

Martin, B.M. *et al. PNAS* (USA) 44: 15645 (2004).

Martin, W., Muller, M. *Nature* 392: 37-41 (1998).

Martin, W., Russell, M.J. *Phil. Trans. Roy. Soc. London* 358B: 59-83 (2003).

Martinez, A. *et al. Appl. Environ. Microbiol.* 70: 3452-2463 (2004).

Martiny, J.B. *et al. Nature Rev. Microbiol.* 4: 102 (2006).

Massana, R., DeLong, E.F., Pedros-Alio, C. *Appl. Environ. Microbiol.* 66: 1777 (2000).

Masse, E., Gottesman, S. *PNAS* (USA) 99: 4620-4625 (2002).

Matias, V.R.F. *et al. J. Bacteriol.* 185: 6112-6118 (2003).

Matsunaga, F. *et al. PNAS* (USA) 98: 11152-11157 (2001).

Matsushima, P., Baltz, R.H. In: *Manual of Industrial Microbiology and Biotechnology*. Pp. 170-183. Demain, A.L., Solomon, N.A. (eds.). American Society of Microbiology, Washington, DC (1986).

Maurelli, A.T. *et al. PNAS* (USA) 95: 3943-3948 (1998).

Maymo-Gatell, X. *et al. Science* 276 1568 (1997).

Mazel, D., Davies, J. *CMLS (Cell. Mol. Life Sci.)* 56: 742-754 (1999).

Mazel, D., Dychinco, B., Webb, V.A., Davies, J. *Science* 280: 605-608 (1998).

McArthur, H.A.I. In: *Developments in Industrial Microbiology–BMP'97*, pp. 43-48, Hutchinson, C.R., McAlpine, J. (eds.) (1998).

McCullers, J.A., Saito, T., Iverson, A.R. *J. Virol.* 78: 12817-12828 (2004).

McCutchen, B.F. *et al. Bio/Technology* 9: 848–852 (1991).

McDonald, I.R. *et al. Appl. Environ. Microbiol.* 74: 1305-1315 (2008).

McFadden, G.I. *J. Eukaryotic Microbiol.* 46: 339-346 (1999).

McFadden, G.I. *J. Phycol.* 37: 951-959 (2001).

McGuire, A.M., Hughes, J.D., Church, G.M. *Genome Res.* 10, 744-757 (2000).

McInerney, J.D., Pisani, D. *Science* 318: 1390-1391 (2007).

McKenny, D., Brown, K.E., Allison, D.G. *J. Bacteriol.* 177: 6989-6992 (1995).

Mehta, M.P., Baross, J.A. *Science* 314: 1783-1786 (2006).

Meinhart, A. *et al. PNAS* (USA) 100: 1661 (2003).

Menard, R. *Cell. Microbiol* 3: 63-73 (2001).

Mergeay, M. *et al. Mol. Gen. Genet.* 209: 61–70 (1987).

Miao, E.A., Miller, S.I. *PNAS (USA)* 96: 9452-9595 (1999).

Michalakis, Y., Roze, D. *Science* 306: 1492-1493 (2004).

Michels, M., Bakker, E.P. *J. Bacteriol.* 161: 231-237 (1985).

Milkman, R., Bridges, M.M. *Genetics* 126: 505-517 (1990).

Miller, M.B. *et al. Cell* 110: 303-314 (2002).

Min, B. *et al. Water Res.* 39: 4961-4968 (2005).

Min, B. *et al. Water Res.* 39: 1675-1686 (2005a).

Mincer, T.J. *et al. Appl. Environ. Microbiol.* 68: 5505-5011 (2002).

Mindich, L. *et al. Virology* 75: 218-223 (1976).

Mino, T. *Biochemistry* (Moscow) 65: 341 (2000).

Minois, N. *et al. PNAS* (USA) 102: 402-406 (2005).

Mira, A. *et al. Trends Genet.* 17: 589-596 (2001).

Miranda, E. *et al. Corros. Sci.* 48: 455-482 (2005).

Mitsushashi, S. *J. Int. Med. Res.* 21: 1-14 (1993).

Miyajima, A. *et al. Gene* 58: 273–282 (1987).

Miyanohara, A., Friedmann, T. *Nature Biotechnol.* 18: 1244-1245 (2000).

Modrich, P. *Ann. Rev. Genet.* 25: 229-253 (1991).

Moll, I. *et al. RNA* 9: 1308-1314 (2003).

Moore, F.B.-G., Rozen, D.E., Lenski, R.E. *Proc. R. Soc. Lond.* B. 267: 515-522 (2000).

Moore, P., Clayton, J. *Nature* 426: 725-731 (2003).

Moran, N.A. *Cell* 108: 583-586 (2002).

Moran, N.C. *PNAS* (USA) 98: 1338-1340 (2001).

Moreira, D. *et al. Environ. Microbiol.* 6: 959-969 (2004).

Morel, F.M.M., Hering, J.G. *Principles and Applications of Aquatic Chemistry*. pp. 244-246, Wiley, New York (1993).

Morris, R.M. *et al. Nature* 420: 806 (2002).

Mosch, H.U. *Contrib. Microbiol.* 5: 185 (2000).

Moss, S.J. *et al. Chem. Commun.* 22: 2341-2343 (2006).

Mota, L.J., Cornelis, G.R. *Ann. Med.* 37: 234-249 (2006).

Montagna, E., Torres, B.B. *Biochem. Mol. Biol. Edu.* 36(2): 99-105 (2008).

Mrazek, J., Guo, X., Shah, A. *PNAS* (USA) 104: 8472-8477 (2007).

MSA. *Geomicrobiology. Interactions Between Microbes and Minerals.* pp. 161-180. Mineralogical Society of America, Washington, D.C. (1997).

Mulkidjanian, A.Y., Junge, W. *Photosynth. Res.* 51: 27-42 (1997).

Mulkidjanian, A.Y. *et al. PNAS* 103: 13126-13131 (2006).

Mullard, A. *Nature* 453: 578-580 (2008).

Müller-Wille, S.E.W. *Stud. Hist. Philos. Biol. Biomed. Sci.* 36: 465-487 (2005).

Mulvey, M.A., Hultgren, S.J. *Science* 732-733 (2000).

Munger, K. *et al. J. Virol.* 78: 11451-11460 (2004).

Murgulis, L., Dolan, M.F., Guerrero, R. *PNAS* (USA) 97: 6954-6959 (2000).

Mushegian, A.R., Koonin, E.V. *PNAS* (USA) 93: 10268-10273 (1996).

Muyzer, G. *Curr. Opin. Microbiol.* 2: 317-322 (1999).

Myers, J.T., Tsang, A.W., Swanson, J.A. *J. Immunol.* 171: 5447-5453 (2003).

Nakagawa, T. *et al. Environ. Microbiol.* 8: 37-49 (2006).

Nanby-Wakao, R. *et al. Genes Cells* 5: 155 (2000).

Narayan, S., Young, J.A.T. *PNAS* (USA) 101: 7721-7726 (2004).

Nat. Acad. Sci. Colloquium (NAS). *PNAS* (USA) 93: 11287 (1996).

Nauhaus, K. et al. *Environ. Microbiol.* 9: 187-196 (2007).

Nedelec, F. *J. Cell Biology* 158: 1005-1015 (2002).

Nedelec, F. *et al. Nature* 389: 305-308 (1997).

Nedelec, F., Surrey, T., Karsenti, E. *Curr. Opin. Cell Biol.* 15: 118-124 (2003).

Neil, S.J.D. *et al. PLoS Pathogens* 2: 354-367 (2006).

Neil, S.J.D. *et al. Cell Host Microbe* 2: 193-203 (2007).

Neil, S.J.D., Zang, T., Bieniasz, P.D. *Nature* 451: 425-430 (2008).

Nelson, D.C., Jannasch, H.W. *Arch. Microbiol.* 136: 262-269 (1983).

Nelson, D.C., Jørgensen, B.B., Revsbech. N.P. *Appl. Environ. Microbiol.* 52: 225-233 (1986).

Nelson, K.E. *et al. Nature* 399: 323-329 (1999).

Nercessian, O. *et al. Environ. Microbiol.* 5: 492-502 (2003).

Nerenberg, R., Rittmann, B.E. *Water Sci. Technol.* 49: 223-230 (2004).

Nesbo, C.L., Boucher, Y., Doolittle, W.F. *J. Mol. Evol.* 53: 340 (2001).

Neufeld, J.D., Mohn, W.W. *Appl. Environ. Microbiol.* 71: 5710-5718 (2005).

Nevels, M. *et al. J. Virol.* 75: 3089-3094 (2001).

Newberry, C.J. *et al. Environ. Microbiol.* 6: 274 (2004).

Newman, D.K. *Science* 292: 1312-1313 (2001).

Newman, D.K., Banfield, J.F. *Science* 296: 1071-1076 (2002).

Ng, T.I. *et al. Hepatology* 45: 1413-1421 (2007).

Nicholson, W.L. *et al. Microbiol. Mol. Biol. Rev.* 64: 548-572 (2000).

Nicholson, W.L. *et al. Mutat. Res.* 571: 249-264 (2005).

Nierman, W.C. *et al. Nature* 438: 1151-1156 (2005).

Nishiyama *et al. Science* 320: p. 376 (2008)

Not, F. *et al. Science* 315: 253 (2007).

Nougayrede, J.P. *et al. Trends Microbiol.* 13: 103 (2005).

Nougayrede, J.-P. *et al. Science* 313: 848-851 (2006).

Novick, R.P. *Science* 319: 910-911 (2008).

Nowrouzian, F.L., Wold, A.E., Adlerberth, I. *J. Infect. Dis.* 191: 1078 (2005).

O'Malley, M.A., Dupre, J. *Biol. and Philos.* 22: 155-191 (2007).

O'Malley, M.A., Dupre, J. *Stud. Hist. Phil. Biol. Biomed. Sci.* 28: 775-779 (2007).

O'Shea, C.C. *Oncogene* 24: 7640-7655 (2005).

O'Toole, G.A. *Nature* 432: 680-681 (2004).

Ochman, H., Lawrence, J.G., Groisman, E.A. *Nature* 405: 299-304 (2000).

Ochman, H., Lerat, E., Daubin, V. *PNAS* (USA) 102: 6595-6599 (2005).

Oda, Y. *et al. Appl. Environ. Microbiol.* 68: 3467 (2002).

Ogata, H. *et al. Virol. J.* 2: 8 (2005).

Oh, J.D. *et al. PNAS* (USA) 103: 9999-10004 (2006).

Oh, S-E., Logan, B.E. *Wat. Res.* 39: 4673-4682 (2005).

Okeke, I.N. *Trends Microbiol.* 16: 198-199 (2008).

Okuta, A., Ohnishi, K., Harayama, S. *Gene* 212: 221-228 (1998).

Olano, C. *et al. Chem. Biol.* 11: 87-97 (2004).

Olson, M.E. *et al. Can. J. Vet. Res.* 66: 86-92 (2002).

Oltvai, Z.N., Barabasi, A.L. *Science* 298, 763-764 (2002).

Onodera, S., Sun, Y., Mindich, L. *Virology* 286: 113-118 (2001).

Oprea, T.J., Matter, H. *Curr. Opin. Chem. Biol.* 8: 349-358 (2004).

Oremland, R.S., King, G.M. In: *Microbial Mats*, Pp. 180-190. In: Cohen, Y., Rosenberg, E. (eds.). Amer. Soc. Microbiol., Washington, D.C. (1989).

Ormö, M. *et al. Science* 273: 1392-1395 (1996).

Orskov, F., Orskov, I. *J. Infect. Dis.* 148: 346-357 (1983).

Osterberg, F. *et al. Proteins* 46: 34-40 (2002).

Oswald, E. *et al. PNAS* (USA) 91: 3814 (1994).

Otto, S.P., Lenormand, T. *Nature Rev. Genet.* 3: 252-261 (2002).

Pace, N.R. *et al. Adv. Microb. Ecol.* 9: 1 (1986).

Pace, N.R. *Science* 276: 734-740 (1997).

Paegel, B.M. *et al. Curr. Opin. Biotechnol.* 14: 42-50 (2003).

Paerl, H.W., Bebout, B.M., Prufert, L.E. In: *Microbial Mats*. Cohen, Y., Rosenberg, E. (eds.), pp. 326-341. ASM, Washington, D.C. (1989).

Paerl, H.W., Keller, P.E. In: *Microbial Ecology*, p. 68-75. Loutit, M.W., Miles, J.A.R. (eds.). Springer, Berlin (1978).

Paerl, H.W. et al. *Nitrogen Fixation*, vol. 1. *Ecology*, pp. 193-240. Broughton W.J. (Ed.). Clarendon Press, Oxford (1981).

Pagani, M. *et al. Science* 309: 600 (2005).

Palecek, S.P., Parikh, A.S., Kron, S.J. *Microbiology* 148: 893 (2002).

Palese, P. *PNAS* (USA) 95: 12750-12751 (1998).

Palmer, R.J. *et al. Infect. Immun.* 69: 5794-5804 (2001).

Palys, T., Nakamura, L.K., Cohan, F.M. *Internat. J. System. Bacteriol.* 47: 1145-1156 (1997).

Panicker, M.M., Minkley, *E.G.* Jr. *J. Bacteriol.* 162: 584 (1985).

Papke, R.T., Doolittle, W.F. *Current Biology* 13: R606-R607 (2003).

Papke, R.T. *et al. Environ. Microbiol.* 5: 650 (2003).

Parker, D.L. *et al. Appl. Environ. Microbiol.* 63: 2324-2329 (1997).

Parker, D.L. *et al. Appl. Environ. Microbiol.* 64: 1545-1547 (1998).

Parkes, R.J. *et al. Nature* 371: 410 (1994).

Parkes, R.J. *et al. Nature* 436: 390 (2005).

Parsek, M.R. *Nature* 450: 805-806 (2007).

Parsek, M.R., Fuqua, C. *J. Bacteriol.* 186: 4427-4440 (2004).

Paterson, G.K., Mitchell, T.J. *Trends Microbiol.* 12: 89-95 (2007).

Patterson, G.H. *Nature Biotechnol.* 22: 1524-1525 (2004).

Paul, J.H. *et al. Mol. Ecol.* 1: 37–46 (1992).

Paul, J.H., Jeffrey, W.H., DeFlaun, M.F. *Appl. Environ. Microbiol.* 53: 170–179 (1987).

Pazos, F. *et al. Nucleic Acids Res.* 33 Database Issue, D588-D592 (2005).

Pedersen, K., Christensen, S.K., Gerdes, K. *Mol. Microbiol.* 45: 501 (2002).

Pedros-Alio, C. *Trends Microbiol.* 14: 257 (2006).

Pedros-Alio, C. *Science* 315: 192-193 (2007).

Pedulla, M.L. *et al. Cell* 113: 171-182 (2003).

Pel, *et al. Nature Biotechnol.* 25: 221-231 (2007).

Pennisi, E. *Science* 289: 1869 (2000).

Pennisi, E. *Science* 305: 334-335 (2004).

Perego, M. *et al. PLoS Pathog.* 4: e1000210 (2008).

Pernthaler, J. *Nature Rev. Microbiol.* 3: 537 (2005).

Pethig, R., Markx, G.H. *Trends Biotechnol.* 15: 426-432 (1997).

Petrini, O. In Fokkema, N.J., van den Heuvel, J. (eds.). *Microbiology of the Phyllosphere*, pp. 175-187. Cambridge Univ. Press, Cambridge (1986)

Pham, C.A. *et al. FEMS Microbiol. Lett.* 223: 129-134 (2003).

Philippe, H. *et al. Proc. R. Soc.* London, Ser. B *Biol. Sci.* 267: 1213-1221 (2000).

Phung, N.T. *et al. FEMS Microbiol. Lett.* 233: 77-82 (2004).

Piel, J. *et al. PNAS* (USA) 101: 16222-16227 (2004).

Pielak, G.J. *PNAS* (USA) 102: 5901 (2005).

Piffaretti, J.-C., Frey, J. *CMLS (Cell. Mol. Life Sci.)* 56: 717-718 (1999).

Pimm, S.., Lawton, J.H. *Nature* 268: 329 (1977).

Pisani, D., Cotton, J.A., McInerney, J.O. *Mol. Biol. Evol.* 24: 1752 (2007).

Pitcher, R.S. *et al. Mol. Cell* 23: 743-748 (2006).

Pitman, A.R. *et al. Curr. Biol.* 15: 2230-2235 (2005).

Pitt, W.G., Ross, S.A. *Biotechnol. Prog.* 19: 1038-1044 (2003).

Piza, F.F. *et al. Antonie Van Leeuwenhoek* 86: 317-328 (2004).

Pol, A. *et al. Nature* 450: 874-878 (2007).

Poli, G. *et al. Science* 244: 575-577 (1989).

Poole, A., Jeffares, D., Penny, D. *Bioessays* 21: 880 (1999).

Poole, A., Penny, D. *Nature* 447: 913 (2007).

Poretsky, R.S. *et al. Appl. Environ. Microbiol.* 71: 4121-4126 (2005).

Portnoy, D.A. *et al. Infect. Immun.* 60: 1263-1267 (1992).

Post, J.C., Stoodley, P., Hall-Stoodley, L. *Curr. Opin. Otolaryngol.* 12: 3185-3190 (2004).

Postgate, J. *Nitrogen Fixation.* Cambridge University Press, Cambridge (1998).

Potera, C. *ASM News* 64: 321-322 (1998).

Potter, M.C. *Proc. R. Soc. Lond. B. Biol. Sci.* 84: 260-276(1911).

Powell, S.K. *et al. Nature Biotechnol.* 18: 1279-1282 (2000).

Prasanna, S.S., Doble, M. *J. Indian Inst. Sci.* (Bangalore) 88: 27-35 (2008).

Prasher, D.C. *et al. Gene* 111: 229-233 (1992).

Prechtl, J. *et al. Mol. Biol. Evol.* 21: 1477-1482 (2004).

Prescott, L.M., Harley, J.P., Klein, D.A. *Microbiology.* 3 ed. Wm. C. Brown, Dubuque, Iowa (1996).

Price, P., Sowers, T. *PNAS* (USA) 101: 4631 (2004).

Proksch, P. *et al. Appl. Microbiol. Biotechnol.* 59: 125-134 (2002).

Pucadyil, T.J. *et al. Biochemistry* 43: 15852-15862 (2004).

Qiao, X.Y., Qiao, J., Onodera, S., Mindich, L. *Virology* 275: 218-224 (2000).

Raaphorst, G.P. *et al. Int. J. Cancer* 90: 336-342 (2000).

Rabaey, K. *et al. Appl. Environ. Microbiol.* 70: 5373-5382 (2004).

Rabaey, K. *et al. Environ. Sci. Technol.* 39: 3401-3408 (2005).

Radajewski, S. *et al. Nature* 403: 646-649 (2000).

Radrnan, M., Wagner, R. *Sci. Amer.* 259: 40-46 (1988).

Radtke, A.L. *et al. Cell Microbiol.* 8: 1720-1729 (2006).

Ragsdale, S.W. *Crit. Rev. Biochem. Mol. Biol.* 26: 261 (1991).

Rainey, P.B., Rainey, K. *Nature* 425, 72-74 (2003).

Raju, D. *et al. BMC Microbiol.* 6: 50 (2006).

Ram, R.J. *et al. Science* 308: 1915-1920 (2005).

Randall, G. *et al. PNAS* (USA) 104: 12884-12889 (2007).

Ranea, J.A.G. *et al. J. Mol. Biol.* 336: 871-887 (2004).

Ranea, J.A.G. *et al. Trends Genet.* 21: 21-25 (2005).

Rankin, D.J. *Science* 320: 1591-1592 (2008).

Raoult, D. *et al. Science* 306: 1344-1350 (2004).

Rappe, M.S., Giovannoni, S.J. *Annu. Rev. Microbiol.* 57: 369 (2003).

Rappuoli, R. *PNAS* (USA) 97: 13467-13469 (2000).

Rastegar, L.A., Bahrami, H.H., Alaghebandan, R. *Iran. Burns.* 24: 637-641 (1998).

Ravasz, E. *et al. Science* 297, 1551 (2002).

Raven, P.H., Williams T. (eds.). *Nature and Human Society: The Quest for a Sustainable World.* pp. 130-155. National Academy Press, Washington, D.C. (1999).

Ravin, V. *et al. J. Mol. Biol.* 299: 53-73 (2000).

Ray, L.B., Alder, E.M., Gough, N.C. *Science* 306: 1505 (2004).

Raymond, J. *et al. Science* 298: 1616 (2002).

Raymond, J. *et al. Philos. Trans. R. Soc. London* B358: 223-230 (2003).

Raymond, J., Blankenship, R.E. *PNAS* (USA) 100: 7419-7420 (2003).

Read, T.D. *et al. Science* 296, 2028 (2002).

Read, T.D. *et al. Nature* 423: 81 (2003).

Redman, R.S. *et al. Science* 298: 1581 (2002).

Rees, C.E.D. *et al. Int. J. Food Microbiol.* 28: 263-275 (1995).

Regis, E. *What Is Life? Investigating the Nature of Life in the Age of Synthetic Biology.* Farrar, Straus and Giroux (2008).

Relman, D.A. *Science* 284, 1308 (1999).

Repoila, F. *et al. Mol. Microbiol.* 48: 855-861 (2003).

Reyes, D.R. *et al. Anal. Chem.* 74: 2623-2636 (2002).

Reynolds, C.S. *Ecology of Phytoplankton*. Cambridge University Press (2006).

Reynolds, C.S., Walsby, A.E. *Biol. Rev.* 50: 437-481 (1975).

Reynolds, T.B., Fink, G.R. *Science* 291: 878-881 (2001).

Reysenbach, A.L., Shock, E.L. *Science* 296: 1077-1082 (2002).

Reysenbach, A-L. *et al. Nature* 442: 444-447 (2006).

Rice, G. *et al. PNAS* (USA) 98: 13341-13345 (2001).

Rice, G. *et al. PNAS* (USA) 101: 7716-7720 (2004).

Richards, M.J. *et al. Crit. Care Med.* 27: 887-892 (1999).

Risenfeld, C.S., Goodman, R.M., Handelsman, J. *Environ. Microbiol.* 6: 981 (2004).

Risenfeld, C.S., Schloss, P.D., Handelsman, J. *Annu. Rev. Genet.* 38 525-552 (2004).

Rittmann, B.E. *Trends Biotechnol.* 24: 261-266 (2006).

Rittmann, B.E. *Water Sci. Technol.* 58: 219-225 (2006a).

Rittmann, B.E. *et al. Water Sci. Technol.: Water Supply* 4: 127-133 (2004).

Rittmann, B.E., McCarty, P.L. *Environmental Biotechnology: Principles and Applications.* McGraw-Hill Book Co., New York (2001).

Rivera, M.C. *et al. PNAS* (USA) 95: 6239-6244 (1998).

Rivera, M.C., Lake, J.A. *Nature* 431: 152-155 (2004).

Roa-Rodriguez, C., Nottenburg, C. Agrobacterium-mediated transformation of plants. CAMBIA technology landscape paper (<http://www.bios.net/ Agrobacterium>) (2003).

Robarts, R.D. *Int. J. Environ. Stud.* 25: 167-175 (1985).

Roberts, M., Cohan, F. *Genetics* 134: 401-408 (1993).

Roberts, M.S., Cohan, F.M. *Evolution* 49: 1081 (1995).

Robinson, N.P. *et al. Cell* 116: 25-38 (2004).

Robson, R.L., Postgate, J.R. *Ann. Rev. Microbiol.* 34: 183-207 (1980).

Rocha, E.P. *Genome Res.* 13: 1123-1132 (2003).

Rocha, E.P., Blanchard, A. *Nucleic Acids Res.* 30: 2031-2042 (2002).

Rohankhedkar, M.S. *et al. J. Bacteriol.* 188: 223-230 (2006).

Rohwerder, T. *et al. Appl. Microbiol. Biotechnol.* 63: 239-248 (2003).

Roizman, B. *Science* 288: 2327-2328 (2000).

Rondon, M.R. *et al. Appl. Environ. Microbiol.* 66, 2541-2547 (2000).

Roossinck, M. *Science* 315: 513-515 (2007).

Rosenzweig, M.L. *Species Diversity in Space and Time.* pp. 8-18, 190-284. Camb. Univ. Press, Cambridge (1995).

Rossello-Mora, R., Amann, R. *FEMS Microbiol. Rev.* 25: 39-67 (2001).

Rossmann, M.G., Johnson, J.E. *Annu. Rev. Biochem.* 58: 533-573 (1989).

Roy, C. *et al. Science* 320: 1651-1654 (2008)

Roy, S.W., Gilbert, W. *PNAS* (USA) 102: 5773 (2005).

Ruepp, A. *et al. Nature* 407: 508-513 (2000).

Ruiz-Trillo, I. *et al. Curro BioI.* 14: R946-R947 (2004).

Rutherford, A.W., Faller, P. *Phil. Trans. R. Soc. Lond.* 358B: 245-253 (2003).

Sachs, J.S. *Discover* pp. 36-40 (Nov. 2005).

Sahgal, M., Johri, B.N. *Current Science* 84: 43-48 (2003).

Salemme, F.R. *Pharmacogenomics* 4: 257-267 (2003).

Salomon *et al. Nat. Prod. Rep.* 21: 105-121 (2004).

Salone, B. *et al. Mutat. Res.* 358: 155-160 (1996).

Salyers, A.A. *Revenge of the Microbes: How Bacterial Resistance is Undermining the Antibiotic Miracle.* ASM Press, Washington, D.C. (2005).

Salyers, A.A., Shoemaker, N.B. *FEMS Microbiol. Ecol.* 15: 15–22 (1994).

Salyers, A.A. *et al. Microbiol. Rev.* 59: 579–590 (1995).

Sancar, A. *et al. Annu. Rev. Biochem.* 73: 39-85 (2004).

Sanchez, R.T. *et al. Mater. Corros.* 52: 614-618 (2001).

Sandeep, K. *Curr. Sci.* 90: 631-633 (2006).

Sandman, K., Reeve, J.N. *Curr. Opin. Microbiol.* 8: 656-661 (2005).

Sanjuan, R., Moya, A., Elena, S.F. *PNAS* (USA) 101: 15376 (2004).

Sanjuan, R., Moya, A., Elena, S.F. *PNAS* (USA) 101: 8396 (2004a).

Santier, S., Gilet, R., Malaise, E. *Radiat. Res.* 104: 224-233 (1985).

Sapp, J. *Microbial Phylogeny and Evolution: Concepts and Controversies.* Oxford Univ. Press, New York, (2005).

Sapp, J. *Microb. Mol. Biol. Rev.* 69: 101-115 (2005).

Sari-Minodier, I. *et al. Mutat. Res.* 629: 111-121 (2007).

Sato, N. *Genome Inform.* 13: 173-182 (2002).

Sauer, A.K. *et al. J. Bacteriol.* 184: 1140-1154 (2002).

Saunders, N.J. *et al. Molec. Microbiol.* 37: 207-215 (2000).

Savina, N., Dalivelya, O., Kuzhir, T. *Mutat. Res.* 535: 195-204 (2003).

Sayeed, L. *et al. J. Bacteriol.* 182: 2416 (2000).

Schaechter, M. *J. Biosci.* 28: 149-154 (2003).

Scherings, G. *et al. Eur. J. Biochem.* 135: 591-599 (1983).

Schippers, A. *et al. Nature* 433: 861 (2005).

Schleper, C. *Nature* 456: 712-714 (2008).

Schleper, C., Jurgens, G., Jonuscheit, M. *Nature Rev. Microbiol.* 3: 479-488 (2005).

Schmid, A. *et al. Nature* 409: 258-268 (2001).

Schmidt, T.M., DeLong, E.F., Pace, N.R. *J. Bacteriol.* 173: 4371 (1991).

Schnupf, P., Portnoy, D.A. *Microbes Infect.* 9: 1176-1187 (2007).

Schreiber, S.L. *Chem. Eng. News* 81: 51-61 (2003).

Schulz, H.N. *et al. Science* 284: 493 (1999).

Schulz, H.N., Schulz, H.D. *Science* 307: 416-418 (2005).

Schulze, F.E. *Physik. Abh. Kgl. Akad. Anz.* 6: 92-97 (1883).

Schwartz, R.M., Dayhoff, M.O. *Science* 199: 395-403 (1978).

Science doi: 10.1126/science.1147112 (2007).

Sedgwick, B., Lindahl, T. *Oncogene* 21: 8886-8894 (2002).

Semrau, J.D., DiSpirito, A.A., Murrell, J.C. *Trends Microbiol.* 16: 190-193 (2008).

Service, R.F. *Science* 299: 640 (2003).

Service, R.F. *Science* 322: 522-523 (2008).

Setlow, B., Setlow, P. *J. Bacteriol.* 178: 3486-3495 (1996).

Setlow, B. *et al. J. Bacteriol.* 188: 3740-3747 (2006).

Setlow, P. *J. Appl. Bacteriol.* 76: (Sympos. Suppl.) 49S-60S (1994).

Setlow, P. *J. Appl. Microbiol.* 101: 514-525 (2006).

Setlow, P. *Trends Microbiol.* 15: 172-180 (2007).

Shankar, B., Pandey, R., Sainis, K. *Int. J. Radiat. Biol.* 82: 537-548 (2006).

Shapiro, J.A. *Gene* 345: 91-100 (2005).

Shapiro, J.A. *Annu. Rev. Microbiol.* 52: 81 (1998).

Shapiro, J.A. *Stud. Hist. Phil. Biol. Biomed. Sci.* 38: 807-819 (2007).

Shapiro, J.A. *Sci. Amer.* 256: 82-89 (1988).

Shapiro, J.A. *Ann. Rev. Microbiol.* 52: 81-104 (1998).

Shapiro, J.A. *Res. Microbiol.* 153: 447-453 (2002).

Shapiro, J.A., Sternberg, R.V. *Biol. Rev.* 80: 227-250 (2005).

Shaughnessy, L.M. *et al. Cell Microbiol.* 8: 781-792 (2006).

Shaughnessy, L.M., Swanson, J.A. *Front. Biosci.* 12: 2683-2692 (2007).

Shigenobu, S. *et al. Nature* 407: 81-86 (2000).

Shimomura, O., Johnson, F.H., Saiga, Y. *J. Cell Comp. Physiol.* 59: 223-239 (1962).

Shin, J.S. *et al. Science* 289: 785 (2000).

Shinn, E.A., Griffin, D.W., Seba, D.B. *Arch. Environ. Health* 58: 498-504 (2005).

Shintani, T., Klionsky. D.J. *Science* 306: 990 (2004).

Sibley, C.G., Comstock, J.A., Ahlquist, J.E. *J. Mol. Evol.* 30: 202-236 (1990).

Silander, O.K. *et al. PNAS* (USA) 102: 19009-19014 (2005).

Silver, P.A., Way, J.C. *Clin. Pharmacol. Ther.* 82: 586-590 (2007).

Simon, H.M. *et al. Appl. Environ. Microbiol.* 71: 4751-4760 (2005).

Simonson, A.B. *et al. PNAS* (USA) 102: 6608-6613 (2005).

Simpson, A.G.B., Roger, A.J. *Curr. Biol.* 14(17): R695-R696 (2004).

Simpson, A.G.B., Roger, A.J. In: *Organelles, Genomes, and Eukaryote Phylogeny: An Evolutionary Synthesis in the Age of Genomics*, pp. 27-54. Hirt, R.P., Homer, D.S. (eds.). CRC Press, Boca Raton (2004a).

Simpson, G.G. *Tempo and Mode in Evolution*. Columbia Univ. Press, New York (1944).

Singer, G.A., Hickey, D.A. *Mol. Biol. Evol.* 17: 1581-1588 (2000).

Singh, K.P. *et al. Nature* 417: 552-555 (2002).

Singh, R., Jamienson, A., Cresswell, P. *Nature* 455: 1244-1247 (2008).

Skalka, A.M., Katz, R.A. *Cell Death Differ.* 12 (suppl. 1): 971-978 (2005).

Smallwood, M.F., Calvert, C.M., Bowles, D.J. (Eds.). *Plant Responses to Environmental Stress.* BIOS Scientific Publishers, Oxford (1999).

Smith, G.P. *Science* 228: 1315-1317 (1985).

Smith, K.R. *J. Biotechnol.* 99: 1-22 (2002).

Snel, B. *et al. Annu. Rev. Microbiol.* 59: 191-209 (2005).

Sogin, M.L. *et al. PNAS* (USA) 103: 12115 (2006).

Sokurenko, E.V., Hasty, D.L., Dykhuizen, D.E. *Trends in Microbiol.* 7: 191-195 (1999).

Sonea, S., Mathieu, L.G. *Internat. Microbiol.* 4: 67-71 (2001).

Song, S. *et al. PNAS* (USA) 101: 2112-2116 (2004).

Sorek, R. *et al. Science* 318: 1449 (2007).

Sorensen, S.J. *et al. Nature Rev. Microbiol.* 3: 700-710 (2005).

Soukos, N.S. *et al. Pharm. Res.* 17: 405-409 (2000).

Splitter, L.J. *Philos. Sci.* 55: 323-348 (1988).

Spring, S., Bazylinski, D.A. In: *The Prokaryotes*, pp. 1-20. Springer, New York (2002).

Staal, M., Meysman, F.J.R., Stal, I.J. *Nature* 425: 504-507 (2003).

Stackebrandt, E., Goebel, B.M. *Int. J. Syst. Bacteriol.* 44: 846-849 (1994).

Stackebrandt, E. *et al. Int. J. Syst. Ecol. Microbiol.* 52: 1043-1047 (2002).

Staley, J.T. *Curr. Opin. Biotechnol.* 8: 340-345 (1997).

Staley, J.T. *Microbial Diversity and Bioprospecting*. Am. Soc. Microbiol., Washington, D.C. (2004).

Staley, J.T., Konopka, A. *Annu. Rev. Microbiol.* 39: 321 (1985).

Staley, J.T., Gosink, J.J. *Annu. Rev. Microbiol.* 53: 189 (1999).

Stanier, R.Y. *et al. Bacteriol. Rev.* 35: 171-205 (1971).

Stanier, R.Y., Cohen-Bazire, G. *Annu. Rev. Microbiol.* 31: 225-274 (1977).

Stavrinides, J. *et al. PLoS Pathog.* 2: e104 (2006).

Stechmann, A., Cavalier-Smith, T. *Curr. Biol.* 13: R665-666 (2003).

Steenkamp, E.T., Baldauf, S.L. In: *Organelles, Genomes and Eukaryote Phylogeny*, pp. 109-129. Hirt, P., Horner, D.S. (eds.) The Systematics Association Special Volume, series 68, CRC Press, London (2004).

Steinbuchel, P. *Materials Today* 4(2): 11 (2001).

Stephens, C. *Curr. Biol.* 14, R65-R66 (2004).

Stetter, K.O. *et al. Nature* 365: 743 (1993).

Stevens, C., Hume, I. *Comparative Physiology of the Vertebrate Digestive System.* Cambridge Univ. Press (2004).

Stevenson, B.S. *et al. Appl. Environ. Microbiol.* 70: 4748-4755 (2004).

Stewart, P. *et al.* In: *Plasmid Biology*, pp. 613. Phillips, G., Funnell, B.E. (eds.), ASM Press, Washington, D.C. (2004).

Stewart, P.L. *et al. Cell* 67: 145-154 (1991).

Stewart, P.S. *Int. J. Med. Microbiol.* 292: 107-113 (2002).

Stock, A.M., Robinson, V.L., Goudreau, P.N. *Annu. Rev. Biochem.* 69: 183 (2000).

Stock, J.B., Surette, M. Chemotaxis. In: *Escherichia coli and Salmonella typhimurium: Cellular and Molecular Biology*, pp. 1103-1129. Neidhardt, F.C. (Ed.) Vol. 1, 2nd ed., Amer. Soc. Microbiol., Washington, D.C. (1996).

Stockert, R.J. *Physiol. Rev.* 75: 591 (1995).

Stockwell, B.R. *Nature* 432: 846-854 (2004).

Stockwell, B.R. *Trends Biotechnol.* 18: 449-455 (2000).

Stone, R. *Science* 317: 1022-1024 (2007).

Stoodley, P., de Beer, D., Lewandowski, Z. *Appl. Environ. Microbiol.* 60: 2711-2716 (1994).

Stotzky, G. In: *Gene Transfer in the Environment*, pp. 165-222. Levy, S.B., Miller, R.V. (eds.). McGraw–Hill, New York (1989).

Stouthamer, A.H. *Antonie van Leeuwenhoek* 39: 545-565 (1973).

Strackebrandt, E. *et al. Int. J. Syst. Evol. Microbiol.* 52: 1043-1047 (2002).

Stracker, T.H. *et al. Nature* 418: 348-352 (2002).

Stroscio *et al. Science* 313: 948 (2006).

Sudbery, P., Gow, N., Berman, *J. Trends Microbiol.* 12: 317 (2004).

uhre, K. *J. Virol.* 79: 14095-14101 (2005).

ukumaran, P.V. *Curr. Sci.* 88: 879-885 (2005).

ulakvelidze, A. *et al. Antimicrob. Agents Chemother.* 45, 649-659 (2001).

un, B. *et al., Science* 298, 1023 (2002).

utton, M.D. *et al. Ann. Rev. Genet.* 34: 479-497 (2000).

wartz, T.E. *et al. Science* 317: 1090 (2007).

y, A. *et al. J. Bacteriol.* 183, 214-220 (2001).

yed, T., Schierwater, B. *Vie Mil.* 52: 177-187 (2002).

zathmary, E. *Genetics* 133: 127 (1993).

zurmant, H., Ordal, G.W. *Microbiol. Mol. Biol. Rev.* 68: 301-319 (2004).

Takai, K., Horikoshi, K. *Genetics* 152: 1285-1297 (1999).

Talloozy, Z. *et al. PNAS* (USA) 99: 190 (2002).

Taylor, B.L., Zhulin, I.B. *Microbiol. Mol. Biol. Rev.* 63: 479 (1999).

Taylor, M.R. *Dark Life.* Scribner, New York (1999).

Taylor, R.K. *Nature* 399: 312-313 (1999).

Tennen, R. *et al. J. Appl. Microbiol.* 89: 330-338 (2000).

Teplitski, M., Robinson, J.B., Bauer, W.D. *Mol. Plant-Microbe Interact.* 13: 637-648 (2000).

Tettelin, H. *et al. Science* 293: 498-506 (2001).

Thauer, R.K. *Science* 318: 1732-1733 (2007).

Thomas, C.M. (Ed.) *The Horizontal Gene Pool.* Harwood Academic Publishers, Amsterdam (2000).

Tiedje, J.M. *Nature Biotech.* 20: 1093-1094 (2002).

Tiku, A.B., Kale, R.K. *J. Biosci.* 29: 111-117 (2004).

Tilman, D., Kareiva, P. *Spatial Ecology: The Role of Space in Population Dynamics and Interspecific Interactions.* Princeton Univ. Press, Princeton (1997).

Timasheff, S.N., Andreu, J.M., Na, G.C. *Pharmacol. Ther.* 52: 191-210 (1991).

Timmis, K.N., Pieper, D.H. *Trends Biotechnol.* 17: 4350-4351 (2001).

Torsvik, V. *et al. Appl. Environ. Microbiol.* 56: 776-782 (1990).

Torsvik, V., Ovreas, L., Thingstad, T.F. *Science* 296: 1064-1065 (2002).

Toth, I.K. *et al. Annu. Rev. Phytopathol.* 44: 305-336 (2006).

Trau, D. *et al. Anal. Chem.* 74: 3168-3173 (2002).

Trevors, J.T., van Elsas, J.D. *Nucleic Acids in the Environment: Methods and Applications.* pp. 29-48 Springer, Heidelberg (1995).

Tringe, S.G. *et al. Science* 308: 554 (2005).

Tripathi, A.K., Verma, S.C., Ron, E.Z., *Res. Microbiol.*, 153, 579-584 (2002).

Trotsenko, Y.A., Khmelenina, V.N. *Arch. Microbiol.* 177: 123-131 (2002).

Truckses, D.M., Garrenton, L.S. Thorner, J. *Science* 306: 1509-1511 (2004).

Trucksis, M. *et al. PNAS* (USA) 95: 14464-14469 (1998).

Tsien, R.Y. *Annu. Rev. Biochem.* 67: 509-544 (1998).

Turley, C. *FEMS Microbiol. Ecol.* 33: 89 (2000).

Turnbaugh, P.J. *et al. Nature* 444: 1027-1031 (2006).

Turnbaugh, P.J. *et al. Nature* 449: 804 (2007).

Tyrrell, T. *Nature* 400: 525-531 (1999).

Tyrrell, T., Law, C.S. *Nature* 393: 318 (1998).

Uchiyama, T. et al. *Nature Biotechnol.* 23: 88-93 (2005).

Ulsh, B.A. *et al. J. Environ. Radioact.* 74: 73-81 (2004).

Upton, A.C. *Crit. Rev. Toxicol.* 31: 681-695 (2001).

Ussery, D.W., Hallin, P.F. *Microbiology* 150: 749 (2004).

Vaganay-Juery, S. *et al. Br. J. Cancer* 83: 514-518 (2000).

Valentin-Hansen, P. *et al. Mol. Microbiol.* 51: 1525-1533 (2004).

Van de Putte, P., Cramer, S., Giphart-Gassler, M. *Nature* 286: 218 (1980).

Van den Burg, B. *Curr. Opin. Microbiol.* 6: 213-218 (2003).

Van Der Hulst, R.W. *et al. Aliment. Pharmacol. Ther.* 13, 1171-1177 (1999).

van der Staay, S.Y.M., Wachter, R.D., Vaulot, D. *Nature* 409: 607-610 (2001).

Van Etten, J.L. *Annu. Rev. Genet.* 37: 153-195 (2003).

Vandamme, P. *et al. Microbiol. Rev.* 60: 407-438 (1996).

Varde, N.K., Pack, D.W. *Expert Opin. Biol. Ther.* 4: 35-51 (2004).

Varnova, E., Inze, D., Van Breusegem, F. *J. Exp. Bot.* 53: 1227-1236 (2002).

Varthakavi, V. *et al. PNAS* (USA) 100: 15154-15159 (2003).

Velicer, G.J. *Trends Microbiol.* 11, 330-337 (2003).

Velicer, G.J., Yu, Y.T. *Nature* 425, 75-78 (2003).

Venkat, S. *et al. J. Environ. Pathol. Toxicol. Oncol.* 20: 165-175 (2001).

Venter, J.C. *et al. Science* 304: 66-74 (2004).

Verpoorte, E. *Electrophoresis* 23: 677-712 (2002).

Vetsigian, K., Woese, C., Goldenfeld, N. *PNAS* (USA) 103: 10696-10701 (2006).

Vezzi, A. *et al.. Science* 307: 1459-1461 (2005).

Vidaver, A.K., Koski, R.K., Van Etten, J.L. *J. Virol.* 11: 799-805 (1973).

Villanueva, R.A.-C. *et al. Mater. Corros.* 57: 543-548 (2006).

Vogel, J. *et al. Nucleic Acids Res.* 31: 6435-6443 (2003).

Voigt, O. *et al. Curr. Biol.* 14(22): R944-R945 (2004).

Voinnet, O. *Nature Rev. Genet.* 6: 206-220 (2005).

Vollmer, W., Höltje, J.-V. *J. Bacteriol.* 186: 5978-5987 (2004).

Von Mering, C. *et al. Science* 315: 1126-1130 (2007).

Von Wintzingerode, F., Goebel, U.B., Stackebrandt, E. *FEMS Microbiol. Rev.* 21: 213 (1997).

Wackett, L.P. *Environ. Microbiol.* 8: 566-567 (2006).

Wackett, L.P. *Trends Biotechnol.* 22: 153-154 (2004).

Walch, M. *Microbial Corrosion. Encyclopaedia of Microbiology*, pp. 585-591. Lederberg, J. (Ed.). Academic Press, New York (1992).

Walsh, C.T., Fischbach, M.A. *Sci. Amer.* pp. 44-51 (Jul., 2009).

Walter, J.M. *et al. PNAS* (USA) 104: 2408-2412 (2007).

Walton, A.W. *Geobiology* 6: 351-364 (2008).

Wang, G.J., Cai, L. *Toxicol. Sci.* 53: 369-376 (2000).

Wang, G.Y. *et al. Org. Lett.* 2: 2401-2404 (2000).

Wang, S.T. *et al. J. Mol. Biol.* 358: 16-37 (2006).

Wang, Y., Dohlman, H.G. *Science* 306: 1508 (2004).

Ward, B.B. In: *Encyclopedia of Environmental Microbiology*, pp. 2144-2167. Capone, D.A. (Ed.). Wiley, New York (2002).

Ward, B.B. *PNAS* (USA) 99: 10234 (2002).

Ward, D.M., Weller, R., Bateson, M.M. *Nature* 345: 63-65 (1990).

Ward, D.V., Zambryski, P.C. *PNAS* (USA) 98: 385-386 (2001).

Warren, L.A., Kauffman, M.E. *Science* 299: 1027-1029 (2003).

Watnick, P., Kolter, R. *J. Bacteriol.* 182: 2675-2679 (2000).

Watve, M. *et al. Curr. Sci.* 78: 1535-1542 (2000).

Wayne, L.G. *et al. Int. J. Syst. Bacteriol.* 37: 463-464 (1987).

Weaver, K.E. *et al. J. Bacteriol.* 185: 2169 (2003).

Webre, D.J., Wolanin, P.M., Stock, J.B. *Curr. Biol.* 13: R47-R49 (2003).

Weibel, D.B. *PNAS* (USA) 105: 18075-18076 (2008).

Weinberg, S.R., Stotzky, G. *Soil Biol. Biochem.* 4: 171–180 (1972).

Weissman, K.J. *Trends Biotechnol.* 25: 139-142 (2007).

Weissman, K.J., Leadlay, P.F. *Nature Rev. Microbiol.* 3: 925 (2005).

Weist, S. *et al. J. Am. Chem. Soc.* 126: 7436-7437 (2004).

Weist, S., Süssmuth, R.D. *Appl. Microbiol. Biotechnol.* 68: 141-150 (2005).

Weitzman, M.D. *et al.* I *DNA Repair* (Amst.) 3: 1165-1173 (2004).

Welch, R.A. *et al. PNAS* (USA) 99: 17020 (2002).

Welch, T.J. *et al. J. Bacteriol*. 175: 7170 (1993).

Wenner, M. *Sci. Amer*. p. 91-92 (Jul. 2008).

West, S.A. *et al. Nature Rev. Microbiol*. 4: 597-607 (2006).

West, S.A., Pen, I., Griffin, A.S. *Science* 296: 72-75 (2002).

Westfall, P.J., Ballon, D.R., Thorner, J. *Science* 306: 1511-1512 (2004).

Whitaker, R.J., Grogan, D.W., Taylor, J.W. *Science* 301: 976-978 (2003).

Whitehead, N.A. *et al. FEMS Microbiol. Rev*. 25: 365-404 (2001).

Whitman, W.B. *et al. PNAS* (USA) 95: 6578-6583 (1998).

Whittaker, G., Sui, M., Helenius, A. *Trends Cell Biology* 6: 67-71 (1996).

Whittam, T.S., Ake, S.E. In: *Molecular Paleopopulation Biology*, pp. 223-245. Takahata, N., Clark, A.G. (ed.). Jap. Sci. Soc. Press, Tokyo (1993).

WHO (World Health Organization). *Meningococcal Meningitis Fact Sheet*. World Health Org,, Geneva (2003).

Wiegel, J., Adams, M.W.W. (eds.). *Thermophiles : The Key to Molecular Evolution and the Origin of Life*. Taylor and Francis, Washington, D.C. (1998).

Wileman, T. *Science* 312: 875-878 (2006).

Wilkinson, B. *et al. Bioorg. Med. Chem. Lett*. 16: 5814-5817 (2006).

Wilson, R.W. (Ed.). *Species: New Interdisciplinary Essays*. pp. 93-118. MIT Press, Cambridge, MA (1999).

Winans, S.C. *Nature* 437: 330 (2005).

Woese, C.R., Fox, G.E. *PNAS* (USA) 74: 5088-5090 (1977).

Woese, C.R., Kandler, O., Wheelis, M.L. *PNAS* (USA) 87: 4576-4579 (1990).

Wolf, J.B., Brodie, E.D.I., Wade, M.J. (eds.) *Epistasis and the Evolutionary Process*. Oxford University Press, New York (2000).

Wolf, Y.I. *et al. Trends Genet*. 18: 472-479 (2002).

Wolff, S. *Environ. Health Persp*. 106: S277-283 (1998).

Wolff, S. *et al. Int. J. Radiat. Biol*. 53: 39-47 (1988).

Wolfram, S. *A New Kind of Science*. Wolfram Media, Champaign, Ill(2002).

Wood, EJ.F. *Marine Microbial Ecology*. Chapman and Hall, London (1965).

Wood, T.K. *et al. Appl. Microbiol. Biotechnol*. 72: 361 (2006).

Woolhouse, M.E.J., Taylor, L.H., Haydon, D.T. *Science* 292: 1109-1112 (2001).

Woolley, A.T. *et al. Anal. Chem*. 68: 4081-4086 (1996).

Wright, S. *Evolution* 36: 427-443 (1982).

Wu, C.H., Mulchandani, A., Chen, W. *Trends Microbiol*. 16: 181-188 (2008).

Wu, L.J., Errington. J. *Science* 264: 572-575 (1994).

Wu, L.J., Errington, J. *Nature* 451: 900-901 (2008).

Wu, X. *et al. Genes Dev.* 18: 1305-1316 (2004).

Wuchter, C.C. *et al. PNAS* (USA) 103: 12317-12322 (2006).

Xavier, J.B., Foster, K.R. *PNAS* (USA) 104: 876-881 (2007).

Xavier, K.B. Bassler, B.L. *Curr. Opin. Microbiol.* 6, 191-197 (2003).

Xiao, C. *et al. J. Mol. Biol.* 353: 493-496 (2005).

Xu, P. *et al. Trends Genet.* 20: 617-624 (2004).

Yamada, T. *et al. Nature Biotech.* 21: 885-890 (2003).

Yamamoto, T.A., Gerdes, K., Tunnadiffe, A. *FEBS Lett.* 519: 191 (2002).

Yan, G., Hua, Z., Du, G., Chen, J. *Curr. Microbiol.* 52: 238-242 (2006).

Yan, X. *et al. Nature Struct. Biol.* 7: 101-103 (2000).

Yanagi, M., Yamasato, K. *FEMS Microbiol. Lett.* 107, 15-120 (1993).

Yang, S., Doolittle, R.F., Bourne, P.E. *PNAS* (USA) 102: 373 (2005).

Yao, X.D., Evans, D.H. *J. Virol.* 75: 6923-6932 (2001).

Yarmolinsky, M.B. *Science* 267: 836 (1995).

Yin, X., Stotzky, G. *Adv. Appl. Microbiol.* 45: 153–212 (Neidleman, S.L., Laskin, A.I. eds.), Academic Press, San Diego (1997).

Yuan *et al. Nature Immunol.* 7: 835 (2006).

Zakeri, F., Assaei, R.G. *Mutat. Res.* 562: 1-9 (2004).

Zengler, K. *et al. Methods Enzymol.* 397: 124 (2005).

Zhang, C.T., Zhang, R. *Nature Biotechnol.* 22, 1079-1079 (2004).

Zhang, Y. *et al. Nature Genet.* 20: 123-128 (1998).

Zhou, H. *et al. Adv. Space Res.* 34: 1368-1372 (2004).

Zhu, J. *et al. PNAS* (USA) 99: 3129-3134 (2002).

Zhu, L. *et al. Lab Chip.* 4: 337-341 (2004).

Zhu, L. *et al. Appl. Environ. Microbiol.* 70: 597-598 (2004).

Ziebuhr, W. *et al. CMLS (Cell. Mol. Life Sci.)* 56: 719-728 (1999).

Zierenbeg, R.A., Adams, M.W.W., Arp, A.J. *PNAS* (USA) 97: 12961-12962 (2000).

Zimmer, C. *Microcosm: E. coli and the New Science of Life*. Pantheon, New York (2008).

Zimmerman, E.S. *et al. J. Virol.* 80: 10407-10418 (2006).

Zinn, M., Witholt, B., Egli, T. *Adv. Drug Deliv. Rev.* 53, 5-21 (2001).

Zohary, T. In: *Microbial Mats,* pp. 52-63. Cohen, Y., Rosenberg, E. (eds.). Amer. Soc. Microbiol., Washington, D.C. (1989).

Zohary, T. *J. Plankton Res.* 7: 399-409 (1985).

Zuberman *et al. PLOS Biol.* 6: e114 (2008).

Zuckerkandl, E., Pauling, L. *J. Theor. Biol.* 8: 357-366 (1965).

Zuo, Y. *et al. Energy Fuels* 20: 1716-1721 (2005).

Zychlinsky, E., Matin, A. *J. Cen. Microbiol.* 156: 1352-1355 (1983).

Index

H

I

K

L